AF379519

Calf and Heifer Feeding and Management

Calf and Heifer Feeding and Management

Editors

Zhijun Cao
Michael Van Amburgh

MDPI • Basel • Beijing • Wuhan • Barcelona • Belgrade • Manchester • Tokyo • Cluj • Tianjin

Editors
Zhijun Cao
China Agricultural University
China

Michael Van Amburgh
Cornell University
USA

Editorial Office
MDPI
St. Alban-Anlage 66
4052 Basel, Switzerland

This is a reprint of articles from the Special Issue published online in the open access journal *Animals* (ISSN 2076-2615) (available at: https://www.mdpi.com/journal/animals/special_issues/Calf_and_Heifer).

For citation purposes, cite each article independently as indicated on the article page online and as indicated below:

LastName, A.A.; LastName, B.B.; LastName, C.C. Article Title. *Journal Name* **Year**, *Article Number*, Page Range.

ISBN 978-3-03943-661-3 (Hbk)
ISBN 978-3-03943-662-0 (PDF)

Cover image courtesy of Hoard's Dairyman.

Contents

About the Editors

Zhijun Cao is Professor and Associate Dean of College of Animal Science & Technology, China Agricultural University (CAU). His bachelor's degree was from Heilongjiang Bayi Agirucltural University and his Ph.D. degree was from CAU. He spent 5 months in South Dakota State University and University of Wisconsin–Madison as an exchange student in 2006, and 15 months in Cornell University as Visiting Professor in 2010–2011. His research program over the last 10 years has been focused on the nutrient requirements of dairy calves and heifers, and the control of energy metabolism in transition dairy cows. As the first author or corresponding author, he has published 62 peer-reviewed articles (11 in Journal of Dairy Science) and is the recipient of several awards, including the National Special Support Program for High-Level Personnel Recruitment, the Beijing NOVA Program Award, and the DBN Award for Young Scientists. He serves as the Committee Member of Nutrient Requirements of Dairy Cattle in China, and Deputy General Secretary of China Cattle Science Association, Chinese Association of Animal Science and Veterinary Medicine. He is also Editor-in-Chief of Hoard's Dairyman China and an editorial board member of *Animals*.

Michael Van Amburgh (Professor, Cornell University) is Professor at the Department of Animal Science, Cornell University. His research is devoted to the discovery and problem solving of contemporary issues in nutrition, physiology, and management in the dairy and livestock industries, and disseminating new knowledge to students and the industry. He is currently leading the development of Cornell's net carbohydrate and protein system (CNCPS). The development and use of CNCPS has greatly improved the efficiency of ruminant nutrition utilization, with improvements for the environment and animal production performance, which has resulted in significant contributions to ruminant research. With the increasing pressure on the dairy industry to reduce the environmental impact of cattle, Van Amburgh and his team are currently working on developing a basic understanding of whole animal nitrogen metabolism and the efficient use of absorbed amino acids. In support of the CNCPS, new tools are also being developed to describe how NDF digestion occurs in various feeds. Van Amburgh is recipient of the Most-Cited Paper Award in 2018 for the Nutrition, Feeding and Calves section of the Journal of Dairy Science. He has won the American Dairy Science Association Stephen H. Weiss Presidential Fellow Award (2016) and the Cornell University Donald C. Burgett Distinguished Advisor Award (2012).

Preface to "Calf and Heifer Feeding and Management"

From birth to first calving, the replacement heifer undergoes tremendous changes anatomically as well as in feeding and management practices. The calf changes from being a pseudo-monogastric to a full ruminant within a period of two months. During the same period, the calf is fed colostrum, milk, or milk replacer, and starter with or without hay. Notably, the lifetime milk production and health of a dairy cow is highly dependent on early life nutrition and management of the calf and, subsequently, the heifer. Hence, animal scientists continue to investigate critical areas such as colostrum feeding, the level of liquid feeding, gut microbial succession, energy and protein levels, housing, health management, and their interactions with the animal in an effort to help dairy producers raise successful and sustainable dairy enterprises. Emerging research techniques have opened new frontiers to better understanding the whole animal and how its diet and environment might influence its microbial, endocrinal, immunity, and metabolic systems. The integration of existing and current knowledge will help refine replacement heifer feeding and management practices. The aim of this Special Issue is to publish current and relevant information related to the nutrition, metabolism, housing, and health of replacement heifers. All papers are open access and can be freely accessed at the Special Issue website: https://www.mdpi.com/journal/animals/special_issues/Calf_and_Heifer.

Zhijun Cao, Michael Van Amburgh
Editors

Review

Review: How Forage Feeding Early in Life Influences the Growth Rate, Ruminal Environment, and the Establishment of Feeding Behavior in Pre-Weaned Calves

Jianxin Xiao [1,†], Gibson Maswayi Alugongo [1,†], Jinghui Li [2], Yajing Wang [1], Shengli Li [1] and Zhijun Cao [3,*]

[1] State Key Laboratory of Animal Nutrition, Beijing Engineering Technology Research Center of Raw Milk Quality and Safety Control, College of Animal Science and Technology, China Agricultural University, Beijing 100193, China; dairyxiao@gmail.com (J.X.); maswayi@yahoo.com (G.M.A.); yajingwang_cau@163.com (Y.W.); lisheng0677@163.com (S.L.)

[2] Department of Animal Science, University of California, Davis, CA 95616, USA; lgreyhui@hotmail.com

[3] State Key Laboratory of Animal Nutrition, College of Animal Science and Technology, China Agricultural University, Beijing 100193, China

* Correspondence: caozhijun@cau.edu.cn; Tel.: +86-10-6273-3746

† These authors contributed equally to this review.

Received: 29 November 2019; Accepted: 19 January 2020; Published: 22 January 2020

Simple Summary: Under natural grazing systems, calves are likely to consume forage in early life. However, forage inclusion in the diet of pre-weaned calves has long been a controversial issue due to it possibly being associated with negative calf performance. Recent published literature seems to confound previous research. This review aims to understand the factors that may influence forage inclusion in the ration of pre-weaned calves. We have explored research related to the effect of feeding forage on rumen and behavioral development to better understand whether forage should be fed to the young calf. Based on the findings, it is concluded that a small amount of good quality forage is recommended for calves to improve their behavioral expression and rumen environment, which may further improve calf performance.

Abstract: The provision of forage to pre-weaned calves has been continuously researched and discussed by scientists, though results associated with calf growth and performance have remained inconsistent. Multiple factors, including forage type, intake level, physical form, and feeding method of both solid and liquid feed, can influence the outcomes of forage inclusion on calf performance. In the current review, we summarized published literature in order to get a comprehensive understanding of how early forage inclusion in diets affects calf growth performance, rumen fermentation, microbiota composition, and the development of feeding behavior. A small amount of good quality forage, such as alfalfa hay, supplemented in the diet, is likely to improve calf feed intake and growth rate. Provision of forage early in life may result in greater chewing (eating and ruminating) activity. Moreover, forage supplementation decreases non-nutritive oral and feed sorting behaviors, which can help to maintain rumen fluid pH and increase the number of cellulolytic bacteria in the rumen. This review argues that forage provision early in life has the potential to affect the rumen environment and the development of feeding behavior in dairy calves. Continued research is required to further understand the long-term effects of forage supplementation in pre-weaned calves, because animal-related factors, such as feed selection and sorting, early in life may persist until later in adult life.

Keywords: calves; forage; performance; rumen fermentation; behavior

1. Introduction

As early as 1897, researchers began to evaluate hay feeding in young calves [1]. Since then, more calf related studies involving various aspects such as genetics, nutrition, health, and welfare have been completed [2]. Likewise, over the last hundred years, the use of forage in pre-weaned calves has remained one of the most key concerns in calf nutrition.

Before the 1950s, forage feeding was generally encouraged in pre-weaned calves, as it was believed to reduce abnormal behavior (e.g., eating bedding material) [3], lower diarrhea [4], and improve rumen development [5,6]. However, new research emerged challenging the fact that forage feeding could improve rumen development to the same degree as calf starter [7]. Volatile fatty acids (VFA) were considered to play a more critical role in stimulating rumen epithelial development rather than the physical form of diets [8]. Specifically, forage ration resulted in a higher proportion of acetate [9], which did not stimulate the growth of rumen papillae to the same extent as butyrate and propionate [10,11]. Concentrates, high in rapidly fermentable carbohydrates, produced more butyrate and propionate [12]. Therefore, higher proportions of concentrates could enhance the development of rumen papillae [13]. Furthermore, as fiber had lower digestibility than starch and sugar, many studies claimed that roughage increased gut fill because of low ruminal fermentation rate, thus curbing the consumption of starter feed which had higher energy density [5,14,15]. Therefore, some dairy farms provided calves with ad libitum access to concentrate feed, with no forage until after weaning [16]. More recently, in the 2000s and 2010s, more studies have investigated the effects of forage feeding in pre-weaned calves, yet results were inconsistent. Some of the studies reported a decrease [15,17], an increase [18–23], or no differences [24–27] in solid dry matter intake (DMI) and average daily gain (ADG) when forage was added in the calf diets. As the solid DMI and growth rate of the pre-weaned calves are important factors that drive rumen development [15] and subsequent milk production in the first lactation [28], it is vital to understand the factors that influence feed consumption and calf growth when forage is added to their diet.

On the other hand, over the last ten years, more research has explored the effects of forage inclusion in diets of young calves not only on calf performance and rumen morphological development but also on ruminal fermentation metabolites, bacterial composition, and feeding behavior [2], giving us a more comprehensive and better understanding of this topic. Therefore, the aims of this review are: (1) to discuss the factors that contribute to the inconsistent results in performance in calves with forage inclusion; (2) to summarize and evaluate the latest literature on the role of forage in rumen fermentation and the establishment of feeding behavior of calves.

2. Factors that Affect Calf Performance with Forage Inclusion

Generally, under natural grazing systems, adult dairy cows spend 7 to 13 h eating grass every day [29]. Young calves acquire nutrients from both milk and fresh grasses and begin to graze as early as week 2 of age [30,31]. The grazing time usually lasts for a short period, around 20 min at 10 days, which increases rapidly to 360 min at 100 days of age, equivalent to 70 percent of the grazing time in adult cattle [30,32].

On some commercial dairy farms, calves are offered free access to starter feed before weaning without forage [33], which is contrary to natural grazing. Forage inclusion in the pre-weaned calf diet has long been discouraged due to its negative effect on the growth rate [15,17]. However, recent research has shown that several factors need to be taken into account when evaluating the impact of forage provision on calf performance [8,9,34]. These factors include the source, amount, particle size, physical form, offering time, and feeding method of forage and concentrate, as well as the amount of milk offered and milk feeding method. A summary of these studies is presented in Tables 1–6.

2.1. Forage Factors

Forage provision in pre-weaned calf remains a controversial topic, hence the proliferation of related research. In earlier studies, it was believed that forage was the main component in calf diets that played an essential role in rumen development [5,6]. In two different studies offering a high proportion of roughage (80% vs. 60% vs. 40% and 67% vs. 50%) to calves as hay to grain respectively, Hibbs et al. [5] and Conard and Hibbs [35] found that DMI and ADG increased as the proportion of concentrate in the ration increased. Stobo et al. [36] obtained similar results with calves provided a maximum daily allowance of concentrate at 0.45, 0.91, 1.36, 1.81, or 2.27 kg/d along with ad libitum access to grass hay (crude protein (CP) = 7.5%, crude fiber (CF) = 28.7%). These authors reported that as the concentrate intake increased, the hay intake decreased linearly [5,36], while live and empty body weight (BW) and rumen development were improved [36]. This is probably because concentrates result in more VFA production, especially propionate and butyrate [12], which enhances papillae development [13,36]. Collectively, these early studies suggested that in a high roughage feeding system, the addition of a bulky forage in the diet might decrease the consumption of energy-dense concentrates, leading to less rumen fermentation and lower degradation rates, and subsequently lower total nutrient intake and calf growth.

In the early 2000s, several studies began to investigate the effect of lower levels of forage inclusion in the diet on pre-weaned calf performance. Most of them included a proportion of forage ranging between 5 and 25% of total solid feed intake [15,19,22,27,37–40]. In contrast to previous studies, most studies either reported an increase [19,22,38,40] or a lack of differences [27,37] in DMI, ADG, and BW, indicative of multifactorial effects in these studies.

Table 1. A summary of studies feeding different levels of forage in pre-weaned dairy calves.

Objectives Forage (%)	Trt [1]	Calf/Trt	Weaning Age (d)	Forage Feeding Age (d)	Forage Source	Cutting Length/ Processing [2]	Solid Feed Offering Method	Concentrate Physical Form	Amount of Milk Fed [3]	Outcomes [4] DMI	Outcomes [4] ADG	Reference
0, 40, 60, 80%	3	7	49	4	Alfalfa and Timothy hay	-	TMR	-	-	N **	N **	*Hibbs* et al., 1956 [5]
50, 67%	4	10	49	3	Grass legume silage	-	TMR	Coarse	-	-	-	*Conard* et al., 1956 [35]
5 to 60%	10	4	-	56	Barley, Rye, Wheat straw	-	-	-	-	-	-	*Jahn* et al., 1970 [41]
20–70%	2	6	-	7	Alfalfa hay	-	TMR	—	-	-	-	*Žitnan* et al., 1998 [42]
0, 7.5, 15%	4	16	31	-	Bromegrass hay	Chopped 8 to 19 mm	TMR	Coarse, Ground	-	*p* *	*p* *	*Coverdale* et al., 2004 [19]
0, 30, 60%	8	8	70	10	Corn silage, Straw, Grass, Corn silage	-	TMR	Pellet starter	-	N **	NS	*Suárez* et al., 2007 [43]
0, 16%	4	16	28	3	Beet pulp	-	TMR	Pellet	80 L (Around 7%)	NS	NS	*Porter* et al., 2007 [37]
0, 5%	2	24	31–32	3–4	Cottonseed hull	GMPL: around 2 mm	TMR	Texture	100 L (Around 10%)	*p* *	NS	*Hill* et al., 2008 (*Trail 1*) [15]
0, 5, 10%	4	12	28	3–4	Cottonseed hull, Timothy hay	GMPL: around 2.2 mm	TMR	Texture	120 L (Around 10%)	N **	N **	*Hill* et al., 2008 (*Trail 2*) [15]
0, 2.5, 5%	3	16	28	3–4	Timothy hay	GMPL: around 2.2 mm	Free Choice	Texture	120 L (Around 10%)	N **	N **	*Hill* et al., 2008 (*Trail 3*) [15]
0, 5, 10%	6	7	53	3	Alfalfa hay	GMPL: 2.6 mm	TMR	Finely ground	Around 10%	*p* **	*p* **	*Beiranvand* et al., 2014 [22]
8, 16%	5	10	51	16	Alfalfa hay	GMPL: 2.92 vs. 5.04 mm	TMR	Ground	190 L (Around 10%)	-	-	*Mirzaei* et al., 2015 [27]
0, 12.5, 25%	4	15	51	3	Alfalfa hay	GMPL: 3 mm	TMR	Finely ground	204 L (Around 10%)	*p* **	*p* **	*Nemati* et al., 2016 [38]
0, 75%, 100%	3	15	56	1	Corn silage	-	TMR	Texture	416 L (Around 18%)	NS	NS	*Kehoe* et al., 2019 [44]

[1] Trt = Treatment. [2] Geometric mean particle length (GMPL) refers to geometric mean particle length, which was determined using American Society of Agricultural and Biological Engineers (ASABE) forage sieve methods (ANSI/ASAE S424.1) [45]. [3] Amount of milk fed is the total amount fed during the whole trial period, while milk feeding rate equals the average daily milk allowance/birth body weight (%). [4] Outcomes = effect of forage inclusion, ** indicates a significant effect ($p < 0.05$), * indicates a tendency ($p < 0.1$), N indicates a negative effect, *p* indicates positive effect, and NS shows no significant effect. DMI and ADG were evaluated by comparing calves fed with or without forage.

2.1.1. Forage Level and Source

Coverdale et al. [19] conducted two experiments in which starter supplemented with relatively low level (7.5 and 15%) of bromegrass hay appeared to improve DMI, ADG, and feed efficiency (FE). In experiment 1, limited amounts of mixed feed (concentrate and forage) were offered before weaning, followed by ad libitum feeding post-weaning. Calves receiving coarse starter with either 7.5 and 15% of bromegrass hay (8 to 19 mm) were heavier and had greater ADG and FE than calves receiving only coarse starter, while calves fed 7.5% of hay tended to have the highest ADG and FE [19]. In experiment 2, calves were offered diets ad libitum and weaned according to intake. The concentrate and total DMI tended to be higher in calves fed 7.5 and 15% of bromegrass hay when compared with the non-forage group [19]. Similarly, Hosseini et al. [40] recently reported that compared to non or 15% straw, feeding 7.5% of chopped wheat straw tended to improve the overall total solid feed intake (659, 685, and 826 g/d, respectively) and ADG (519, 553, and 620 g/d, respectively) when calves were offered 4 L of whole milk per day. Feeding alfalfa hay at 10% of total solid feed increased the overall DMI, ADG, and final BW, and thereby shortened the time to weaning at a target DMI of starter (1 kg for 3 consecutive days), compared with feeding 0 or 5% of alfalfa hay [22]. Nemati et al. [38] also observed a linear increase in total DMI and ADG during the postweaning (d 52 to 70) and overall periods (d 3 to 70) of calves supplemented with chopped alfalfa hay at 0, 12.5, and 25% on dry matter (DM) basis. However, gut fill could be a confounding factor when evaluating the effect of forage feeding on improving ADG in dairy calves. It is commonly believed that an increase in ADG and BW in calves fed forage could be due to greater gut fill [15,20,27]. Therefore, there is a need for further investigation of the relationship between gut fill and ADG.

Poor performance has also been commonly observed when including forage in the diet [15,17]. Hill et al. [15] reported that feeding either 2.5 and 5% of chopped timothy hay linearly reduced starter intake, ADG, empty body weight ADG, and FE. The quality of forage significantly influences the digestibility and the palatability of the diet [46]. Ülger et al. [47] compared two calf total mixed ration (TMR) diets with either 20% of a high-quality alfalfa hay (CP = 18.1%, acid detergent fiber (ADF) = 36.1% and neutral detergent fiber (NDF) = 44.4% on DM basis, relative feed value (RFV) = 127.2) or lower quality wheat straw (CP = 3.7%, ADF = 52.4% and NDF = 80.1% on DM basis, RFV = 55.9) and found that the high-quality roughage improved FE and numerically increased ADG during the preweaning period. In a more recent study, Hill et al. [48] found that moderate to low-quality grass hay (5.4% CP and 62.8% NDF on DM basis) reduced the digestibility of DM, OM, and CP in young calves consuming a textured starter. It is noteworthy that the type of hay was not specified in this study, and we speculated that timothy or mixed hay was included based on other studies at the same period by the authors [49]. On the contrary, Castells et al. [21] and Hosseini et al. [40] reported that a low-quality straw (CP = 4.2%, and NDF = 74.0% on DM basis) could also improve DMI and ADG. The inconsistency in results on calf performance when providing low-quality hay was likely due to the different amount of milk offered, which may affect the solid feed consumption and forage preference as discussed below.

Table 2. A summary of studies feeding different forage sources in pre-weaned dairy calves.

Objectives Forage Source	Trt [1]	Calf/Trt	Weaning Age (d)	Forage Feeding Age (d)	Forage (%)	Cutting Length/ Processing [2]	Solid Feed Offering Method	Concentrate Physical Form	Amount of Milk Fed [3]	Outcomes [4] DMI	ADG	Reference
Alfalfa hay, Cottonseed	3	24	-	7	25% of Cottonseed, Ad libitum of alfalfa hay	Chopped to 10 cm	Free choice	-	Around 7%	NS	NS	Anderson et al., 1982 [50]
Beet pulp, Soybean hulls, Corn grits	5	32	84	-	0, 30.3, 46.4, 91.3%	-	TMR	Pellet	608 L (around 18%)	p **	p **	Suárez et al., 2006 [51]
Straw; Corn silage, Dried grass	8	8	70	10	-	Chopped	TMR	Pellet starter;	-	N **	NS	Suárez et al., 2007 [43]
Beet pulp	4	16	28	3	0, 16%	-	TMR	Pellet	80 L (Around 7%)	NS	NS	Porter et al., 2007 [37]
Cottonseed hull	2	24	31–32	3–4	0, 5%	GMPL: around 2 mm	TMR	Texture	100 L (Around 10%)	p *	NS	Hill et al., 2008 (Trail 1) [15]
Cottonseed hull, Timothy hay	4	12	28	3–4	0, 5, 10%	GMPL: around 2.2 mm	TMR	Texture	120 L (Around 10%)	N **	N **	Hill et al., 2008 (Trail 2) [15]
Timothy hay	3	16	28	3–4	0, 2.5, 5%	GMPL: around 2.2 mm	Free Choice	Texture	120 L (Around 10%)	N **	N **	Hill et al., 2008 (Trail 3) [15]
Alfalfa hay	6	7	53	3	0, 5, 10%	GMPL: 2.6 mm	TMR	Finely ground	Around 10%	p **	p **	Beiranvand et al., 2014 [22]
Alfalfa hay	4	15	51	3	0, 12.5, 25%	GMPL: 3 mm	TMR	Finely ground	204 L (Around 10%)	p **	p **	Nemati et al., 2016 [38]
Alfalfa hay, Ryegrass hay	3	20	57	14.1 ± 4.2	Ad libitum	Chopped	Free choice	Pellet	214 L (Around 9.6%)	NS	NS	Castells et al., 2012 (Trail 1) [21]
Oat hay, Barley straw,	3	20	57	14.1 ± 4.2	Ad libitum	Chopped	Free choice	Pellet	214 L (Around 9.6%)	p **	p **	Castells et al., 2012 (Trail 2) [21]
Triticale silage, Corn silage	3	20	57	14.1 ± 4.2	Ad libitum	Chopped	Free choice	Pellet	214 L (Around 9.6%)	p **	p **	Castells et al., 2012 (Trail 3) [21]

Table 2. *Cont.*

| Objectives | | | | | | | | | Outcomes [4] | | |
Forage Source	Trt [1]	Calf/Trt	Weaning Age (d)	Forage Feeding Age (d)	Forage (%)	Cutting Length/ Processing [2]	Solid Feed Offering Method	Concentrate Physical Form	Amount of Milk Fed [3]	DMI	ADG	Reference
Alfalfa hay; Oat hay	3	5	56	3	Ad libitum	Chopped	Free choice	Pellet	214 L (Around 10%)	NS	NS	*Castells* et al., 2013 [52]
Oat Hay	4	16	51	9 ± 4.4	Ad libitum (4, 5%)	Chopped	Free choice	Pellet	152 L (Around 10%)	p **	p **	*Terré* et al., 2013 [53]
Orchard hay, Timothy hay	2	8	56	42	0, 20%	-	-	-	88 L (Around 4%)	NS	-	*Kim* et al., 2016 [54]
Wheat straw, Alfalfa hay	2	15	56	14	20%	Chopped to 1–2 cm	TMR	Pellet	212 L (Around 10%)	-	-	*Ülger* et al., 2017 [47]
Alfalfa hay, Beet pulp	2	13	50	4	0, 10, 20%	-	TMR	Texture	228 L (Around 11%)	p *	p **	*Maktabi* et al., 2016 [39]
Alfalfa hay, Corn silage	6	10	49	3	0, 15%	GMPL: 2.9 and 12.07 mm	TMR	Fine ground	196 L (Around 10%)	p **	p **	*Mirzaei* et al., 2017 [55]
Fresh Ryegrass	4	6	49	7–10	Ad libitum	Chopped to approximately 4 cm	Free choice	Pellet	-	NS	NS	*Phillips* et al., 2004 [56]
Coastal Bermuda grass hay	2	16	56	17 ± 3	Ad libitum (15%)	Chopped to 5 cm	Free choice	Pellet	426 L (Around 19%)	NS	NS	*Horvath* et al., 2019 [57]
Grass hay	4	4	42	3	Ad libitum	Long (without details)	Free choice	Texture	182 L (Around 10%)	NS	NS	*Hill* et al., 2019a [48]
Corn silage, Reconstituted alfalfa, Reconstituted beet pulp	3	18	49	3	10%	GMPL: alfalfa, 5 mm and corn silage, 12–15 mm	TMR	Ground	283 L (Around 14.5%)	-	-	*Kagar* et al., 2019 [58,59]

[1] Trt = Treatment. [2] GMPL refers to geometric mean particle length, which was determined using ASABE forage sieve methods (ANSI/ASAE S424.1) [45]. [3] Amount of milk fed is the total amount fed during the whole trial period, while milk feeding rate equals the average daily milk allowance/birth body weight (%). [4] Outcomes = effect of forage inclusion, ** indicates a significant effect ($p < 0.05$), * indicates a tendency ($p < 0.1$), N indicates a negative effect, p indicates positive effect, and NS shows no significant effect. DMI and ADG were evaluated by comparing calves fed with or without forage.

Castells et al. [21] evaluated ad libitum provision of different types of forages in the diets from 2 weeks of age and found that the inclusion of alfalfa hay (CP = 16.6%, ADF = 30.2%, and NDF = 40.2% on DM basis), rye-grass hay (CP = 6.8%, ADF = 35.1%, and NDF = 59.3% on DM basis), oat hay (CP = 8.4%, ADF = 31.8%, and NDF = 59.6% on DM basis), barley straw (CP = 4.2%, ADF = 42.5%, and NDF = 74.0% on DM basis), corn silage (CP = 8.6%, ADF = 25.2%, and NDF = 41.9% on DM basis), or triticale silage (CP = 7.5%, ADF = 42.3%, and NDF = 64.7% on DM basis) resulted in similar or increased intake and gains without impairing FE and nutrient digestibility. Increased DMI and ADG were observed when oat hay, barley straw, or triticale silage were offered, and the inclusion of alfalfa hay did not exhibit similar benefits, probably due to the preference for this high-quality and palatable forage [21]. Indeed, forage intake was highest in calves fed alfalfa hay (14% of total solid DM) compared with the other treatments (oat hay: 8% rye-grass hay: 4%, barley straw: 5%, corn silage: 5%, and triticale silage: 4%). The higher proportion of forage to concentrate ratio may limit the DM digestibility and hence restrict the DMI and ADG, as we have already discussed [5,35]. More recently, a meta-analysis by Imani et al. [34] evaluating the effect of forage provision on growth performance of dairy calves using 27 published studies from 1998 to 2016 revealed that concentrate DMI was higher in calves offered alfalfa hay compared with those offered other types of forages.

2.1.2. Forage Physical Forms and Processing

The focus is not only on the forage source and level of feeding, but also on the physical form and processing of forage. Mirzaei et al. [27] evaluated the effects of particle size (short at 2.92 mm vs. long at 5.04 mm as geometrical means) of alfalfa hay on growth performance of dairy calves at two different inclusion rates (low at 8% vs. high at 16% on DM basis). The authors observed no differences in growth rates between calves fed with or without hay, but greater DMI and weaning BW were found in calves fed low levels of alfalfa with a long particle size (8% and 5.04 mm) and high levels with a short particle size (16% and 2.92 mm) compared with calves fed low levels with a short particle size (8% and 2.92 mm) and high levels with a long particle size (16% and 5.04 mm). The short particle size at a low level of alfalfa might not have the potential to increase the capacity, motility, and development of the rumen [11,60], while the negative effect on performance by the long particle size at a high level might be attributed to the lower digestibility rate of long particles compared with short particles [61]. Montoro et al. [62] also found that when calves were supplemented with 10% of long chopped (3 to 4 cm) ryegrass hay, DMI, ADG, and FE were greater than those fed 10% finely ground (2 mm) grass hay. Longer particle size improved performance [8], probably because of the increased rumination time of calves, which increased saliva production, and consequently improved buffering effect on the ruminal environment [63,64]. However, inconsistent results have been obtained by Omidi-Mirzaei [65] and Suárez-Mena [66]. Omidi-Mirzaei et al. [65] reported that when calves were fed forage with different particle size (alfalfa hay: short = 1.96 mm or long = 3.93 mm; and wheat straw: short = 2.03 mm or long = 4.10 mm as geometric mean), rumination time increased in calves fed forage with long particle size, but concentrate DMI, ADG, and FE were not affected. Suárez-Mena et al. [66] compared four different particle sizes (0.82, 3.04, 7.10, and 12.7 mm as geometric mean) of low-quality forage (5% straw) mixed in the diet and observed no effect on DMI, growth performance, and minimal changes in rumen fermentation and pH among treatments. In summary, these results imply that interactions may exist among forage source, level, and particle size, and the optimal inclusion level of forage should be determined based on the forage source and particle size.

In recent years, attempts have been made on alternative ways to increase solid feed consumption of dairy calves, such as using non-forage fiber [39], silage based feed [55,67], moisturized starter [68,69], and reconstituted hay [58,59,70]. Beet pulp is a common source included in the diet as a non-effective fiber source. Maktabi et al. [39] observed that 10% of beet pulp in the diet improved DMI and ADG compared to a control group (no fiber inclusion), but growth was not enhanced when 20% of beet pulp was used. Inclusion of corn silage early in life of dairy calves has recently gained more interest [55,67]. In an experiment that compared supplementing 15 against 0% of corn silage, DMI,

ADG, and BW increased probably because of the higher moisture content of corn silage that contributed to reduced dustiness and increased palatability of the feed [55,67]. However, feeding a high level of corn silage (30 or 60%; 75 or 100%) offered no benefits compared with feeding concentrate alone [43,44]. Suárez et al. [43] reported that substitution of 30 or 60% of the concentrate by corn silage did not affect DMI and ADG but feeding 30% of straw reduced DMI. Kehoe et al. [44] also found no differences in DMI and ADG when including 0, 75, or 100% of corn silage in pre-weaned calf diet but fed solely corn silage diet stunted the growth of rumen papillae and tended to impair intestinal morphology. Hence, it is possible that corn silage can be used to partly replace the concentrate with little harmful effects on the growth and development of the calf.

It has been documented that moisturizing the concentrate starter feed by adding water to change the DM from 90 to 50% increased DMI, ADG, and VFA production in dairy calves [68,69]. More recently, hay processing by reconstituting with water was evaluated in a series of studies published by Kargar et al. [58,59,70,71]. Hay was soaked in water for 24 h and mixed every 6 h to obtain a theoretical DM content of 20% [71]. This method has been used previously in the diets of mature cows to increase fecal consistency [72] and reduce digestion lag time in the rumen as a result of a higher fiber digestibility [73]. Kargar et al. [70,71] replaced dry alfalfa hay (10%) with a similar amount of reconstituted alfalfa hay, resulting in similar DMI and ADG, but higher NDF digestibility during the pre-weaned period. Furthermore, a greater improvement in health status (fecal score and general appearance score) was obtained with reconstituted hay, possibly due to decreased dustiness, similar to corn silage [70]. Therefore, corn silage, reconstituted hay, and beet pulp can be used interchangeably in dairy calf diets based on availability and the relative feed price [58,59].

Table 3. A summary of studies related to forage particle size and methods of processing forage in pre-weaned dairy calves.

| Objectives | | | | | | | | | Outcomes [4] | | | |
Cutting Length/ Processing [1]	Trt [2]	Calf/Trt	Weaning Age (d)	Forage Feeding Age (d)	Forage Source	Forage (%)	Solid Feed Offering Method	Concentrate Physical Form	Amount of Milk Fed [3]	DMI	ADG	Reference
Chopped (GMPL, 5.4 mm) vs. Pelleted (GMPL, 5.8 mm)	3	11	76	3	Alfalfa hay	0, 10%	TMR	Semi-texture	500 L (17%)	NS	NS	*Jahani-Moghadam* et al., 2015 [26]
GMPL: 2.92 vs. 5.04 mm	5	10	51	16	Alfalfa hay	8, 16%	TMR	Ground	190 L (Around 10%)	-	-	*Mirzaei* et al., 2015 [27]
GMPL: 0.82, 3.04, 7.0, 12.7 mm	4	10	56	1	Straw	5%	TMR	Pellet	Around 12%	-	-	*Suárez -Mena* et al., 2015 [66]
GMPL: Alfalfa (1.96 and 3.93 mm) vs. Wheat straw (2.03 and 4.10 mm)	4	12	49	1	Alfalfa; Wheat straw	Ad libitum	Free choice	Texture	279 L (Around 14%)	-	-	*Omidi-Mirzaei* et al., 2018 [65]
Chopped to 20 to 40 cm	2	24	42	2–3	Timothy hay (*Phleum pratense*)	Ad libitum	Free choice	Texture	178 L (Around 10%)	NS	NS	*Hill* et al., 2019b [49]
Chopped 3 to 4 cm vs. ground to 2 mm	2	10	49	5	Ryegrass hay	10%	TMR	Crumb	200 L (Around 9%)	-	-	*Montoro* et al., 2013 [62]
Non-forage fiber	4	16	28	3	Beet pulp	0, 16%	TMR	Pellet, Mash	80 L (Around 7%)	NS	NS	*Porter* et al., 2007 [37]
Non-forage fiber	2	13	50	4	Alfalfa hay, Beet pulp	0, 10, 20%	TMR	Texture	228 L (Around 11%)	*p* *	*p* **	*Maktabi* et al., 2016 [39]
Silage based feed	6	10	49	3	Alfalfa hay, Corn silage;	0, 15%	TMR	Fine ground	196 L (Around 10%)	*p* **	*p* **	*Mirzaei* et al., 2017 [55]
Silage based feed	4	12	56	3	Corn silage	0, 15%	TMR	Mash, Texture	291 L (Around 13%)	*p* **	*p* **	*Mirzaei* et al., 2016 [67]

Table 3. *Cont.*

| Objectives | | | | | | | | | | Outcomes [4] | | |
Cutting Length/ Processing [1]	Trt [2]	Calf/Trt	Weaning Age (d)	Forage Feeding Age (d)	Forage Source	Forage (%)	Solid Feed Offering Method	Concentrate Physical Form	Amount of Milk Fed [3]	DMI	ADG	Reference
Silage based feed	3	15	56	1	Corn silage	0, 75%, 100%	TMR	Texture	416 L (Around 18%)	NS	NS	*Dill-McFarland* et al., 2019 [74]
Silage based feed	3	15	56	1	Corn silage	0, 75%, 100%	TMR	Texture	416 L (Around 18%)	NS	NS	*Kehoe* et al., 2019 [44]
Reconstituted hay	3	18	49	3	Corn silage, Reconstituted alfalfa, Reconstituted beet pulp	10%	TMR	Ground	283 L (Around 14.5%)	-	-	*Kagar* et al., 2019 [58,59]

[1] GMPL refers to geometric mean particle length, which was determined using ASABE forage sieve methods (ANSI/ASAE S424.1) [45]. [2] Trt = Treatment. [3] Amount of milk fed is the total amount fed during the whole trial period, while milk feeding rate equals the average daily milk allowance/birth body weight (%). [4] Outcomes = effect of forage inclusion, ** indicates a significant effect ($p < 0.05$), * indicates a tendency ($p < 0.1$), N indicates a negative effect, p indicates a positive effect, and NS shows no significant effect. DMI and ADG were evaluated by comparing calves fed with or without forage.

2.1.3. Time and Method of Offering Forage

Time [24,75,76] and method [25,77,78] of offering forage are essential factors that can influence how dairy calves utilize forage. While Wu et al. [24] found no differences in DMI, ADG, and rumen development in calves fed alfalfa hay or oat hay either at day 3 or 15 of age, different outcomes were shown in two other studies, investigating the effect of age at which alfalfa hay [75] and oat hay [76] were introduced to calves. Both studies observed improved DMI and growth performance with forage provision, and the greatest growth performance and rumen development were obtained in calves offered hay from the 2nd week rather than the 4th or 6th week of age [75,76]. Based on these studies, suggestions may be made to include alfalfa or oat hay in diets of calves as early as week 2 or even right after birth in order to improve DMI and ADG in dairy calves [24,75].

Forage have been provided as a mixture with concentrate in previous studies, while recent studies investigated calf preference to different feeds by providing forage and concentrate separately. Castell et al. [21] observed a greater DMI and ADG in calves provided forage rather than without forage. In this study, calves consumed around 5% of forage when it was offered ad libitum and separately from the concentrate. However, several studies did not observe a positive effect on DMI when part of the concentrate was substituted for forage before weaning [43,47]. In research by Ülger et al. [47], who mixed forage at 20% with the concentrate and Suárez et al. [43] at either 30 or 60%, calves consumed at a predetermined fixed forage to concentrate ratio which was much greater compared to calves fed free choice. Therefore, the greater forage proportion in the mixture diet (i.e., containing 30 or 60% forage) [43] might mask the positive effect on DMI and growth performance compared with forage consumed voluntarily (around 5% forage) by calves [21]. A meta-analysis study has proven this inference; when forage was offered separately, starter feed intake and ADG increased compared to a mixed ration diet [34]. On the contrary, some studies claimed that DMI and growth performance were not different between two feeding methods (mixed vs. separate) [77,78], possibly due to the low proportion of forage (15 or 10%) in mixed ration, which was similar to that (11 or 10%) with separate forage provision in studies by Overvest [77] and EbnAli [78]. Moreover, the high level of milk feeding (around 13 or 26% of birth weight) might have decreased the solid feed consumption [77,78]. Although the forage feeding method may not always lead to better performance, it certainly affects the expression and development of dairy calf behavior [25,79], as discussed below.

Table 4. A summary of studies differing in methods and time of offering forage in pre-weaned dairy calves.

| Objectives | | | | | | | | | Outcomes [4] | | | |
Forage Feeding Method/Time	Trt [1]	Calf/Trt	Weaning Age (d)	Forage Feeding Age (d)	Forage Source	Forage (%)	Cutting Length/ Processing [2]	Concentrate Physical Form	Amount of Milk Fed [3]	DMI	ADG	Reference
Sole vs. TMR Free choice-	4	12	50	1	Grass hay	TMR (0, 15%), Free choice	Chopped <2.5 cm	Texture	534 L (Around 26%)	NS	NS	*Overvest* et al., 2015 [77]
Sole vs. TMR, Free choice	3	15	57	3	Alfalfa hay	TMR (0, 10%), Free choice	GMPL: 3 mm	Finely ground	262 L (Around 11%)	*p* **	NS	*EbnAli* et al., 2016 [78]
Sole vs. Free choice	3	60	56	4	Alfalfa hay, Oats hay	Free choice	Chopped to approximately 2.5 cm	Pellet	376 L (Around 17%)	NS	NS	*Xiao* et al., 2018 [25]
Day 14, 28, 42	4	10	57	-	Alfalfa hay	TMR (0, 15%)	GMPL: 3 mm	Ground	Around 10%	*p* **	*p* **	*Hosseini* et al., 2015 [75]
Day 3, 15	5	8	56	-	Alfalfa hay, Oats hay	Free choice	Chopped	Pellet	358 L (Around 16%)	NS	NS	*Wu* et al., 2018 [24]
Day 14, 42	3	6	63	-	Oat hay	Free choice	-	-	252 L (Around 10%)	*p* **	*p* **	*Lin* et al., 2018 [60,76]

[1] Trt = Treatment. [2] GMPL refers to geometric mean particle length, which was determined using ASABE forage sieve methods (ANSI/ASAE S424.1) [45]. [3] Amount of milk fed is the total amount fed during the whole trial period, while milk feeding rate equals the average daily milk allowance/birth body weight (%). [4] Outcomes = effect of forage inclusion, ** indicates a significant effect ($p < 0.05$), N indicates a negative effect, p indicates positive effect, and NS shows no significant effect. DMI and ADG were evaluated by comparing calves fed with or without forage.

2.2. Concentrate and Milk Factors

2.2.1. The Physical Form of Concentrate Feed

There are different forms and types of calf starters. Porter et al. [37] reported that whether forage was included or not, calves on coarse mash (average particle size: 2014 µm) ate and gained more than those on pelleted diets (average particle size: 888 µm). Moreover, rumination was initiated earlier. Hence, up to 8 weeks of age, calves raised on a complete concentrate diet without forage did not experience a significant depression in growth performance, which might be due to the long particle size of coarse mash feed initiating rumination early and preventing bloat and parakeratosis in the rumen [65]. Two consecutive experiments were conducted by Terré et al. in 2005 to evaluate the influence of the physical form of concentrate feed (textured or pelleted) with or without forage inclusion on the performance of young calves. Calves receiving pelleted concentrate feed with straw exhibited a greater solid intake and higher rumen fluid pH compared with those receiving a pelleted concentrate feed without straw. Calves that received the texturized (containing whole corn) starter feed had equivalent rumen fluid pH to those fed a pelleted concentrate with straw. However, rumen fluid pH and performance were not improved when another texturized (containing rolled mixed grains) concentrate feed was offered [80]. These results show that the physical forms of concentrate feed may affect the calf performance and rumen environment differently. For example, calves fed a texturized concentrate feed containing whole corn had a greater rumen fluid pH than steam-flaked corn, dried-rolled corn, and roasted-rolled corn [81], likely because the calves spent a longer time chewing the whole corn feed, which increased saliva production, hence neutralizing the rumen pH and acids. In another study involving a mashed (with or without corn silage) and a textured concentrate (with or without corn silage), regardless of the physical form of concentrate feed, forage inclusion resulted in greater DMI, ADG, and final BW than non-included calves [67]. It was evident that forage provision had more effect on the growth performance than the physical form of the feed, whereby steam-flaked grains were the main component in the concentrate. In agreement, Mojahedi et al. [82] reported that including alfalfa hay could improve DMI and ADG of calves fed steam-flaked corn, as opposed to a cracked corn-based diet, probably because of higher amounts of gelatinized starch in the steam-flaked corn (44.1 vs. 12.5% of total starch, respectively). Possibly, forage inclusion enhanced starch fermentability of the steam-flaked corn through the provision of effective fiber. Collectively, a decrease in solid feed consumption in calves fed finely ground or pelleted starter on commercial farms compared with those fed textured concentrate [83] warrants forage provision to improve the solid feed intake, growth performance, and rumen environment to a greater extent [80,82].

Table 5. A summary of studies on forage inclusion in dairy calves based on different physical forms of concentrate.

| Objectives | | | | | | | | | | Outcomes [4] | | |
Physical Form of Concentrate	Trt [1]	Calf/Trt	Weaning Age (d)	Forage Feeding Age (d)	Forage Source	Forage (%)	Forage Cutting Length/ Processing [2]	Solid Feed Offering Method	Amount of Milk Fed [3]	DMI	ADG	Reference
Pellet, Mash	4	16	28	3	Beet pulp	0, 16%	-	TMR	80 L (Around 7%)	NS	NS	Porter et al., 2007 [37]
Pellet, Texture	3	11	49	7	Ryegrass hay	Ad libitum (0, 6.8, 11.6%)	-	Free choice	274 L (Around 16%)	NS	NS	Terré et al., 2015 (Trail 1) [80]
Pellet, Texture	3	20	52	8	Ryegrass hay	Ad libitum (0, 4.3%)	-	Free choice	233 L (Around 13%)	NS	NS	Terré et al., 2015 (Trail 2) [80]
Mash, Texture	4	12	56	3	Corn silage	0, 15%	GMPL: 0.5, 1.1, 3.0, and 4.0 mm	TMR	291 L (Around 13%)	*p* **	*p* **	Mirzaei et al., 2016 [67]

[1] Trt = Treatment. [2] GMPL refers to geometric mean particle length, which was determined using ASABE forage sieve methods (ANSI/ASAE S424.1) [45]. [3] Amount of milk fed is the total amount fed during the whole trial period, while milk feeding rate equals the average daily milk allowance/birth body weight (%). [4] Outcomes = effect of forage inclusion, ** indicates a significant effect ($p < 0.05$), N indicates a negative effect, p indicates positive effect, and NS shows no significant effect. DMI and ADG were evaluated by comparing calves fed with or without forage.

Table 6. A summary of studies on forage inclusion in dairy calves based on different volumes and methods of milk feeding.

Objectives										Outcomes [4]		
Milk Feeding Amount/ Method [1]	Trt [2]	Calf/Trt	Weaning Age (d)	Forage Feeding Age (d)	Forage Source	Forage (%)	Cutting Length/ Processing [3]	Solid Feed Offering Method	Concentrate Physical Form	DMI	ADG	Reference
359 L (Around 16%), 221 L (Around 10%)	4	8	56	1	Beet pulp	Beet pulp (0%, 18%)	Chopped	Free choice	Pellet	NS	N *	Kosiorowska et al., 2010 [84]
350 L (Around 20%)	2	15	56	3	Orchard grass hay	Ad libitum	Chopped	-	-	NS	NS	Khan et al., 2011 [20]
534 L (Around 26%)	4	12	50	1	Grass hay	TMR (0, 15%), Free choice-ad libitum	Chopped < 2.5 cm	Free choice; TMR	Texture	NS	NS	Overvest et al., 2015 [77]
212 L (Around 9%), 338 L (Around 15%)	6	10	56	4	Wheat straw	0, 7.5, 15%	-	TMR	Ground	p *	p *	Hosseini et al., 2019 [40]
Step Down vs. Conventional; 313 L (Around 13%,)	4	20	60	3	Alfalfa hay	0, 15%	-	TMR	Finely ground	p **	p **	Daneshvar et al., 2015 [23]
Teat vs. Bucket; 241 L (Around 13%)	3	10	45 ± 2	1–3	Timothy hay	Ad libitum	Chopped around 5 cm	-	Pellet	NS	NS	Horvath et al., 2017 [85]

[1] The amount of milk fed is the total amount fed during the whole trial period, while the milk feeding rate equals the average daily milk allowance/birth body weight (%). [2] Trt = Treatment. [3] GMPL refers to geometric mean particle length, which was determined using ASABE forage sieve methods (ANSI/ASAE S424.1) [45]. [4] Outcomes = effect of forage inclusion, ** indicates a significant effect ($p < 0.05$), * indicates a tendency ($p < 0.1$), N indicates a negative effect, p indicates positive effect, and NS shows no significant effect. DMI and ADG were evaluated by comparing calves fed with or without forage.

2.2.2. The Amount and Method of Milk Feeding

Most studies that suggested exclusive concentrate feeding were conducted with calves fed low amounts of milk [15,36]. For example, in the Hill et al. [15] study, only 120 L of milk was fed to calves before 28 days of life (weaning date), averaging around 4 L/d of milk (approximately 10% of birth body weight) which is insufficient for optimal growth. This low milk feeding rate might stimulate greater concentrate intake in calves to make up for the deficit in energy requirements. Indeed, a strong negative correlation between liquid and concentrate feed intake has been elucidated in a meta-analysis that shows calves fed high milk or milk replacer resulted in limited daily starter intakes [28].

As discussed earlier, compared to concentrates, forage are bulkier and are less digestible and have lower fermentation rates [5], which can lead to a low voluntary intake when low energy forage is offered separately or as a mixture with the concentrates [8,86]. Castells et al. [21] claimed that when calves were offered different forages (hay, straw, and silages, respectively) with concentrates ad libitum and separately, forage consumption was only 4–6% of the total solid feed intake. Interestingly, the proportion of hay consumed across studies seems to range from 3 to 45% of total solid feed intake [20,21,25,52,87]. The difference in the proportion of forage consumed across multiple studies may also depend upon milk feeding amounts. In two studies feeding different amounts of milk, Castells et al. [52] reported that calves consumed 3% of total solid feed as forage at a low level of milk feeding (214 L from d 0 to d 57, averaged 4 L/d, around 10% of birth body weight), while Xiao et al. [52] reported a greater ratio of forage to total solid feed intake, approximately 45% when a high amount of milk was offered (376 L from d 0 to d 56, averaged 6.8 L/d, around 17% of birth body weight) [25]. Milk contains a high content of fat and sugar, which provide the energy required by the calves, and greater milk amounts might alter concentrate requirements [25]. This speculation concurs with another study in which calves fed low amounts of milk consumed more concentrates, resulting in a lower ratio of forage to total solid feed intake in a low compared to a high milk feeding group (13.2% vs. 18.6%) [84].

Feeding patterns and methods could also affect forage intake in calves. When investigating the effect of either step down (fed at around 15% of birth body weight per day) or conventional (approximately 10% of birth body weight per day) feeding patterns in dairy calves, Khan et al. [88] found that the former had better performance. In a different study, Daneshvar et al. [23] reported that when similar amounts of milk were fed using different feeding patterns (step down vs. conventional), solid feed intake did not differ between treatments. Horvath et al. [85] showed that the feeding method (bucket vs. teat feeding) did not alter the forage and concentrate intake. Hence, milk allowance might have a greater impact on solid feed consumed by calves as opposed to the milk feeding pattern or method.

Limited studies directly investigating the relationship between milk allowance and forage consumption in pre-weaned calves are available, which calls for scientists to turn their attention to this area, especially with more farms leaning towards high milk volume feeding. Forage inclusion can promote total solid feed consumption and BW gain in calves, but factors such as the amount of forage, forage sources, forage feeding method, physical form of forage and concentrate, and milk allowance might confound these benefits. Calves should be slowly introduced to relatively low levels of forage while guarding against the use of low digestible forage (i.e., straw), which may depress total DMI and BW gain. Moreover, forage should be available free-choice and in separate containers from concentrate feed.

Table 7 shows a summary of selected studies that determined the effects of forage inclusion on performance, rumen fermentation and development, and expressive behavior in dairy calves. While the effects of feeding forage on performance, such as DMI and ADG, were controversial, relatively consistent results were obtained in other parameters, like rumen fermentation and expressive behavior.

Table 7. Summary of selected studies on forage inclusion and their effect on performance, rumen fermentation, and expressive behavior compared to calves fed only concentrates [1].

Parameters [2]	Studies with Positive Impact	Studies with Negative Impact	Studies with No Effect
Total solid DMI	[15,18–23,38–40,43,50,51,55,67,75,76,78,80]	[5,15,17,39,43]	[21,24–26,37,44,48–50,52,54,56,57,77,80,84,86]
ADG	[18–23,38–40,50,51,53,55,67,75,76,80]	[5,15,17,39,84]	[15,21,24–26,37,43,44,48–50,52,56,57,77,78,80,85]
DM digestibility	/	[23,37,41,48]	[21,39,52,75,78]
Feed to Gain ratio	[19,39]	[5,15,22,51]	[21,23,24,26,27,37–40,44,49,52,53,55,56,67,75,78,80,85]
Structural growth	[55]	[15]	[20,23,25,27,38–40,44,49,57,67,76,80]
Fecal score	/	[27,37]	[19,26,39,44,49,75]
Rumen fluid pH	[20,22,23,25,27,37–42,52–55,67,75,76,78,80]	/	[5,24,39,43]
Total VFA	[22]	[19,23,25,35,40,52–54,67]	[24,27,38,39,42,43,76,78]
Acetate	[22,23,40,43,52,53,67]	/	[19,24,27,38,39,76,78]
Propionate	[53]	[25,35,40,43,54,67]	[19,22–24,27,38,39,52,76,78]
Butyrate	/	[23,25,27,39,40,53]	[19,22,24,38,39,43,52,54,67,76,78]
Valerate	/	[23,25,52,53,67]	[40,76]
Acetate to Propionate ratio	[27,37,38,40,42,54,67]	[53]	[19,22–24,39,52]
Lactate	/	[51]	[43]
NH3	[42]	[43]	[24]
Rumen papillae length	/	[42,44,52]	[20,22,27,37,43,76,84]
Rumen plaque formation	/	[22,43]	[42]
Rumen weight	[11,20,50]	[27,43]	[24,52,76,84]
Rumen volume	[11,42,52]	/	[24,76]
Ruminating	[21,39,40,53,55,56,75–78]	/	[39]
Total eating behavior	[39,40,57,76,77]	/	[39,55,78,85]
Concentrate eating behavior	[57]	[56,85]	[21,53,75]
Drinking behavior	[76]	[56]	/
Non-nutritive oral behavior/Abnormal behavior	/	[21,40,53,55–57,76,85]	[75,78]
Lying behavior	/	[21,53,75,76]	[39,40,55,56,77,78]
Standing behavior	[76]	/	[21,39,40,53,55,56,75,78]
Satisfaction behavior	[56]	/	[57]
Urination and Defecation behavior	/	/	[56]
Sorting behavior	/	[25,87]	[77]

[1] Forages included dry hay, silage, straw, and by-products (e.g., cottonseed hulls) and reconstituted hay. Positive, negative, and no effect on a parameter was determined by adding forage in the diet compared with no forage inclusion in those studies. "Positive effect" represent an increase or improved effect ($p < 0.05$), "Negative effect" represent a decreased effect in the related parameter ($p < 0.05$), "/" means no studies were found to affect this parameter in the current review ($p > 0.05$). [2] Parameters were measured in dairy calves within 3 months of age. Non-nutritive oral behavior/abnormal behavior included tongue rolling, licking buckets, pen or surface, sniffing, vocalizing, and eating beddings; Satisfaction behavior included tail swishing, self-grooming, and rubbing.

3. Rumen Environment

3.1. Rumen Fluid pH and Fermentation

The rumen is the largest and most crucial compartment of the digestive system in adult ruminants, as it is vital for acquiring metabolic substrates through microbial fermentation. Although young calves have an undeveloped rumen, fermentation begins at a very early age [89] and may affect the development and health of the rumen. Prolonged low rumen fluid pH may cause subacute ruminal acidosis (SARA) in adult cows, which is well defined in beef feedlot cattle (pH < 5.8 for 3 h/d) [90] and dairy cows (as periods of moderately depressed pH, from about 5.0 to 5.5) [91]. However, in young calves, ruminal acidosis has not been clearly defined. Previous studies reported that rumen fluid pH in young calves is often well below 5.8 [89,92]. Some researchers believe that, as in mature cows, dairy calves can experience ruminal acidosis [34,89], probably due to the high amount of concentrate fed [92] in artificially rearing systems and the relatively low saliva [93] secreted at a young age.

Concentrate feed, high in rapidly fermentable carbohydrates, such as sugar and starch, provides energy for optimal growth, but the fermentation rate tends to generate lots of VFA and lactic acid, resulting in low rumen fluid pH [51]. Forage, high in fiber, may play a role in mitigating this challenge. Most of the studies (21 studies, accounting for 84% of summarized studies) explored in this review showed a positive effect of forage inclusion on rumen fluid pH in dairy calves, while very few reported no difference (four studies, accounting for 16% of summarized studies) or negative impact (Table 7). In agreement with a previous meta-analysis, our literature search showed that forage could improve the rumen fluid pH when supplemented to calves, though it might be dependent on the forage source [38,40]. Alfalfa hay is more likely to modulate rumen fluid pH during the milk-feeding period than other types of forages [33]. Maktabi et al. [39] reported that increasing fiber content by adding beet pulp (10% and 20%) in the concentrate diet failed to improve rumen fluid pH, while supplementing alfalfa hay (10%) resulted in a significant improvement in this parameter by providing more effective NDF. Terré et al. [53] also demonstrated that increasing NDF content (18.2 vs. 26.7%) by adding soybean hulls in the pelleted starter could not alter rumen fluid pH, but adding chopped oat hay, containing more effective fiber, could improve the ruminating behavior resulting in a higher pH. In agreement, Laarman et al. [94] reported a positive relationship between forage intake and rumen fluid pH, while SARA (rumen fluid pH below 5.8) could be exacerbated when calves are fed less than 0.08 kg/d [64], suggesting that even small amounts of forage consumption (timothy hay, 0.08 kg/d) can reduce rumen acidosis in calves.

The possible reasons for the increased rumen fluid pH when adding forage in the diet are multifactorial. On the one hand, forage is bulkier and has lower digestibility compared to the concentrate. The higher forage consumption leads to increased intake of effective fiber, which in turn stimulates the chewing (ruminating and eating) activity of calves [56,95], and subsequently improves the saliva production and rumen buffering [38,63]. On the other hand, the rapidly fermentable carbohydrates generate abundant VFA that may exceed and overwhelm the absorptive capacity of the undeveloped rumen [96]. Feeding forage in pre-weaned calves could reduce the concentration of VFA [19,23,25,35,40,52–54,67] and decrease rumen plaque formation [42,52], increasing the absorptive surface area of the rumen epithelium and hence reducing the accumulation of VFA and maintaining the appropriate rumen fluid pH. In addition, an increased passage rate in the rumen was observed in calves fed forage compared to those fed concentrate only, which lowered the feed retention time in the gastrointestinal tract (28.4 h for concentrate feed vs. 18.8 h for oat hay group), fermentation time and VFA concentration [52]. In the same study, calves fed forage tended to have a higher expression of monocarboxylate transporter-1 in the rumen wall [52], which plays a central role in transporting acetate, lactate, and protons from the rumen lumen to the bloodstream [97,98], hence alleviating VFA accumulation in the rumen as well.

A greater acetate [22,23,40,43,52,53,67], and a lower propionate [25,34,40,43,54,67], butyrate [23,25,27,39,40,53], and valerate [23,25,52,53,67] concentration/proportions have been reported in calves fed with forage than those fed only concentrate. These dynamics in fermentation patterns are probably related to the changes in the rumen microbial ecosystem. For example, cellulolytic microbes, such as *Ruminococcus flavefaciens* and *Ruminococcus albus*, are more prevalent in animals fed high forage diets, which increase fiber degradation and elevates the proportion of acetate in the rumen [42,52]. Both propionate and butyrate stimulate and enhance rumen epithelial development [12,13,99], with butyrate serving as the preferred energy source as well as modulating the gene expression in the rumen epithelium [96]. A low proportion of these two VFA may limit the growth of rumen papillae [42,52]. Due to its relatively low proportions, the valerate has received little attention. It has been suggested that cellulolytic microbes utilize valerate in the rumen [100], which might explain its decrease in calves supplemented with forage in their diets.

Lactic acid decreases in the rumen when forage is included in the diet [51], which might end up positively altering rumen fluid pH. Terré et al. [53] found an interesting relationship between rumen VFA and rumen fluid pH. When rumen fluid pH was above 5.1, total VFA and rumen fluid pH were linearly correlated; however, when it fell below 5.1, the correlation disappeared. The implications are that lactic acid (a much stronger acid than VFA) may alter rumen fluid pH at a pH below 5.1 [53,101]. In adult cows, an acute ruminal acidosis was observed, with excessive consumption of concentrate feed leading to a sudden and uncompensated drop in rumen fluid pH (below about 5.0). Owens et al. [60] showed that lactic acid concentrations increased with a decline in rumen fluid pH. However, when rumen fluid pH (around 5.0–5.5) was moderately depressed, lactic acid accumulation was inconsistent [102] or transiently fluctuated [103]. Hence, although moderate depression in rumen fluid pH may cause SARA in dairy cows, it is not because of the lactic acid accumulation, but might be due to the accumulation of VFA alone [102].

Collectively, previous studies demonstrate that forage provision has a positive effect on rumen fluid pH and alters rumen fermentation in calves. However, the majority of these studies assessed rumen fluid pH and VFA at only a single time point. Further research is encouraged to test the dynamic changes in rumen fluid pH and VFA when different types, amounts, timing, and particle sizes of forage are supplemented in the diet, which may help us define SARA in young calves more accurately.

3.2. Rumen Microbes

A developed rumen is full of microorganisms that ferment and degrade multiple nutritional fractions (sugar, starch, fiber, protein, fat, and so on) and provides necessary metabolic substrates and nutrients to the dairy cattle. The microbial ecosystem differs between young calves and adult cows [104]. At birth, young calves possess no anaerobic microorganisms in the rumen [105], with recent evidence suggesting that colonization occurs immediately after birth [104,106]. Dominant microbes that are involved in normal rumen function of mature cows are present as early as one day of age [106]. This colonization of microorganisms and the presence of substrates trigger fermentation activity, which then provides indispensable nutrients for rumen development. It might take as long as a year for the rumen to mature and for calves to establish a stable rumen microbiota system with many factors involved [104].

Both liquid and solid feed appear to affect the microbial community in young calves [107]. In this review, our discussion is restricted to the effect of forage feeding on the microbiota of young calves. Castells et al. [52] reported that forage supplementation (alfalfa) numerically increased cellulolytic microbes (*Ruminococcus albus*) compared with calves fed only concentrate feeds. Similarly, Kim et al. [54] observed significantly higher copy numbers of cellulolytic bacteria (*Ruminococcus flavefaciens* and *Ruminococcus albus*) in calves supplemented with forages (orchard and timothy hay). Early studies also evaluated the effects of the physical form (finely ground, 1 mm theoretical particle size vs. unground, 0.64 cm theoretical particle size) of diet on rumen microbiota with two identical diets (25% alfalfa hay and 75% grain) that varied only in particle size. Calves offered the ground diet had a relatively lower rumen fluid pH and lower number of cellulolytic bacteria than calves fed the unground diet [89].

These results revealed that effective fiber might play a crucial role in changing the rumen environment, hence altering the microbial populations.

The next-generation sequencing (NGS) analysis has revealed that the major phyla in the rumen are *Firmicutes* (around 43%), *Bacteroidetes* (around 21%), *Actinobacteria* (around 18%), and *Proteobacteria* (around 4%) [108]. Relatively higher abundance of *Bacteroidetes* and lower abundance of *Actinobacteria* was observed in calves supplemented with forage compared to those fed only concentrate [54]. *Bacteroidetes*, the second most dominant phyla in calves' rumen, may stimulate the development of the digestive tract [109]. In adult cows fed concentrate feed, the relative abundance of *Bacteroidetes* dropped and cows were more susceptible to SARA [110]. The lower level of *Bacteroidetes* in calves fed only concentrate could be partly explained by greater feed intakes resulting in low rumen pH. Furthermore, Kim et al. [54] found that the most dominant genus in *Bacteroidetes* phylum was *Prevotella*, a highly active hemicellulolytic and starch-degraders [111] that mainly produce acetate. A relatively higher abundance of *Prevotella* may be related to the greater acetate proportion in calves offered forage [54]. Jami et al. [104] reported that *Prevotella* was the predominant genus in animals fed high-fiber diets rather than high-caloric diets. *Olsenella* is an important lactic acid-producing bacterium under the phylum *Actinobacteria* [112]. Forage inclusion decreases *Olsenella* (hay, 3.9% vs. concentrate, 13.2%) significantly, which contributes to the lower abundance of *Actinobacteria* (hay, 4.7% vs. concentrate, 13.9%) [54]. Thus, it can be speculated that forage inclusion in the diet might affect the growth of lactic acid-producing bacteria (such as *Olsenella*) by limiting the proportion of rapidly fermentable substrates (e.g., starch) replaced by fiber. In contrast, although numerical differences were observed in *Bacteroidetes* and *Actinobacteria* when evaluating the effect of forage supplementation, Lin et al. [108] reported that neither alpha nor beta diversity indices and microbiota were significantly different among the dietary groups [108], possibly because of the volume of milk (Lin, 252 L, 10% vs. Kim, 88 L, 4% of birth bodyweight) fed to calves, resulting in varying solid feed consumed. Alternatively, different forage sources (Lin, Oat hay vs. Kim, Timothy) and feeding levels (Forage/Total Solid: Lin, 6% vs. Kim, 20%) (Tables 2 and 4) might have led to insufficient forage consumption causing the changes in the composition of rumen microbiota in the Lin et al. study. These results indicate that the population of most predominant microbiota (e.g., *Bacteroidetes* and *Actinobacteria*) in the rumen is closely related to the amount and type of solid feed consumed. Further studies need to focus on these major groups to illustrate the relationship between rumen microbiota, fermentation and feed consumption in dairy calves.

It is important to note that the effect of various nutritional (sugars, starch, rumen degradable protein, NDF, or ADF) and physical (effective forage fiber or non-forage fiber) fractions on rumen microorganisms in young calves is not clearly defined in literature. Furthermore, most of the studies evaluating microbial and molecular changes in young calves have only studied a small number of microbes of interest. With the rapid development of the NGS technologies in the past 20 years, we recommend the exploration of the global changes in microbial abundance to better understand substrate fermentation and absorption and epithelium development in the rumen as forage are supplemented to calves in early life.

4. Feeding Behavior

4.1. Ruminating and Eating Behavior

Feed experiences and behavior development in early life might affect the behavioral expression of adult ruminants [79]. In the last decade, researchers have increased their attention on the development of calf behaviors, such as eating, standing, lying, and ruminating when forage is included in the diet. Forage inclusion in the diet undoubtedly increases chewing in calves even before weaning when only a small amount of solid feed is consumed. Increased chewing activity may be as a result of higher rumination [21,39,40,53,55,56,75–78] or the total time spent eating [39,40,57,76,77] when calves are fed forage.

In newborn ruminants, rumination is initially absent and emerge a few weeks after birth [113]. Providing forage to young calves can accelerate the development of rumination behavior [114,115]. van Ackeren et al. [116] observed that chewing time was lower in calves receiving a low NDF diet (26.2%) compared with those receiving a high NDF diet (31.3%). Porter et al. [37] claimed that calves began ruminating by week 4 of age when fed a more physically effective solid feed, while those fed a finely pelleted feed began ruminating from week 6. Rumination is crucial in ruminants helping maintain the rumen fluid pH by stimulating saliva production that neutralize VFA and lactic acid in the rumen and thus to maintain a healthy rumen environment [105].

Meal feeding patterns (meal size, frequency, and duration) can also impact the rumen environment. Generally, rumen fluid pH declines rapidly after feed ingestion, and the rate of decrease is associated with meal size and feeding frequency [117]. Large meal sizes and infrequent meals may result in a greater drop in rumen fluid pH post-ingestion. Horvath et al. [57] illustrated that the provision of forage not only increased the total eating time but also influenced the solid feed meal patterns. An improved meal frequency and duration were observed in their study, which leads to relatively slower post-prandial drops in rumen fluid pH, potentially decreasing the risk of SARA [63].

4.2. Sorting Behavior

Feed sorting is well demonstrated in adult cows, since they are highly sensitive to sweet taste [118]. Probably, the preference for sweetness reflects the inclination towards higher energy demands, hence the tendency to sort out for concentrates (sweet, high energy-density) in a total mixed ration [119]. The sorting out of the mixed ration can result in an unbalanced nutrient intake, whereby cows sort out for the rapidly fermentable cereals as opposed to forage, leading to a drop in rumen fluid pH, and hence inducing SARA [120]. Ingesting excessive fermentable concentrate feed can result in rumen acidosis [63]. In turn, the sorting behavior is altered further, leading the animals to choose the part of diets with longer particle size and slower fermentable rate [121]. These results suggest that ruminants develop feed preferences based on post-ingestive feedback [122] and they may be biased towards choosing certain nutrients as the situation demands.

Feed sorting is also seen in the early life of calves. When feeding concentrates and forage free-choice, variation in the proportion of forage to total solid intake was observed (ranging from 5 and 45%) [21,25,87]. The changes in dietary selection across multiple studies may depend on forage related factors and milk feeding allowance, as has been discussed above. It is interesting to note that feed preference and sorting can be established early and persist later on in life. Miller-Cushon et al. [123] reported that calves fed either concentrate or forage before weaning were likely to consume the feed that they were already familiar with, even when switched to a mixed diet after weaning. Similarly, our research group found that calves are likely to eat feed they were originally introduced to and familiar with even after switching to a free-choice diet, though this effect only lasted for a short period. However, after switching the diet at weaning, the provision of both concentrate and hay separately early in life led to a greater hay intake ratio (35.6%) than providing concentrate (17.7%) or hay (16.5%) solely before weaning. Furthermore, exposure to a diet of both concentrate and hay early in life could numerically improve the calves' ability to sort for long particles 6 months later [25]. Therefore, these results suggest that early exposure to feed experience can affect the feed preference immediately after switching diets and may have a long-lasting effect. The feeding method may also play an essential role in influencing the learning of feed sorting behavior. When we compared three different feeding methods (solely concentrate, separated concentrate and forage, mixed concentrate and forage for the first month; data unpublished) in 2 month old calves, the lowest sorting activity was observed in calves fed concentrate and forage separately. Hence, calves exposed early to a diet of concentrate or mixed ration are likely to sort for fine-grain particles, probably because these calves have already established their sorting behavior, which can last even after changing to a new mixed diet (data unpublished). Similarly, the provision of solid feed in pre-weaned calves as separate components reduced the extent of feed sorting after weaning compared to offering the diet as a mixed ration [87]. As already stated, feed sorting is likely to influence rumen fluid pH and may lead to SARA. Separately feeding different solid feed components at the same time may avert sorting for fine particles when compared to feeding solely concentrate or mixed diets and might lead to a more stable rumen fluid pH and a healthier rumen in the calf. However, we cannot ignore the fact that the effect of forage inclusion on sorting behavior is dependent on a myriad of other factors (e.g., forage source, level, physical form, dry matter, and milk allowance).

4.3. Other Behaviors

Access to forage by dairy calves may also reduce the occurrence of other non-nutritive oral and abnormal behaviors [21,40,53,55–57,76,85], such as tongue rolling, licking of buckets, pen or surface, sniffing, vocalizing, and eating the bedding material. Horvath et al. [85] demonstrated that providing forage decreased the non-nutritive oral behaviors, and when combined with feeding milk by teat, the effects were more significant compared with bucket feeding. These results further buttress the fact that liquid and solid feeding can influence the development of pre-weaned calf behaviors. Furthermore, supplementation of good quality forage increased other behaviors that may indicate satisfaction (tail swishing, self-grooming, and rubbing) [56]. Worthy to note is that the decline in non-nutritive oral behavior may have also reduced the formation of hair and fiber balls in the rumen [4], which have been associated with poor health and growth of calves. Further research is encouraged to explore whether forage inclusion early in life would have a long-term effect on sorting and other behaviors.

5. Conclusions

Understanding factors that influence responses to forage inclusion in pre-weaned calves is of significant importance from a management point of view because the effect of offering forage on calf feed intake and growth rate has been inconsistent. In recent studies, a small amount of good quality forage such as alfalfa supplemented in the diet is likely to improve the DMI and ADG. However, these performances are dependent on the type of concentrate and the amount of milk offered. Although controversy remains on whether forage improves growth rate, it has been well documented that its inclusion early in life can help with the establishment of feeding behavior, leading to greater rumination and eating behavior as well as lowering the abnormal feed sorting behavior. All these positive effects can result in a higher rumen fluid pH and a more stable rumen environment, with a corresponding positive effect on rumen microbiota and fermentation. Further research is required to understand the long-term effects of offering forage to pre-weaned calves, since animal-related factors, such as feed selection and sorting, established early in life may persist later on in life.

Author Contributions: The paper was mainly conceived and designed by Z.C. Research articles were collected and analyzed by J.X. and G.M.A. The manuscript was mainly written by J.X. and G.M.A. and edited by J.L., Y.W., S.L., and Z.C. All authors have read and agreed to the published version of the manuscript.

Funding: This research received no external funding.

Conflicts of Interest: The authors declare no conflict of interest.

References

1. Davenport, E. On the importance of the physiological requirements of the animal body; results of an attempt to grow cattle without coarse feed. *III Agric. Exp. Sta. Bul.* **1897**, *46*, 362.
2. Kertz, A.F.; Hill, T.M.; Heinrichs, A.J.; Linn, J.G.; Drackley, J.K. A 100-Year Review: Calf nutrition and management. *J. Dairy Sci.* **2017**, *100*, 10151. [CrossRef] [PubMed]
3. Mccandlish, A.C. Studies in the Growth and Nutrition of Dairy Calves: VI. The Addition of Hay and Grain to a Milk Ration for Calves. *J. Dairy Sci.* **1923**, *6*, 500–508. [CrossRef]
4. Pounden, W.D.; Hibbs, J.W.; Cole, C.R. Observations on the relation of diet to diarrhea in young dairy calves. *J. Am. Vet. Med. Assoc.* **1951**, *118*, 400.

5. Hibbs, J.W.; Conrad, H.R.; Pounden, W.D.; Frank, N. A High Roughage System for Raising Calves Based on Early Development of Rumen Function. VI. Influence of Hay to Grain Ratio on Calf Performance, Rumen Development, and Certain Blood Changes. *J. Dairy Sci.* **1956**, *39*, 171–179. [CrossRef]

6. Hibbs, J.W.; Pounden, W.D.; Conrad, H.R. A high roughage system for raising calves based on the early development of rumen function. 1. Effect of variations in the ration on growth, feed consumption, and utilization. *J. Dairy Sci.* **1953**, *36*, 717–727. [CrossRef]

7. Warner, R.G.; Flatt, W.P.; Loosli, J.K. Dietary factors influencing the development of the ruminant stomach. *J. Agric. Food Chem.* **1956**, *4*, 788–792. [CrossRef]

8. Suarez-Mena, F.X.; Hill, T.M.; Jones, C.M.; Heinrichs, A.J. Review: Effect of forage provision on feed intake in dairy calves. *Prof. Anim. Sci.* **2016**, *32*, 383–388. [CrossRef]

9. Davis, C.L.; Drackley, J.K. *The Development, Nutrition, and Management of the Young Calf*; Iowa State University Press: Ames, IA, USA, 1998.

10. Sander, E.G.; Warner, R.G.; Harrison, H.N.; Loosli, J.K. The Stimulatory Effect of Sodium Butyrate and Sodium Propionate on the Development of Rumen Mucosa in the Young Calf. *J. Dairy Sci.* **1959**, *42*, 1600–1605. [CrossRef]

11. Tamate, H.; Mcgilliard, A.D.; Jacobson, N.L.; Getty, R. Effect of Various Dietaries on the Anatomical Development of the Stomach in the Calf 1. *J. Dairy Sci.* **1962**, *45*, 408–420. [CrossRef]

12. Carroll, E.; Hungate, R. The magnitude of the microbial fermentation in the bovine rumen. *Appl. Microbiol.* **1954**, *2*, 205. [CrossRef] [PubMed]

13. Stobo, I.J.F.; Roy, J.H.B.; Gaston, H.J. Rumen development in the calf: 2.* The effect of diets containing different proportions of concentrates to hay on digestive efficiency. *Br. J. Nutr.* **1966**, *20*, 189. [CrossRef] [PubMed]

14. Leibholz, J. Ground roughage in the diet of the early-weaned calf. *Anim. Prod.* **1975**, *20*, 93–100. [CrossRef]

15. Hill, T.M.; Ii, H.G.B.; Aldrich, J.M.; Schlotterbeck, R.L. Effects of the Amount of Chopped Hay or Cottonseed Hulls in a Textured Calf Starter on Young Calf Performance. *J. Dairy Sci.* **2008**, *91*, 2684–2693. [CrossRef]

16. National Academies Press. *Nutrient Requirements of Dairy Cattle*; National Academies Press: Washington, WA, USA, 2001.

17. Hill, T.M.; Ii, H.G.B.; Pas, J.M.A.; Schlotterbeck, R.L. Roughage for Diets Fed to Weaned Dairy Calves. *Prof. Anim. Sci.* **2009**, *25*, 283–288. [CrossRef]

18. Castells, L.; Bach, A.; Terré, M. Short- and long-term effects of forage supplementation of calves during the preweaning period on performance, reproduction, and milk yield at first lactation. *J. Dairy Sci.* **2015**, *98*, 4748–4753. [CrossRef]

19. Coverdale, J.A.; Tyler, H.D.; Brumm, J.A. Effect of various levels of forage and form of diet on rumen development and growth in calves. *J. Dairy Sci.* **2004**, *87*, 2554–2562. [CrossRef]

20. Khan, M.A.; Weary, D.M.; Keyserlingk, M.A.G.V. Hay intake improves performance and rumen development of calves fed higher quantities of milk. *J. Dairy Sci.* **2011**, *94*, 3547–3553. [CrossRef]

21. Castells, L.; Bach, A.; Araujo, G.; Montoro, C.; Terré, M. Effect of different forage sources on performance and feeding behavior of Holstein calves. *J. Dairy Sci.* **2012**, *95*, 286–293. [CrossRef]

22. Beiranvand, H.; Ghorbani, G.R.; Khorvash, M.; Nabipour, A.; Dehghanbanadaky, M.; Homayouni, A.; Kargar, S. Interactions of alfalfa hay and sodium propionate on dairy calf performance and rumen development. *J. Dairy Sci.* **2014**, *97*, 2270–2280. [CrossRef]

23. Daneshvar, D.; Khorvash, M.; Ghasemi, E.; Mahdavi, A.H.; Moshiri, B.; Mirzaei, M.; Pezeshki, A.; Ghaffari, M.H. The effect of restricted milk feeding through conventional or step-down methods with or without forage provision in starter feed on performance of Holstein bull calves. *J. Anim. Sci.* **2015**, *93*, 3979–3989. [CrossRef] [PubMed]

24. Wu, Z.; Azarfar, A.; Simayi, A.; Li, S.; Jonker, A.; Cao, Z. Effects of forage type and age at which forage provision is started on growth performance, rumen fermentation, blood metabolites and intestinal enzymes in Holstein calves. *Anim. Prod. Sci.* **2018**, *58*, 2288–2299. [CrossRef]

25. Xiao, J.; Guo, L.; Alugongo, G.; Wang, Y.; Cao, Z.; Li, S. Effects of different feed type exposure in early life on performance, rumen fermentation, and feed preference of dairy calves. *J. Dairy Sci.* **2018**, *101*, 8169–8181. [CrossRef] [PubMed]

26. Jahanimoghadam, M.; Mahjoubi, E.; Yazdi, M.H.; Cardoso, F.C.; Drackley, J.K. Effects of alfalfa hay and its physical form (chopped versus pelleted) on performance of Holstein calves. *J. Dairy Sci.* **2015**, *98*, 4055–4061. [CrossRef]

27. Mirzaei, M.; Khorvash, M.; Ghorbani, G.R.; Kazemi-Bonchenari, M.; Riasi, A.; Nabipour, A.; Borne, J.J.G.C. Effects of supplementation level and particle size of alfalfa hay on growth characteristics and rumen development in dairy calves. *J. Anim. Physiol. Anim. Nutr.* **2015**, *99*, 553–564. [CrossRef]

28. Gelsinger, S.L.; Heinrichs, A.J.; Jones, C.M. A meta-analysis of the effects of preweaned calf nutrition and growth on first-lactation performance1. *J. Dairy Sci.* **2016**, *99*, 6206–6214. [CrossRef]

29. Kilgour, R.J. In pursuit of "normal": A review of the behaviour of cattle at pasture. *Appl. Anim. Behav. Sci.* **2012**, *138*, 1–11. [CrossRef]

30. Nicol, A.M.; Sharafeldin, M.A. Observations on the behaviour of single-suckled calves from birth to 120 days. *Proc NZ Soc. Anim. Prod.* **1975**.

31. Tedeschi, L.O.; Fox, D.G. Predicting milk and forage intake of nursing calves. *J. Anim. Sci.* **2009**, *87*, 3380–3391. [CrossRef]

32. Miller-Cushon, E. Expression and Development of Feeding Behaviour in Dairy Calves. Ph.D. Thesis, University of Guelph, Guelph, ON, Canada, 6 May 2014.

33. Drackley, J.K. Calf Nutrition from Birth to Breeding. *Vet. Clin. North Am. Food Anim. Pract.* **2008**, *24*, 55–86. [CrossRef]

34. Imani, M.; Mirzaei, M.; Baghbanzadehnobari, B.; Ghaffari, M.H. Effects of forage provision to dairy calves on growth performance and rumen fermentation: A meta-analysis and meta-regression. *J. Dairy Sci.* **2017**, *100*, 1136–1150. [CrossRef] [PubMed]

35. Conrad, H.R.; Hibbs, J.W. A High Roughage System for Raising Calves Based on the Early Development of Rumen Function. VII. Utilization of Grass Silage, Pasture, and Pelleted Alfalfa Meal. *J. Dairy Sci.* **1956**, *39*, 1170–1179. [CrossRef]

36. Stobo, I.J.F.; Roy, J.H.B.; Gaston, H.J. Rumen development in the calf. 1. The effect of diets containing different proportions of concentrates to hay on rumen development. *Br. J. Nutr.* **1966**, *20*, 171–188. [CrossRef]

37. Porter, J.C.; Warner, R.G.; Pas, A.F.K. Effect of Fiber Level and Physical Form of Starter on Growth and Development of Dairy Calves Fed No Forage. *Prof. Anim. Sci.* **2007**, *23*, 395–400. [CrossRef]

38. Nemati, M.; Amanlou, H.; Khorvash, M.; Mirzaei, M.; Moshiri, B.; Ghaffari, M.H. Effect of different alfalfa hay levels on growth performance, rumen fermentation, and structural growth of Holstein dairy calves. *J. Anim. Sci.* **2016**, *94*, 1141. [CrossRef] [PubMed]

39. Maktabi, H.; Ghasemi, E.; Khorvash, M. Effects of substituting grain with forage or nonforage fiber source on growth performance, rumen fermentation, and chewing activity of dairy calves. *Anim. Feed Sci. Technol.* **2016**, *221*, 70–78. [CrossRef]

40. Hosseini, S.; Mirzaei-Alamouti, H.; Vazirigohar, M.; Mahjoubi, E.; Rezamand, P. Effects of whole milk feeding rate and straw level of starter feed on performance, rumen fermentation, blood metabolites, structural growth, and feeding behavior of Holstein calves. *Anim. Feed Sci. Technol.* **2019**, *255*, 114238. [CrossRef]

41. Jahn, E.; Chandler, P.T.; Polan, C.E. Effects of fiber and ratio of starch to sugar on performance of ruminating calves. *J. Dairy Sci.* **1970**, *53*, 466–474. [CrossRef]

42. Zitnan, R.; Voigt, J.; Schönhusen, U.; Wegner, J.; Kokardová, M.; Hagemeister, H.; Levkut, M.; Kuhla, S.; Sommer, A. Influence of dietary concentrate to forage ratio on the development of rumen mucosa in calves. *Arch. Anim. Nutr.* **1998**, *51*, 279–291.

43. Suárez, B.; Van Reenen, C.; Stockhofe, N.; Dijkstra, J.; Gerrits, W. Effect of roughage source and roughage to concentrate ratio on animal performance and rumen development in veal calves. *J. Dairy Sci.* **2007**, *90*, 2390–2403. [CrossRef]

44. Kehoe, S.I.; Dill-McFarland, K.A.; Breaker, J.D.; Suen, G. Effects of corn silage inclusion in preweaning calf diets. *J. Dairy Sci.* **2019**, *102*, 4131–4137. [CrossRef] [PubMed]

45. ASAE. *Method of Determining and Expressing Particle Size of Chopped Forage Material by Screening*; ASAE: St. Joseph, MI, USA, 1998.

46. Booth, J.A. *Effect of Forage Addition to the Diet on Rumen Development in Calves*; Retrospective Theses and Dissertations; Iowa State University: Ames, IA, USA, 2003; p. 564.

47. ÜLGER, İ.; Kaliber, M.; BEYZİ, S.B.; Konca, Y. Effects of Different Quality Roughage Supply on Performance of Holstein Calves during Preweaning Period. *Tarim Bilimleri Dergisi J. Agric. Sci.* **2017**, *23*, 386–394.

48. Hill, T.; Dennis, T.; Suárez-Mena, F.; Quigley, J.; Aragona, K.; Schlotterbeck, R. Effects of free-choice hay and straw bedding on digestion of nutrients in 7-week-old Holstein calves. *Appl. Anim. Sci.* **2019**, *35*, 312–317. [CrossRef]

49. Hill, T.; Suárez-Mena, F.; Dennis, T.; Quigley, J.; Schlotterbeck, R. Effects of free-choice hay on intake and growth of Holstein calves fed a textured starter to 2 months of age. *Appl. Anim. Sci.* **2019**, *35*, 161–168. [CrossRef]

50. Anderson, M.J.; Khoyloo, M.; Walters, J.L. Effect of Feeding Whole Cottonseed on Intake, Body Weight, and Reticulorumen Development of Young Holstein Calves. *J. Dairy Sci.* **1982**, *65*, 764–772. [CrossRef]

51. Suárez, B.J.; Van Reenen, C.G.; Beldman, G.; Van, D.J.; Dijkstra, J.; Gerrits, W.J. Effects of supplementing concentrates differing in carbohydrate composition in veal calf diets: I. Animal performance and rumen fermentation characteristics. *J. Dairy Sci.* **2006**, *89*, 4365–4375. [CrossRef]

52. Castells, L.; Bach, A.; Aris, A.; Terré, M. Effects of forage provision to young calves on rumen fermentation and development of the gastrointestinal tract. *J. Dairy Sci.* **2013**, *96*, 5226–5236. [CrossRef]

53. Terré, M.; Pedrals, E.; Dalmau, A.; Bach, A. What do preweaned and weaned calves need in the diet: A high fiber content or a forage source? *J. Dairy Sci.* **2013**, *96*, 5217–5225. [CrossRef]

54. Kim, Y.-H.; Nagata, R.; Ohtani, N.; Ichijo, T.; Ikuta, K.; Sato, S. Effects of dietary forage and calf starter diet on ruminal pH and bacteria in Holstein calves during weaning transition. *Front. Microbiol.* **2016**, *7*, 1575. [CrossRef]

55. Mirzaei, M.; Khorvash, M.; Ghorbani, G.R.; Kazemi-Bonchenari, M.; Ghaffari, M.H. Growth performance, feeding behavior, and selected blood metabolites of Holstein dairy calves fed restricted amounts of milk: No interactions between sources of finely ground grain and forage provision. *J. Dairy Sci.* **2017**, *100*, 1086–1094. [CrossRef]

56. Phillips, C.J. The effects of forage provision and group size on the behavior of calves. *J. Dairy Sci.* **2004**, *87*, 1380–1388. [CrossRef]

57. Horvath, K.; Miller-Cushon, E. Evaluating effects of providing hay on behavioral development and performance of group-housed dairy calves. *J. Dairy Sci.* **2019**, *102*, 10411–10422. [CrossRef] [PubMed]

58. Kargar, S.; Kanani, M.; Albenzio, M.; Caroprese, M. Substituting corn silage with reconstituted forage or nonforage fiber sources in the starter diets of Holstein calves: Effects on performance, ruminal fermentation, and blood metabolites. *J. Anim. Sci.* **2019**, *97*, 3046–3055. [CrossRef]

59. Kargar, S.; Kanani, M. Substituting corn silage with reconstituted forage or nonforage fiber sources in the starter feed diets of Holstein calves: Effects on intake, meal pattern, sorting, and health. *J. Dairy Sci.* **2019**, *102*, 7168–7178. [CrossRef]

60. Owens, F.N.; Secrist, D.S.; Hill, W.J.; Gill, D.R. Acidosis in cattle: A review. *J. Anim. Sci.* **1998**, *76*, 275–286. [CrossRef] [PubMed]

61. Kononoff, P.; Heinrichs, A.J. The effect of reducing alfalfa haylage particle size on cows in early lactation. *J. Dairy Sci.* **2003**, *86*, 1445–1457. [CrossRef]

62. Montoro, C.; Miller-Cushon, E.; DeVries, T.; Bach, A. Effect of physical form of forage on performance, feeding behavior, and digestibility of Holstein calves. *J. Dairy Sci.* **2013**, *96*, 1117–1124. [CrossRef]

63. Krause, K.M.; Oetzel, G.R. Understanding and preventing subacute ruminal acidosis in dairy herds: A review. *Anim. Feed Sci. Technol.* **2006**, *126*, 215–236. [CrossRef]

64. Laarman, A.; Oba, M. Effect of calf starter on rumen pH of Holstein dairy calves at weaning. *J. Dairy Sci.* **2011**, *94*, 5661–5664. [CrossRef]

65. Omidi-Mirzaei, H.; Azarfar, A.; Mirzaei, M.; Kiani, A.; Ghaffari, M. Effects of forage source and forage particle size as a free-choice provision on growth performance, rumen fermentation, and behavior of dairy calves fed texturized starters. *J. Dairy Sci.* **2018**, *101*, 4143–4157. [CrossRef]

66. Suárez-Mena, F.X.; Heinrichs, A.J.; Jones, C.M.; Hill, T.M.; Quigley, J.D. Straw particle size in calf starters: Effects on digestive system development and rumen fermentation. *J. Dairy Sci.* **2015**, *99*, 341–353. [CrossRef] [PubMed]

67. Mirzaei, M.; Khorvash, M.; Ghorbani, G.R.; Kazemi-Bonchenari, M.; Riasi, A.; Soltani, A.; Moshiri, B.; Ghaffari, M.H. Interactions between the physical form of starter (mashed versus textured) and corn silage provision on performance, rumen fermentation, and structural growth of Holstein calves. *J. Anim. Sci.* **2016**, *94*, 678–686. [CrossRef] [PubMed]

68. Beiranvand, H.; Khani, M.; Omidian, S.; Ariana, M.; Rezvani, R.; Ghaffari, M. Does adding water to dry calf starter improve performance during summer? *J. Dairy Sci.* **2016**, *99*, 1903–1911. [CrossRef] [PubMed]

69. Beiranvand, H.; Khani, M.; Ahmadi, F.; Omidi-Mirzaei, H.; Ariana, M.; Bayat, A. Does adding water to a dry starter diet improve calf performance during winter? *Animal* **2019**, *13*, 959–967. [CrossRef] [PubMed]

70. Kargar, S.; Kanani, M. Reconstituted versus dry alfalfa hay in starter feed diets of Holstein dairy calves: Effects on feed intake, feeding and chewing behavior, feed preference, and health criteria. *J. Dairy Sci.* **2019**, *102*, 4061–4071. [CrossRef]

71. Kargar, S.; Kanani, M. Reconstituted versus dry alfalfa hay in starter feed diets of Holstein dairy calves: Effects on growth performance, nutrient digestibility, and metabolic indications of rumen development. *J. Dairy Sci.* **2019**, *102*, 4051–4060. [CrossRef]

72. Van Soest, P.J. *Nutritional Ecology of the Ruminant*; Cornell university press: Ithaca, NY, USA, 1994; Volume 44, pp. 2552–2561.

73. Pasha, T.N.; Prigge, E.C.; Russell, R.W.; Bryan, W.B. Influence of moisture content of forage diets on intake and digestion by sheep. *J. Anim. Sci.* **1994**, *72*, 2455–2463. [CrossRef]

74. Dill-McFarland, K.A.; Weimer, P.J.; Breaker, J.D.; Suen, G. Diet influences early microbiota development in dairy calves without long-term impacts on milk production. *Appl. Environ. Microbiol.* **2019**, *85*, e02141-18. [CrossRef]

75. Hosseini, S.M.; Ghorbani, G.R.; Khorvash, P.R.M. Determining optimum age of Holstein dairy calves when adding chopped alfalfa hay to meal starter diets based on measures of growth and performance. *Anim. Int. J. Anim. Biosci.* **2015**, *10*, 607. [CrossRef]

76. Lin, X.; Wang, Y.; Wang, J.; Hou, Q.; Hu, Z.; Shi, K.; Yan, Z.; Wang, Z. Effect of initial time of forage supply on growth and rumen development in preweaning calves. *Anim. Prod. Sci.* **2018**, *58*, 2224–2232. [CrossRef]

77. Overvest, M.A.; Bergeron, R.; Haley, D.B.; Devries, T.J. Effect of feed type and method of presentation on feeding behavior, intake, and growth of dairy calves fed a high level of milk. *J. Dairy Sci.* **2015**, *99*, 317–327. [CrossRef] [PubMed]

78. Ebnali, A.; Khorvash, M.; Ghorbani, G.R.; Mahdavi, A.H.; Malekkhahi, M.; Mirzaei, M.; Pezeshki, A.; Ghaffari, M.H. Effects of forage offering method on performance, rumen fermentation, nutrient digestibility and nutritional behaviour in Holstein dairy calves. *J. Anim. Physiol. A Anim. Nutr.* **2016**, *100*, 820–827. [CrossRef] [PubMed]

79. Miller-Cushon, E.K.; DeVries, T.J. Feed sorting in dairy cattle: Causes, consequences, and management. *J. Dairy Sci.* **2017**, *100*, 4172–4183. [CrossRef] [PubMed]

80. Terrã, M.; Castells, L.; Khan, M.A.; Bach, A. Interaction between the physical form of the starter feed and straw provision on growth performance of Holstein calves. *J. Dairy Sci.* **2015**, *98*, 1101–1109. [CrossRef]

81. Lesmeister, K.; Heinrichs, A.J. Effects of corn processing on growth characteristics, rumen development, and rumen parameters in neonatal dairy calves. *J. Dairy Sci.* **2004**, *87*, 3439–3450. [CrossRef]

82. Mojahedi, S.; Khorvash, M.; Ghorbani, G.R.; Ghasemi, E.; Mirzaei, M.; Hashemzadeh-Cigari, F. Performance, nutritional behavior, and metabolic responses of calves supplemented with forage depend on starch fermentability. *J. Dairy Sci.* **2018**, *101*, 7061–7072. [CrossRef]

83. Bateman Ii, H.; Hill, T.; Aldrich, J.; Schlotterbeck, R. Effects of corn processing, particle size, and diet form on performance of calves in bedded pens. *J. Dairy Sci.* **2009**, *92*, 782–789. [CrossRef]

84. Kosiorowska, A.; Puggaard, L.; Hedemann, M.S.; Sehested, J.; Jensen, S.K.; Kristensen, N.B.; Kuropka, P.; Marycz, K.; Vestergaard, M. Gastrointestinal development of dairy calves fed low- or high-starch concentrate at two milk allowances. *Animal* **2011**, *5*, 211–219. [CrossRef]

85. Horvath, K.; Miller-Cushon, E. The effect of milk-feeding method and hay provision on the development of feeding behavior and non-nutritive oral behavior of dairy calves. *J. Dairy Sci.* **2017**, *100*, 3949–3957. [CrossRef]

86. Thomas, D.B.; Hinks, C.E. The effect of changing the physical form of roughage on the performance of the early-weaned calf. *Anim. Sci.* **1982**, *35*, 375–384. [CrossRef]

87. Millercushon, E.K.; Bergeron, R.; Leslie, K.E.; Mason, G.J.; Devries, T.J. Effect of early exposure to different feed presentations on feed sorting of dairy calves. *J. Dairy Sci.* **2013**, *96*, 4624–4633. [CrossRef] [PubMed]

88. Khan, M.A.; Lee, H.J.; Lee, W.S. Structural growth, rumen development, and metabolic and immune responses of Holstein male calves fed milk through step-down and conventional methods. *J. Dairy Sci.* **2007**, *90*, 3376–3387. [CrossRef] [PubMed]

89. Beharka, A.A.; Nagaraja, T.G.; Morrill, J.L.; Kennedy, G.A.; Klemm, R.D. Effects of form of the diet on anatomical, microbial, and fermentative development of the rumen of neonatal calves. *J. Dairy Sci.* **1998**, *81*, 1946. [CrossRef]

90. Schwartzkopf-Genswein, K.S.; Beauchemin, K.A.; Gibb, D.J.; Crews, D.H.; Mcallister, T.A. Effect of bunk management on feeding behavior, ruminal acidosis and performance of feedlot cattle: A review. *J. Anim. Sci.* **2003**, *81*, E149–E158.

91. Garrett, E.F.; Pereira, M.N.; Nordlund, K.V. Diagnostic Methods for the Detection of Subacute Ruminal Acidosis in Dairy Cows. *J. Dairy Sci.* **1999**, *82*, 1170–1178. [CrossRef]

92. Quigley, J.D.; Steen, T.M.; Boehms, S.I. Postprandial Changes of Selected Blood and Ruminal Metabolites in Ruminating Calves Fed Diets with or Without Hay. *J. Dairy Sci.* **1992**, *75*, 228–235. [CrossRef]

93. Kay, R.N.B. The rate of flow and composition of various salivary secretions in sheep and calves. *J. Physiol.* **1960**, *150*, 515–537. [CrossRef]

94. Laarman, A.H.; Ruiz-Sanchez, A.L.; Sugino, T.; Guan, L.L.; Oba, M. Effects of feeding a calf starter on molecular adaptations in the ruminal epithelium and liver of Holstein dairy calves. *J. Dairy Sci.* **2012**, *95*, 2585–2594. [CrossRef]

95. Hodgson, J. The development of solid food intake in calves. 1. The effect of previous experience of solid food, and the physical form of the diet, on the development of food intake after weaning. *Anim. Sci.* **1971**, *13*, 15–24. [CrossRef]

96. Baldwin, R.L.V.; Mcleod, K.R.; Klotz, J.L.; Heitmann, R.N. Rumen development, intestinal growth and hepatic metabolism in the pre- and postweaning ruminant. *J. Dairy Sci.* **2004**, *87*, E55–E65. [CrossRef]

97. Kirat, D.; Masuoka, J.; Hayashi, H.; Iwano, H.; Yokota, H.; Taniyama, H.; Kato, S. Monocarboxylate transporter 1 (MCT1) plays a direct role in short-chain fatty acids absorption in caprine rumen. *J. Physiol.* **2006**, *576*, 635–647. [CrossRef] [PubMed]

98. Graham, C.; Gatherar, I.; Haslam, I.; Glanville, M.; Simmons, N.L. Expression and localization of monocarboxylate transporters and sodium/proton exchangers in bovine rumen epithelium. *AJP Regul. Integr. Comp. Physiol.* **2007**, *292*, R997–R1007. [CrossRef] [PubMed]

99. Flatt, W.P.; Warner, R.G.; Loosli, J.K. Influence of Purified Materials on the Development of the Ruminant Stomach. *J. Dairy Sci.* **1958**, *41*, 1593–1600. [CrossRef]

100. Cline, J.H.; Hershberger, T.V.; Bentley, O.G. Utilization and/or Synthesis of Valeric Acid during the Digestion of Glucose, Starch and Cellulose by Rumen Micro-Organisms. *J. Anim. Sci.* **1958**, *17*, 284–292. [CrossRef]

101. Briggs, P.; Hogan, J.; Reid, R. Effect of volatile fatty acids, lactic acid and ammonia on rumen pH in sheep. *Aust. J. Agric. Res.* **1957**, *8*, 674. [CrossRef]

102. Oetzel, G.R.; Nordlund, K.V.; Garrett, E.F. Effect of ruminal pH and stage of lactation on ruminal lactate concentration in dairy cows. *J. Dairy Sci.* **1999**, *82*, 38.

103. Kennelly, J.J.; Robinson, B.; Khorasani, G.R. Influence of carbohydrate source and buffer on rumen fermentation characteristics, milk yield, and milk composition in early-lactation Holstein cows. *J. Dairy Sci.* **1999**, *82*, 2486–2496. [CrossRef]

104. Jami, E.; Israel, A.; Kotser, A.; Mizrahi, I. Exploring the bovine rumen bacterial community from birth to adulthood. *ISME J.* **2013**, *7*, 1069–1079. [CrossRef]

105. Khan, M.A.; Bach, A.; Weary, D.M.; von Keyserlingk, M.A.G. Invited review: Transitioning from milk to solid feed in dairy heifers. *J. Dairy Sci.* **2016**, *99*, 885–902. [CrossRef]

106. Guzman, C.E.; Bereza-Malcolm, L.T.; De Groef, B.; Franks, A.E. Presence of Selected Methanogens, Fibrolytic Bacteria, and Proteobacteria in the Gastrointestinal Tract of Neonatal Dairy Calves from Birth to 72 Hours. *PLoS ONE* **2015**, *10*, e0133048. [CrossRef] [PubMed]

107. Rey, M.; Enjalbert, F.; Combes, S.; Cauquil, L.; Bouchez, O.; Monteils, V. Establishment of ruminal bacterial community in dairy calves from birth to weaning is sequential. *J. Appl. Microbiol.* **2014**, *116*, 245–257. [CrossRef] [PubMed]

108. Lin, X.; Wang, j.; Hou, Q.; Wang, Y.; Hu, Z.; Shi, K.; Yan, Z.; Wang, Z. Effect of hay supplementation timing on rumen microbiota in suckling calves. *MicrobiologyOpen* **2018**, *7*, e00430. [CrossRef] [PubMed]

109. Thomas, F.; Hehemann, J.H.; Rebuffet, E.; Czjzek, M.; Michel, G. Environmental and gut bacteroidetes: The food connection. *Front. Microbiol.* **2011**, *2*, 93. [CrossRef] [PubMed]

110. Sato, S. Pathophysiological evaluation of subacute ruminal acidosis (SARA) by continuous ruminal pH monitoring. *Anim. Sci. J.* **2016**, *87*, 168–177. [CrossRef]

111. Matsui, H.; Ogata, K.; Tajima, K.; Nakamura, M.; Nagamine, T.; Aminov, R.I.; Benno, Y. Phenotypic Characterization of Polysaccharidases Produced by FourPrevotellaType Strains. *Curr. Microbiol.* **2000**, *41*, 45–49. [CrossRef]

112. Hobson, P.N.; Stewart, C.S. *The Rumen Microbial Ecosystem*; Springer Science & Business Media: Berlin/Heidelberg, Germany, 2012.

113. Morisse, J.P.; Huonnic, D.; Cotte, J.P.; Martrenchar, A. The effect of four fibrous feed supplementations on different welfare traits in veal calves. *Anim. Feed Sci. Technol.* **2000**, *84*, 129–136. [CrossRef]

114. Costa, J.H.C.; Daros, R.R.; von Keyserlingk, M.A.G.; Weary, D.M. Complex social housing reduces food neophobia in dairy calves. *J. Dairy Sci.* **2014**, *97*, 7804–7810. [CrossRef]

115. Khan, M.A.; Weary, D.M.; Veira, D.M.; Von Keyserlingk, M.A.G. Postweaning performance of heifers provided hay during the milk feeding period. *J. Dairy Sci.* **2012**, *95*, 3970–3976. [CrossRef]

116. Van Ackeren, C.; Steingaß, H.; Hartung, K.; Funk, R.; Drochner, W. Effect of roughage level in a total mixed ration on feed intake, ruminal fermentation patterns and chewing activity of early-weaned calves with ad libitum access to grass hay. *Anim. Feed Sci. Technol.* **2009**, *153*, 48–59. [CrossRef]

117. Allen, M.S. Relationship Between Fermentation Acid Production in the Rumen and the Requirement for Physically Effective Fiber. *J. Dairy Sci.* **1997**, *80*, 1447–1451. [CrossRef]

118. Ginane, C.; Baumont, R.; Favreaupeigné, A. Perception and hedonic value of basic tastes in domestic ruminants. *Physiol. Behav.* **2011**, *104*, 666–674. [CrossRef]

119. Leonardi, C.; Armentano, L.E. Effect of quantity, quality, and length of alfalfa hay on selective consumption by dairy cows. *J. Dairy Sci.* **2003**, *86*, 557–564. [CrossRef]

120. Gao, X.; Oba, M. Relationship of severity of subacute ruminal acidosis to rumen fermentation, chewing activities, sorting behavior, and milk production in lactating dairy cows fed a high-grain diet. *J. Dairy Sci.* **2014**, *97*, 3006–3016. [CrossRef] [PubMed]

121. Maulfair, D.D.; McIntyre, K.K.; Heinrichs, A.J. Subacute ruminal acidosis and total mixed ration preference in lactating dairy cows. *J. Dairy Sci.* **2013**, *96*, 6610–6620. [CrossRef] [PubMed]

122. Burritt, E.A.; Provenza, F.D. Lambs form preferences for nonnutritive flavors paired with glucose. *J. Anim. Sci.* **1992**, *70*, 1133. [CrossRef] [PubMed]

123. Miller-Cushon, E.K.; Devries, T.J. Effect of early feed type exposure on diet-selection behavior of dairy calves. *J. Dairy Sci.* **2011**, *94*, 342. [CrossRef]

Review

Review of Strategies to Promote Rumen Development in Calves

Qiyu Diao [1,*,†], **Rong Zhang** [1,2,†] **and Tong Fu** [3]

[1] Key Laboratory of Feed Biotechnology of the Ministry of Agriculture, Feed Research Institute, Chinese Academy of Agricultural Sciences, Beijing 100081, China

[2] Precision Livestock and Nutrition Unit, Gembloux Agro-Bio Tech, University of Liège, Passage des Déportés, 2, 5030 Gembloux, Belgium

[3] College of Animal Science and Veterinary Medicine, Henan Agricultural University, Zhengzhou 450002, China

* Correspondence: diaoqiyu@caas.cn; Tel.: +86-10-8210-6055; Fax: +86-10-8210-6216-9105

† These authors contributed equally to this work.

Received: 22 May 2019; Accepted: 23 July 2019; Published: 26 July 2019

Simple Summary: The rumen is an important digestive organ that plays a key role in the growth, production performance and health of ruminants. Promoting rumen development has always been a key target of calf nutrition. Current research reveals that an early feeding regime and nutrition have effects on rumen development and the establishment of rumen microbiota. The effects may persist for a long time, and consequently, impact the lifetime productive performance and health of adult ruminants. The most sensitive window for rumen manipulation may exist in the postnatal and weaning period. Thus, the early feeding regime and nutrition of calves deserve further research. The establishment of the rumen bacterial community is a mysterious and complex process. The development of microbial 16S rDNA gene sequencing and metagenome analysis enables us to learn more about the establishment of rumen microbes and their interactions in host gastrointestinal (GI) tract development.

Abstract: Digestive tract development in calves presents a uniquely organized system. Specifically, as the rumen develops and becomes colonized by microorganisms, a calf physiologically transitions from a pseudo-monogastric animal to a functioning ruminant. Importantly, the development of rumen in calves can directly affect the intake of feed, nutrient digestibility and overall growth. Even minor changes in the early feeding regime and nutrition can drastically influence rumen development, resulting in long-term effects on growth, health, and milk yields in adult cattle. Rumen development in newborn calves is one of the most important and interesting areas of calf nutrition. This paper presents a comprehensive review of recent studies of the gastrointestinal (GI) tract development in calves. Moreover, we also describe the effect of the environment in shaping the GI tract, including diet, feed additives and feeding management, as well as discuss the strategies to promote the physiological and microbiological development of rumen.

Keywords: calves; rumen; epithelium; microbiota; diet; feed additives; feeding management

---

1. Introduction

Rearing healthy calves is very important as it can have a significant impact on their growth and milk production performance in adult life. Adequate calf development is therefore crucially important for the entire dairy industry. Calves are challenged by a series of stress factors after they are born, including changes in their surroundings. Specifically, the living environment changes from the sterile uterus to natural outside conditions, in addition to changes in nutrition from that provided by the mother to the digestion and absorption of feed by calves themselves. However, due to the poor immunity and the incomplete development of the digestive system in young calves, any interference from the external environment or changes to the nutrition can drastically affect the development of calves [1]. Some of the problems include diarrhea and slow weight gain, as well as respiratory tract disease, which can lead to high levels of morbidity and mortality, and pose significant challenges to breeding.

2. Rumen Development

Compared with monogastric animals, the forestomaches of ruminants have a specialized structure and function, which results in differences in digestion and physiology between ruminants and monogastric animals. Moreover, calves have additional unique systems that are present in their digestive tract during their development. At birth, the rumen is not completely developed, and significant changes in rumen development have to occur first before the calves can digest dry feed to guarantee their own growth needs. The specific changes include the development of the rumen organ and rumen epithelium, and the establishment of rumen microbiota. Understanding rumen development in newborn calves is one of the most important focus areas of calf nutrition.

2.1. Rumen Organ Development

The digestive system of young ruminants begins to develop during the embryonic period. For example, the stomach chambers are visible by day 56 in bovine embryos [2]. At birth, the weights of reticulorumen, omasum, and abomasum account for 38%, 13%, and 49% of the entire stomach weight, respectively [3]. By eight weeks of age, these proportions change to 61.23%, 13.40%, and 25.37% of the stomach weight, respectively [1]. Finally, at 12–16 weeks of age, they reach 67%, 18%, and 15% of the stomach weight, respectively [1,3] (Table 1). The esophageal groove, namely the rumoreticuler groove, is one of the unique features inside the gastrointestinal (GI) tract of calves. The majority of the liquid feed, such as colostrum, whole milk and milk replacer (MR), can bypass the rumen, reticulum and omasum, and flow directly into the abomasum as a result of the reflex closure of the esophageal groove. The abomasum of newborn calves is the only fully developed and functional stomach, and is also the most important digestive organ for calves at birth. The digestion of fat, carbohydrates, and protein is predominantly dependent on the digestive enzymes secreted by the abomasum and small intestine, which is similar to the digestive system in monogastric animals. Over time, with the increase in dry feed intake, the rumen begins to develop and starts to play more important digestive roles.

Table 1. The development of the forestomach.

Items [1]	0 w	8 w	12–16 w
Reticulorumen %	38	61.23	67
Omasum %	13	13.4	18
Abomasum %	49	25.37	15
Total	100	100	100

[1] Each stomach compartment is expressed as a percentage of the total weight of the forestomach.

2.2. Rumen Epithelium

The ruminal epithelium performs many important functions and plays the key role in rumen development, including absorption, transportation, short-chain fatty acid metabolism, and protection. The proliferation and growth of the rumen squamous epithelium promotes the growth of papillae length and width, and increases the thickness of the interior rumen wall [4]. Work by Lesmeister and coworkers (2004) [5] considered the papillae length of the rumen as the most important factor for the evaluation of rumen development, followed by the papillae width and rumen wall thickness. However, papillae per square centimeter is not used as an indicator of rumen development.

Newborn calves have a smooth epithelium with no prominent papillae. Calves fed solely with liquid feed have been shown to have limited rumen development characterized by decrease in rumen weight, papillary growth, degree of keratinization, pigmentation and musculature development [6,7]. Of note, increased intake of solid feed contributes to the rapid development of ruminal fermentation. As calves consume more starter feed, rumen digesta pH decreases, whereas volatile fatty acid (VFA) concentration gradually increases during the first two months. The molar proportion of acetate decreases during the first two months, and then starts to increase until nine months of age as forage intake increases [1,8]. The presence and absorption of VFAs in the rumen provides chemical stimuli required for the proliferation of rumen epithelium [6,9]. Importantly, intraruminal administration of acetate, propionate, and butyrate can stimulate the growth of rumen epithelium in young ruminants, with the effect of butyrate being the most prominent, followed by propionate [4,6]. Studies suggest that rumen papilla proliferation is associated with increased blood flow through the rumen wall [10,11] and a direct effect of butyrate and propionate on gene expression [12].

Despite many studies indicating that VFA can promote the development of rumen epithelium in vivo, the in vitro results suggest the opposite. For example, butyrate treatment decreases DNA synthesis of rumen epithelial cells in culture [13], while the proliferation of rumen epithelial cells is inhibited by rumen fluid in vitro [14]. The divergent in vivo and in vitro response may be linked with an indirect hormonal response to VFA metabolites. Several hormones, such as insulin, pentagastrin, and glucagon, have been implicated as possible VFA mediators that stimulate rumen epithelial proliferation [12,15]. A previous study by Baldwin (1999) reported that proliferation rates of rumen epithelial cells induced by insulin, epidermal growth factor, and insulin-like growth factor (IGF-1) were 75%, 97% and 96%, respectively [16]. Importantly, other studies also suggested that insulin, epidermal growth factor, and IGF-1 can overcome the inhibitory effect of butyrate [16,17].

2.3. Ruminal Microbiota

At birth, the GI tract of young ruminants is sterile. During the first hours of life, the forestomach becomes rapidly colonized with an abundant bacterial population. The neonates acquire bacteria from the dam, partners, feed, housing and environment. The early gut microbes of suckled lambs were mainly derived from the mother's teats (43%) and ambient air (28%), whereas those of bottle-fed lambs were dominated by bacteria from the mother's vagina (46%), ambient air (31%), and the sheep pen floor (12%) (Bi et al., 2019) [18]. Facultative anaerobes such as *Streptococcus* and *Enterococcus* are the early colonizers of rumen, which convert rumen to a fully anaerobic environment to promote the rapid establishment of strictly anaerobic bacteria [19]. By two days of age, the rumen microflora reaches 10^9 cells/mL with strictly anaerobic bacteria being predominantly found in the rumen of lambs [20]. The aerobic and facultatively anaerobic bacteria were 10- to 100-fold lower than the strictly anaerobic bacterial count observed during the first week, which continued to decrease afterwards [20].

Compared to older animals, the abundance of phylum Bacteroidetes was significantly lower in one-day-old calves and was mainly composed of the genus *Bacteroides*, whereas older animals were mainly colonized with *Prevotella* [19]. Work by Malmuthuge and coworkers (2014) [21] reported that the rumen contents of three-week-old calves contained a similar level of *Bacteroides* (15.8%) and *Prevotella* (15.1%), which may suggest that starter feed can propel rumen microbiome development to more mature status. The presence of cellulolytic and methanogenic bacteria was observed in lambs at three–four days of age, and the population of these bacteria reached a level similar to that observed in mature sheep within seven days of age [20]. Study by Jami and colleagues (2013) [19] reported that cellulolytic bacteria and other bacterial species important to rumen function can be detected as early as one day after birth. Thus, the establishment of these rumen bacteria occurs long before young ruminants have access to concentrated feed or forage. Dill-McFarland and coworkers (2017) [22] indicated that calves sampled a few days after weaning had a more diverse rumen community compared to calves sampled during weaning. Several fungal operational taxonomic units (OTUs) observed in weaned calves are also present in adults. As fungi mainly colonize fibrous solids, this may suggest an introduction of forage allows previously low-abundant or transient fungi to persist and multiply.

The rumen bacterial population of two-week-old calves fed milk replacer (MR) was reported to contain 45 bacterial genera belonging to 15 phyla [23]. Similarly, 47 bacterial genera belonging to 13 phyla were observed in the three-week-old calves [21]. Interestingly, the rumen microbiota of the two-week-old calves has more heterogeneous microbiota and harbors more abundant yet transient bacterial species and genera compared to calves at 42 days of age [23]. Another study suggested that the diversity and intra-group similarity of rumen microbiota increases with age, suggesting a transition from a heterogeneous and less distinct community to a more homogeneous and diverse mature bacterial population [19]. This is further supported by a recent study, where gut communities showed higher alpha-diversity but lower beta-diversity with age [22]. Co-habitation facilitates individuals to acquire a shared microbiota [24]. The rumen microbiota was similar in weaned and adult goats that were co-housed pre-weaning [25]. This may also contribute to a convergence toward a similar microbiota in the adult animals.

The composition of the rumen bacterial community varied significantly among individual calves, suggesting a strong host-microbiota specificity in the rumen [19,23]. Similarly, the communities of archaea and fungi in rumen varied considerably among individuals [22,26]. This may suggest that the composition of the rumen microbial community is associated with the physiological condition of the host [19]. Moreover, work by Mayer and coworkers (2012) [27] found that fecal microbial composition was more similar between twin calves than between siblings, implying that host genetics partly define individual gut microbial composition.

Additionally, the bacterial composition was different among the gastrointestinal tract regions and between mucosa- and digesta-associated communities [21]. Colonization of calf rumen starts early in life with a distinct segregation of bacteria between digesta and epithelial surfaces. Similarly, the methanogen community also varies along the gastrointestinal tract [26]. This indicates that previous studies on fecal samples cannot adequately represent the complexity of the gut microbiome. Future studies should focus on both mucosa- and digesta-associated communities in rumen directly.

3. Strategies to Promote Rumen Development

Strategies to promote morphological structure and metabolic function of rumen in pre-ruminants are an ongoing issue which greatly attracts a lot of attention from the scientific community. Numerous studies and approaches attempt to modulate rumen fermentation and the microbial community in young ruminants to accelerate rumen development. These approaches include alteration of diet composition and physical forms, addition of new types of feed additives, and introduction of variables in the feeding management.

3.1. Diet

3.1.1. Liquid Feed

Liquid feed may affect plasma concentration of hormones and growth factors, such as insulin and IGF-1, which play important roles in stimulating proliferation of rumen epithelial cells [16,28]. Colostrum contains many biologically active substances, mainly polypeptide growth factors and steroid hormones, including insulin, IGF-1, and transforming growth factor (TGF). Intake of colostrum has been associated with the development, digestion, and absorption ability of the GI tract in the newborn calves [16,28]. Moreover, a whole milk calf diet was shown to have a positive effect on milk yield during the first lactation of the adults compared to calves fed an MR diet. These results highlight the importance of biologically active milk-borne factors [29].

Soybean protein can be used as an alternative to milk protein in formulating MR [30]. Previous studies suggested that MR formulated with soy proteins can negatively affect the development of the small intestine [31,32]. The abomasal pH declines more slowly and pH is higher in calves fed MR containing soy flour compared to calves given whole milk [33–35]. Decreasing the pH of MR emulsion by addition of an acidifier reduces the pH of digesta pH in the rumen, reticulum, and omasum. Specifically, pH reduction of MR emulsion was found to be beneficial for the development of ruminal epithelium [36]. Work by Górka and coworkers (2011a) [37] reported a shorter papillae length of the cranial dorsal sac in calves fed MR compared to calves fed whole milk, and noted positive relationships between reticulorumen weight and small intestine weight, or with brush border enzyme activities. There is a close relationship between the development of the rumen and the small intestine. Importantly, different types of liquid feed affect the development of the small intestine, the intake of solid feed later in life, as well as the growth and metabolic status of calves, thereby indirectly affecting the development of forestomaches [37]. Enhancing the nutrition level of MR in calves induces changes in the expression of genes coding for proteins directly influencing rumen epithelial growth [38]. Moreover, liquid feed may flow into the rumen due to the closure of the esophageal groove. This can occur even in calves that are not clinically defined as rumen drinkers. Specifically, in veal calves that received large amounts of milk, the amount of leakage liquid was approximately 14–35%, which may induce ruminal and metabolic acidosis in a clinical case [39,40].

3.1.2. Starter Feed

Feeding readily fermentable carbohydrates to calves increases VFA production in the rumen, which is necessary to stimulate the development of rumen epithelium [41,42]. Calves fed milk-only diet during the first three weeks present with a different microbial community in their GI tract and feces compared to calves given milk and solid feed [43]. Diets differing in carbohydrate composition lead to differences in rumen fermentation patterns and VFA profiles which may have a variable effect on rumen development [44,45]. For example, high concentrations of ruminal ammonia, acetate, propionate, and butyrate were detected in calves fed corn- and wheat-based diets compared to calves fed barley- and oat-based diets. Moreover, the forestomach weight and papillae growth were greater in calves fed corn- and wheat-based diets [46]. The mucosal thickness was greater in veal calves fed starch- and pectin-based diets compared to calves on neutral detergent fiber (NDF)-based diets, however, a higher incidence of poorly developed mucosa was observed in calves fed starch-based diet than in animals fed pectin- and NDF-based diets [40]. It was reported that the stimulatory effects of VFAs are different, with butyrate being most stimulatory followed by propionate and then acetate [4,6]. Butyrate provides energy required for rumen wall thickening, formation of papillae and stimulating capillary development [47]. Butyrate can also increase the blood flow during nutrient absorption and metabolism and can directly affect gene expression in the ruminal epithelium [4].

Rumen development can also be affected by the dietary nutrient level. Interestingly, lambs fed a high protein diet had a higher concentration of ammonia nitrogen (NH_3–N) but a lower proportion of total VFA and propionate [48]. Moreover, study by Shen and coworkers (2004) [49] identified that a high energy diet lead to rumen papillae proliferation, which was associated with IGF-1 receptors and increased plasma IGF-1 levels in baby goats. However, excessive consumption of rapidly fermentable starter feed may predispose calves to rumen acidosis. Specifically, it can reduce ruminal pH, decrease rumen motility, and result in keratinization of papillae, causing a decreased in VFA absorption [42,50,51].

3.1.3. Forage

Forage is less energy-intensive than starter feed. The low digestibility of forage in the rumen increases gut fill and decreases voluntary intake of concentrated feed by calves, which results in insufficient levels of VFAs required to stimulate rumen growth [52]. However, forage consumption is associated with positive effects of fiber on rumination and salivation in the GI tract [53,54]. The inclusion of forage in the diet increases rumen pH in both pre-weaning and post-weaning calves [55,56]. Importantly, intake of forage was negatively correlated with the severity of subacute ruminal acidosis (SARA), suggesting that a small quantity of consumed forage (0.080 kg/day) can alleviate rumen acidosis in calves [57]. The empty rumen weight was greater in calves supplemented with hay compared to calves fed a hay-free diet [54,56]. During weaning transition, feeding dietary forage in calves mitigates ruminal acidosis and induces changes in ruminal bacterial diversity and abundance [58]. Thus, two completely opposite opinions exist as to whether to feed forage to calves before weaning. To address this issue, several studies have been conducted to compare the effect of different initial time of forage provision on growth and rumen development in calves. Lin and colleagues (2017) [59] indicated that supplementation of oat hay to pre-weaned calves increased starter feed intake, ruminal pH, and reduced non-nutritive oral behaviors. Calves with hay supplementation initiated at two weeks of age showed the best productivity. Another study found that feeding forage to calves, either from 3 or 15 days of age, had no effect on growth rate, feed intake and rumen fermentation parameters compared to calves fed no forage, which also justified the supply of forage to young calves [60]. Inclusion of forage in the starter feed was positively linked with muscular development of the rumen [61,62] and morphological appearances of rumen epithelial cells, and caused decreased plaque formation [40,61]. Replacing 50% barley or corn with corn silage in the diet given to 10- or 90-day-old calves improved the thickness of the rumen wall, but had no significant effect on the papillae [63].

Different forage sources have different effects on stimulating chewing activity and saliva production [64]. Supplementation with NDF from alfalfa hay in the starter diet was shown to be more effective than beet pulp in increasing rumen pH and stimulating chewing activity [65]. A recent meta-analysis indicated that forage consumption can affect starter feed intake and performance in calves, which was modulated by forage level, sources, and physical forms of the starter [66].

3.1.4. Physical Form

The physical form and particle size distribution of the diet exert significant influence on the anatomical and microbial development of the rumen [50,54]. For example, calves fed a ground diet had shorter papillae with a smaller surface area compared to calves fed the unground diet. Moreover, a decrease in cellulolytic bacteria and an increase in amylolytic bacteria were detected in calves fed the ground diet [50]. Consumption of finely ground diets can reduce ruminal pH [57] and lead to rumen parakeratosis [50,67]. Given these considerations, 75% of the particles in the starter feed should exceed 1190 μm in diameter [68]. Work by Lesmeister and Heinrichs (2004) [69] reported that calves fed texturized starter feed containing whole corn had higher ruminal pH compared to calves fed diet with dry-rolling corn, roasted-rolling corn, or steam-flaked corn. Increasing particle score of alfalfa hay from 1 mm to 3 mm can affect non-nutritive oral behaviours in calves fed finely ground starter feed [70]. However, research by Suarez-Mena and coworkers (2015, 2016) [8,71] suggested that increasing particle size of the starter diet by adding whole oat or straw of different lengths had no effect on rumen fermentation and calf development. Moreover, chopping of hay grass (~50% particles > 1.9 cm) decreased chewing time of calves [72], meanwhile, the richness and diversity of rectal microflora was reduced [73]. Provision of rations containing finely ground hay (2 mm) to calves may increase feed sorting and result in imbalanced intake of nutrients after weaning [74]. Increasing length of chopped hay from 2 mm to 3–4 cm reduced non-nutritive oral behaviors and improved nutrient digestibility [75]. The effect of the physical form and shape of the diet on calves is closely related to the inclusion rate, source, nutrient matrix and processing method of each ingredient. Importantly, the optimal calf diet specification designed specifically to promote rumen development has not been yet defined.

3.2. Feed Additives

3.2.1. Probiotics

Probiotics are viable and beneficial microorganisms that help maintain GI microbial balance and promote rumen development. Feeding probiotics to calves around weaning age may facilitate the development of rumen bacterial communities and help calves with a transition from liquid feed to dry feed and forage [76,77]. Fermentation products of *Saccharomyces cerevisiae* have been shown to positively influence ruminal microbiota and improve ruminal morphology [78,79]. Specifically, effects of *Bacillus licheniformis*, *Saccharomyces cerevisiae* and their compounds can increase nitrogen utilization of the rumen microbial community and affect the fermentation pattern which was shown to be beneficial for growth of fattening lamb [80]. An oral dose of *Megasphaera elsdenii* NCIMB 41125 given to calves at 14 days of age increased ruminal butyrate, reticulorumen weight and papillae growth, suggesting an improvement in epithelial metabolism [81]. Supplementation of *Bacillus subtilis natto* in starter feed was shown to aid the development of rumen bacterial communities by increasing the growth of cellulolytic bacteria in calves after weaning [82].

However, feeding probiotics to calves has not always been shown to exert positive effects on the development of cellulolytic bacteria. For example, adding a mixture of *Lactobacillus plantarum* and *Bacillus subtilis* to MR and starter feed affected the denatured gradient gel electrophoresis (DGGE) fingerprint of the 16S ribosomal RNA genes, and reduced the number of *Ruminococcus albus* in calves [83]. In contrast, other studies reported that pH and enzymatic activities of rumen fluid were unaffected by three kinds of probiotic feeding in newborn calves [84]. Supplementation of *Candida tropicalis* in MR had no effect on the morphology of the forestomach and enzymatic activities of ruminal digesta [85]. Rumen and papillae measurements of Holstein bull calves were not affected by inclusion of *Aspergillus oryzae* fermentation extract in MR and starter feed [86]. Overall, the effects of probiotics on rumen development in calves are inconclusive, and frequently driven by differences in viable probiotic bacterial numbers, probiotics species, administration methods, and health status of animals.

3.2.2. Effects of VFAs

VFAs are the primary products of rumen fermentation and contribute to rumen epithelium development in calves. Previous studies suggested that infusion of sodium propionate or sodium butyrate greatly promotes the development of the rumen papillae in calves [6,42,87]. Supplementation of MR with sodium butyrate was associated with increased reticulorumen weight and increased length and width of papillae [37,88,89]. Another study showed that calves receiving a blend of short and medium chain fatty acids as monoglycerides (0.2%) in MR had less degenerative tissue accumulation and a higher number of cytoplasmic protrusions on the exposed horn surfaces [90].

Branched-chain VFAs (BCVFA), such as isobutyrate, isovalerate and 2-methylbutyrate, are naturally derived from the catabolism of branched-chain amino acids. Adequate levels of BCVFA are essential for the growth of some cellulolytic bacteria and digestion of structural carbohydrates in the rumen [91–93]. Supplementation of isobutyrate and isovalerate in milk and concentrate feed can accelerate the growth of calves by improving ruminal fermentation, rumen enzyme activities and growth of cellulolytic bacteria [94,95]. Administration of VFAs have been proved to be effective in promoting rumen development in calves, however, the optimal inclusion rate of VFAs and BCVFA in feed deserved further researches.

3.2.3. Plant Extracts

There are many studies focused on evaluating the potential of plant extracts as alternatives to feed antibiotics and growth promoters in ruminant nutrition. Plant extracts have been shown to favorably affect rumen microbiota [96] and modulate ruminal fermentation in ruminants [97–99]. However, studies evaluating how plant extracts affect rumen development in young ruminants are limited. A recent study revealed that adding Aloe barbadensis to milk was beneficial in increasing total VFA concentration and bacterial count in cross-bred calves [100]. Supplementation of mulberry leaf flavonoids in MR increased α-amylase activity in ruminal digesta and protease activity in abomasal digesta in calves [85]. Supplementation of caraway and garlic in concentrated feed can improve rumen fermentation parameters by increasing total VFAs, increasing rumen pH and decreasing rumen ammonia in growing buffalo calves [101]. Thyme and cinnamon essential oils were shown to decrease the molar proportion of acetate and lower the ratio of acetate to propionate, as well as increase the level of propionate in Holstein calves consuming a high-concentrate diet. Finally, cinnamon essential oil was shown to increase rumen molar concentration of butyrate [102]. Plant extracts are among the most promising alternatives to antibiotics due to their extensive biological effects, and can be used in calf feed to prevent diarrhea. However, the efficacy of plant extracts is subject to a series of factors, including the composition of active components, addition levels, and physiological status of animals. The use of different types of plant extracts at various inclusion rates in the diet deserves further research. Moreover, effects of plant extracts on the colonization of microbial populations remains to be determined in calves.

3.3. Feeding Management

Weaning age can influence the development of rumen in pre-ruminants. For example, calves weaned at six weeks of age had longer and wider papillae compared to calves weaned at nine weeks of age [103]. In early-weaned calves, the ruminal pH, molar proportion of acetate and the ratio of acetate to propionate were lower, but the molar proportion of propionate and butyrate were greater [104]. The β-diversity of ruminal microbiota shifts rapidly in calves weaned at six weeks, while a more gradual shift is observed in calves weaned at eight weeks [105]. The colonization pattern substantially differs between newborn goats reared naturally with the dam and those reared artificially with MR. A higher bacterial diversity was observed in natural-fed goats [106]. Compared with suckling feeding, bottle feeding mode tended to increase the number of potential pathogens and delay the establishment of anaerobic microbes in the gut of lambs [18]. The total rumen bacterial population of lambs grazing at pasture with the nursing mother was lower compared to lambs weaned at 21 or 35 days of age, whereas methanogens and protozoa population were lower in early-weaned lambs compared to grazing lambs [107]. Kehoe and coworkers (2007) [108], however, reported that weaning age had no effect on rumen papillae length, width or rumen wall thickness. Different weaning methods (conventional weaning or concentrate-dependent weaning) result in similar rumen development [109]. The development of ruminal microbiome was not affected by the weaning strategy, and there was no effect of gradual or abrupt weaning [110]. The difference may mainly be associated with rumen development status. Due to the differences in feeding and management during the pre-weaning period, rumen development of calves may vary in different experiments. Calves with a well-developed rumen are able to utilize grains and forage efficiently. The effect of weaning age may only be detected in calves with undeveloped rumen. Additionally, pair-housed calves were shown to consume more solid feed at an earlier age compared to calves housed individually [111,112].

Intensive feeding of milk or MR may decrease starter feed intake, thereby delaying rumen development (Cowles, 2006) [113]. Hence, the amount of milk supplied to calves is normally restricted to promote starter feed intake and rumen development in conventional feeding practices [3,114]. However, calves fed limited amounts of milk had lower growth rates and abnormal behavior due to reduced nutrient intakes [55]. Schäff and colleagues (2018) [115] reported that compared to calves fed MR ad libitum, calves fed a restricted amount of MR had greater density of the rumen papillae in the atrium and ventral blind sac, but lower villus surface area and villus height/crypt depth ratio in the distal jejunum. Enhanced MR feeding increased the concentration of plasma IGF-1 and insulin [116,117], which may be beneficial for gastrointestinal growth in pre-weaning calves [16,28]. Furthermore, increasing nutrient intake from milk or MR resulted in enhanced milk yield in the first lactation [118]. Thus, intensive feeding practices have been widely adopted by producers; however, supporting feeding programs, such as a gradual weaning plan, need to be detailed to ensure optimum rumen development.

4. Conclusions

To summarize, it is beneficial for rumen development for calves to be fed high-quality liquid feed rich in biologically active substances. Minimization of the use of soy protein or appropriate acidification of MR may alleviate gastrointestinal epithelium lesions. Feeding readily fermentable carbohydrates to calves to increase VFA production can stimulate rumen development. A pellet or texturized starter feed is superior to a finely ground meal. Providing calves with high-graded forage, such as alfalfa hay, can reduce the occurrence of rumen acidosis and papillae keratinization. Moreover, additives can be used in calf feed due to their potential advantages in rumen manipulation, however, the types and the optimal inclusion rate deserve further study. More importantly, there is no fixed pattern of calf feed. The diet compositions and nutrient specifications should be matched with the feeding program and management to better promote rumen development.

The rumen is a unique part of the GI tract in ruminants. As the rumen develops and becomes colonized by microorganisms, a calf physiologically transitions from a pseudo-monogastric to a functioning ruminant. The development of rumen in calves can directly affect feed intake, nutrient digestibility and eventual growth of calves. Any changes in the early feeding regime and nutrition can influence rumen development, and thus, lead to long-lasting effects on subsequent growth, health, and milk production performance. Study by Moallem and colleagues (2010) [29] reported higher milk yields during the first lactation in heifers fed whole milk compared to heifers fed MR. Moreover, the same study suggested that MR did not impart any milk-borne effects in calves [29]. Increasing the nutrient intake from milk or MR prior to weaning resulted in an increased milk yield during the first lactation [118–120]. This phenomenon may be associated with epigenetic effects of early nutrition [118].

Additionally, an early feeding regime and nutrition can influence rumen development and rumen microbial composition, ultimately exerting an effect on the lifetime milk yield in cattle. Studies indicated that diets can modify the establishment of the bacterial community in lambs during weaning, which can persist for four to five months [121,122]. The postnatal period is frequently referred to as the most sensitive window for rumen manipulation [123–125], although studies evaluating ruminal imprinting are still limited. The majority of published studies focus on rumen organ development, rumen fermentation parameters, morphology, and changes in the population of cellulolytic bacteria. With the development of microbial 16S rDNA gene sequencing and metagenome analysis, additional studies will likely reveal the interactions between host GI tract development and establishment of rumen bacteria.

Author Contributions: Conceptualization, Q.D.; Writing—original draft, Q.D. and R.Z.; Writing—review & editing, Q.D. and T.F.

Funding: This research was supported by the Earmarked Fund for Beijing Dairy Industry Innovation Consortium of Agriculture Research System (BAIC06-2019) and the Collaborative Innovation Task of the Agricultural Science and Technology Innovation Program in Chinese Academy of Agricultural Sciences-Integration Innovation of Technology in Dairy Environment-Safe Rearing and Breeding.

Conflicts of Interest: The authors declare no conflict of interest.

References

1. Diao, Q.; Zhang, R.; Tu, Y. Current research progresses on calf rearing and nutrition in China. *J. Integr. Agric.* **2017**, *16*, 2805–2814. [CrossRef]
2. Warner, E.D. The organogenesis and early histogenesis of the bovine stomach. *Am. J. Anat.* **1958**, *102*, 33. [CrossRef] [PubMed]
3. Davis, C.L.; Drackley, J.K. *The Development, Nutrition, and Management of the Young Calf*; Iowa State University Press: Iowa City, IA, USA, 1998.
4. Baldwin, R.L.; McLeod, K.R.; Klotz, J.L.; Heitmann, R.N. Rumen development, intestinal growth and hepatic metabolism in the pre- and postweaning ruminant. *J. Dairy Sci.* **2004**, *87* (Suppl. E), E55–E65. [CrossRef]
5. Lesmeister, K.E.; Tozer, P.R.; Heinrichs, A.J. Development and analysis of a rumen tissue sampling procedure. *J. Dairy Sci.* **2004**, *87*, 1336–1344. [CrossRef]
6. Tamate, H.; McGilliard, A.D.; Jacobson, N.L.; Getty, R. Effect of various dietaries on the anatomical development of the stomach in the calf. *J. Dairy Sci.* **1962**, *45*, 408–420. [CrossRef]

7. Gilliland, R.L.; Bush, L.J.; Friend, J.D. Relationship of ration composition to rumen development in early weaned dairy calves with observations on ruminal parakeratosis. *J. Dairy Sci.* **1962**, *45*, 1211–1217. [CrossRef]

8. Suarez-Mena, F.X.; Heinrichs, A.J.; Jones, C.M.; Hill, T.M.; Quigley, J.D. Straw particle size in calf starters: Effects on digestive system development and rumen fermentation. *J. Dairy Sci.* **2016**, *99*, 341–353. [CrossRef]

9. Flatt, W.P.; Warner, R.G.; Loosli, J.K. Influence of purified materials on the development of the ruminant stomach. *J. Dairy Sci.* **1958**, *41*, 1593–1600. [CrossRef]

10. Sander, E.G.; Warner, H.N.; Harrison, H.N.; Loosli, J.K. The stimulatory effect of sodium butyrate and sodium propionate on the development of rumen mucosa in the young calf. *J. Dairy Sci.* **1959**, *42*, 1600–1605. [CrossRef]

11. Glauber, J.G.; Wandersee, N.J.; Little, J.A.; Ginder, G.D. 5′-Flanking sequences mediate butyrate stimulation of embryonic globin gene expressioin in adult erythroid cells. *Mol. Cell. Biol.* **1991**, *11*, 4690–4697. [CrossRef]

12. Galfi, P.; Neogrady, S.; Sakata, T. Effects of volatile fatty acids on the epithelial cell proliferation of the digestive tract and its hormonal mediation. In *Physiological Aspects of Digestion and Metabolism, Proceedings of the Seventh International Symposium on Ruminant Physiology, Sendai, Japan, 28 August–1 September 1989*; Tsuda, T., Sasaki, Y., Kawashima, R., Eds.; Academic Press: Cambridge, MA, USA, 1991; pp. 49–59.

13. Neogrady, S.; Galfi, P.; Kutas, F. Effects of butyrate and insulin and their interaction on the DNA synthesis of rumen epithelial cells in culture. *Experientia* **1989**, *45*, 94–96. [CrossRef] [PubMed]

14. Wang, A.; Jiang, H. Rumen fluid inhibits proliferation and stimulates expression of cyclin-dependent kinase inhibitors 1A and 2A in bovine rumen epithelial cells1. *J. Anim. Sci.* **2010**, *88*, 3226–3232. [CrossRef] [PubMed]

15. Sakata, T.; Hikosaka, K.; Shiomura, Y.; Tamate, H. Stimulatory effect of insulin on ruminal epithelium cell mitosis in adult sheep. *Br. J. Nutr.* **1980**, *44*, 325–331. [CrossRef] [PubMed]

16. Baldwin, R. The proliferative actions of insulin, insulin-like growth factor-I, epidermal growth factor, butyrate and propionate on ruminal epithelial cells in vitro. *Small Rumin. Res.* **1999**, *32*, 261–268. [CrossRef]

17. Galfi, P.; Neogrady, S. Epithelial and non-epithelial cell and tissue culture from the rumen mucosa. *Asian Aust. J. Anim. Sci.* **1989**, *2*, 143–149. [CrossRef]

18. Bi, Y.L.; Cox, M.S.; Zhang, F.; Suen, G.; Zhang, N.F.; Tu, Y.; Diao, Q.Y. Feeding modes shape the acquisition and structure of the initial gut microbiota in newborn lambs. *Environ. Microbiol.* **2019**, *21*, 2333–2346. [CrossRef] [PubMed]

19. Jami, E.; Israel, A.; Kotser, A.; Mizrahi, I. Exploring the bovine rumen bacterial community from birth to adulthood. *ISME J.* **2013**, *7*, 1069–1079. [CrossRef] [PubMed]

20. Fonty, G.; Gouet, P.; Jouany, J.P.; Senaud, J. Establishment of the microflora and anaerobic fungi in the rumen of lambs. *J. Gen. Microbiol.* **1987**, *133*, 1835–1843. [CrossRef]

21. Malmuthuge, N.; Griebel, P.J.; Guan, L.L. Taxonomic Identification of Commensal Bacteria Associated with the Mucosa and Digesta throughout the Gastrointestinal Tracts of Preweaned Calves. *Appl. Environ. Microb.* **2014**, *80*, 2021–2028. [CrossRef]

22. Dill-McFarland, K.A.; Breaker, J.D.; Suen, G. Microbi-al succession in the gastrointestinal tract of dairy cows from 2 weeks to first lactation. *Sci. Rep.* **2017**, *7*, 40864. [CrossRef]

23. Li, R.W.; Connor, E.E.; Li, C.; Baldwin VI, R.L.; Sparks, M.E. Characterization of the rumen microbiota of pre-ruminant calves using metagenomic tools. *Environ. Microbiol.* **2012**, *14*, 129–139. [CrossRef] [PubMed]

24. Song, S.J.; Lauber, C.; Costello, E.K.; Lozupone, C.A.; Humphrey, G.; Berg-Lyons, D.; Caporaso, J.G.; Knights, D.; Clemente, J.C.; Nakielny, S.; et al. Cohabiting familymembers share microbiota with one another and with their dogs. *eLife* **2013**, *2*, e00458. [CrossRef] [PubMed]

25. Wang, L.; Xu, Q.; Kong, F.; Yang, Y.; Wu, D.; Mishra, S.; Li, Y. Exploring the goat rumen microbiome from seven days to two years. *PLoS ONE* **2016**, *11*, e0154354. [CrossRef] [PubMed]

26. Zhou, M.; Chen, Y.; Griebel, P.J.; Guan, L.L. Methano-gen prevalence throughout the gastrointestinal tract of pre-weaned dairy calves. *Gut Microbes* **2014**, *5*, 628–663. [CrossRef]

27. Mayer, M.; Abenthuma, J.M.; Matthesa, D.; Kleebergera, M.J.; Egeb, C.; Holzel, J.; Bauer, J.; Schwaiger, K. Development and genetic influence of the rectal bacterial flora of newborn calves. *Vet. Microbiol.* **2012**, *161*, 179–185. [CrossRef] [PubMed]

28. Žitňan, R.; Kuhla, S.; Sanftleben, P.; Bilska, A.; Schneider, F.; Zupcanova, M.; Voigt, J. Diet induced ruminal papillae development in neonatal calves not correlating with rumen butyrate. *Vet. Med. Praha* **2005**, *50*, 472–479. [CrossRef]

29. Moallem, U.; Werner, D.; Lehrer, H.; Zachut, M.; Livshitz, L.; Yakoby, S.; Shamay, A. Long-term effects of ad libitum whole milk prior to weaning and prepubertal protein supplementation on skeletal growth rate and first-lactation milk production. *J. Dairy Sci.* **2010**, *93*, 2639–2650. [CrossRef] [PubMed]

30. Dawson, D.P.; Morrill, J.L.; Reddy, P.G.; Minocha, H.C. Soy protein concentrate and heated soy flours as protein sources in milk replacer for preruminant calves. *J. Dairy Sci.* **1988**, *71*, 1301–1309. [CrossRef]

31. Seegraber, F.J.; Morrill, J.L. Effect of protein source in calf milk replacers on morphology and absorptive ability of small intestine. *J. Dairy Sci.* **1986**, *69*, 460–469. [CrossRef]

32. Blattler, U.; Hammon, H.M.; Morel, C.; Philipona, C.; Rauprich, A.; Rome, V.; Le Huerou-Luron, I.; Guilloteau, P.; Blum, J.W. Feeding colostrum, its composition and feeding duration variably modify proliferation and morphology of the intestine and digestive enzyme activities of neonatal calves. *J. Nutr.* **2001**, *131*, 1256–1263. [CrossRef]

33. Colvin, B.M.; Lowe, R.A.; Ramsey, H.A. Passage of digesta from the abomasums of a calf fed soy flour milk replacers and whole milk. *J. Dairy Sci.* **1969**, *52*, 687–688. [CrossRef]

34. Smith, R.H.; Sissons, J.W. The effect of different feeds, including those containing soya-bean products, on the passage of digesta from the abomasums of the preruminant calf. *Br. J. Nutr.* **1975**, *33*, 329–349. [CrossRef]

35. Constable, P.D.; Ahmed, A.F.; Misk, N.A. Effect of suckling cow's milk or milk replacer on abomasal luminal pH in dairy calves. *J. Vet. Intern. Med.* **2005**, *19*, 97–102. [CrossRef]

36. Zhang, R.; Diao, Q.Y.; Zhou, Y.; Yun, Q.; Deng, K.D.; Qi, D.; Tu, Y. Decreasing the pH of milk replacer containing soy flour affects nutrient digestibility, digesta pH, and gastrointestinal development of preweaned calves. *J. Dairy Sci.* **2017**, *100*, 236–243. [CrossRef]

37. Górka, P.; Kowalski, Z.; Pietrzak, P.; Kotunia, A.; Jagusiak, W.; Zabielski, R. Is rumen development in newborn calves affected by different liquid feeds and small intestine development? *J. Dairy Sci.* **2011**, *94*, 3002–3013. [CrossRef]

38. Naeem, A.; Drackley, J.K.; Stamey, J.; Loor, J.J. Role of metabolic and cellular proliferation genes in ruminal development in response to enhanced plane of nutrition in neonatal Holstein calves. *J. Dairy Sci.* **2012**, *95*, 1807–1820. [CrossRef]

39. Berends, H.; van Reenen, C.G.; Stockhofe-Zurwieden, N.; Gerrits, W.J.J. Effects of early rumen development and solid feed composition on growth performance and abomasal health in veal calves. *J. Dairy Sci.* **2012**, *95*, 3190–3199. [CrossRef]

40. Suárez, B.J.; Reenen, C.G.V.; Stockhofe, N.; Dijkstra, J.; Gerrits, W.J.J. Effect of roughage source and roughage to concentrate ratio on animal performance and rumen development in veal calves. *J. Dairy Sci.* **2007**, *90*, 2390–2403. [CrossRef]

41. Heinrichs, A.J.; Lesmeister, K.E. Rumen development in the dairy calf. In *Calf and Heifer Rearing*; Garnworthy, P.C., Ed.; Nottingham University Press: Nottingham, UK, 2004; pp. 53–66.

42. Mentschel, J.; Leiser, R.; Mülling, C.; Pfarrer, C.; Claus, R. Butyric acid stimulates rumen mucosa development in the calf mainly by a reduction of apoptosis. *Arch. Für Tierernaehrung.* **2001**, *55*, 85–102. [CrossRef]

43. Guzman, C.E.; Bereza-Malcolm, L.T.; DeGroef, B.; Franks, A.E. Uptake of milk with and without solid feed during the monogastric phase: Effect on fibrolytic and methanogenic microorganisms in the gastrointestinal tract of calves. *Anim. Sci. J.* **2015**, *87*, 378–388. [CrossRef]

44. Bannink, A.; Kogut, J.; Dijkstra, J.; France, J.; Kebreab, E.; Van Vuuren, A.M.; Tamminga, S. Estimation of the stoichiometry of volatile fatty acid production in the rumen of lactating cows. *J. Theor. Biol.* **2006**, *238*, 36–51. [CrossRef] [PubMed]

45. Suárez, B.J.; Van Reenen, C.G.; Beldman, G.; van Delen, J.; Dijkstra, J.; Gerrits, W.J.J. Effects of supplementing concentrates differing in carbohydrate composition in veal calf diets: I. Animal performance and rumen fermentation characteristics. *J. Dairy Sci.* **2006**, *89*, 4365–4375. [CrossRef]

46. Khan, M.A.; Lee, H.J.; Lee, W.S.; Kim, H.S.; Kim, S.B.; Park, S.B.; Ha, J.K.; Choi, Y.J. Starch Source Evaluation in Calf Starter: II. Ruminal Parameters, Rumen Development, Nutrient Digestibilities, and Nitrogen Utilization in Holstein Calves. *J. Dairy Sci.* **2008**, *91*, 1140–1149. [CrossRef] [PubMed]

47. Weigand, E.; Young, J.W.; McGilliard, A.D. Volatile fatty acid metabolism by rumen mucosa from cattle fed hay or grain. *J. Dairy Sci.* **1975**, *58*, 1294–1300. [CrossRef]

48. Yang, C.; Si, B.; Diao, Q.; Jin, H.; Zeng, S.; Tu, Y. Rumen fermentation and bacterial communities in weaned Chahaer lambs on diets with different protein levels. *J. Integr. Agric.* **2016**, *15*, 1564–1574. [CrossRef]

49. Shen, Z.; Seyfert, H.M.; Lohrke, B.; Schneider, F.; Zitnan, R.; Chudy, A.; Kuhla, S.; Hammon, H.M.; Blum, J.W.; Martens, H.; et al. An energy rich diet causes rumen papillae proliferation associated with more IGF type 1 receptors and increased plasma IGF-1 concentrations in young goats. *J. Nutr.* **2004**, *134*, 11–17. [CrossRef] [PubMed]

50. Beharka, A.A.; Nagaraja, T.G.; Morrill, J.L.; Kennedy, G.A.; Klemm, R.D. Effect of form of the diet on anatomical, and fermentative development of rumen of neonatal calves. *J. Dairy Sci.* **1998**, *81*, 1946–1955. [CrossRef]

51. Greenwood, R.H.; Morril, J.L.; Titgemeyer, E.C.; Kennedy, G.A. A new method of measuring diet abrasion and its effect on the development of the forestomach. *J. Dairy Sci.* **1997**, *80*, 2534–2541. [CrossRef]

52. Coverdale, J.A.; Tyler, H.D.; Quigley, J.D.; Brumm, J.A. Effect of various levels of forage and form of diet on rumen development and growth in calves. *J. Dairy Sci.* **2004**, *87*, 2554–2562. [CrossRef]

53. McBurney, M.I.; van Soest, P.J.; Chase, L.E. Cation exchange capacity and buffering capacity of neutral detergent fibers. *J. Sci. Food Agric.* **1983**, *34*, 910–916. [CrossRef]

54. Pazoki, A.; Ghorbani, G.R.; Kargar, S.; Sadeghi-Sefidmazgi, A.; Drackley, J.K.; Ghaffari, M.H. Growth performance, nutrient digestibility, ruminal fermentation, and rumen development of calves during transition from liquid to solid feed: Effects of physical form of starter feed and forage provision. *Anim. Feed Sci. Tech.* **2017**, *234*, 173–185. [CrossRef]

55. Khan, M.A.; Weary, D.M.; von Keyerslingk, M.A.G. Hay intake improves performance and rumen development of calves fed higher quantities of milk. *J. Dairy Sci.* **2011**, *94*, 3547–3553. [CrossRef] [PubMed]

56. Castells, L.; Bach, A.; Arís, A.; Terré, M. Effects of forage provision to young calves on rumen fermentation and development of the gastrointestinal tract. *J. Dairy Sci.* **2013**, *96*, 5226–5236. [CrossRef] [PubMed]

57. Laarman, A.H.; Oba, M. Short communication: Effect of calf starter on rumen pH of Holstein dairy calves at weaning. *J. Dairy Sci.* **2011**, *94*, 5661–5664. [CrossRef] [PubMed]

58. Kim, Y.-H.; Nagata, R.; Ohtani, N.; Ichijo, T.; Ikuta, K.; Sato, S. Effects of Dietary Forage and Calf Starter Diet on Ruminal pH and Bacteria in Holstein Calves during Weaning Transition. *Front. Microbiol.* **2016**, *7*, 1575. [CrossRef] [PubMed]

59. Lin, X.Y.; Wang, Y.; Wang, J.; Hou, Q.L.; Hu, Z.Y.; Shi, K.R.; Yan, Z.G. Effect of initial time of forage supply on growth and rumen development in preweaning calves. *Anim. Prod. Sci.* **2018**, *58*, 2224–2232. [CrossRef]

60. Wu, Z.H.; Azarfar, A.; Simayi, A.; Li, S.L.; Jonker, A.; Cao, Z.J. Effects of forage type and age at which forage provision is started on growth performance, rumen fermentation, blood metabolites and intestinal enzymes in Holstein calves. *Anim. Prod. Sci.* **2018**, *58*, 2288–2299. [CrossRef]

61. Beiranvand, H.; Ghorbani, G.R.; Khorvash, M.; Nabipour, A.; Dehghan-Banadaky, M.; Homayouni, A.; Kargar, S. Interactions of alfalfa hay and sodium propionate on dairy calf performance and rumen development. *J. Dairy Sci.* **2014**, *97*, 2270–2280. [CrossRef]

62. Mirzaei, M.; Khorvash, M.; Ghorbani, G.R.; Kazemi-Bonchenari, M.; Riasi, A.; Nabipour, A.; van-den-Borne, J.J.G.C. Effects of supplementation level and particle size of alfalfa hay on growth characteristics and rumen development in dairy calves. *J. Anim Physiol. Anim. Nutr.* **2015**, *99*, 553–564. [CrossRef]

63. Sosin-Bzducha, E.; Strzetelski, J.; Borowiec, F.; Kowalczyk, J.; Okon, K. Effect of feeding ensiled maize grain on rumen development and calf rearing performance. *J. Anim. Feed. Sci.* **2010**, *19*, 195–210. [CrossRef]

64. Mertens, D.R. Creating a system for meeting the fiber requirements of dairy cows. *J. Dairy Sci.* **1997**, *80*, 1463–1481. [CrossRef]

65. Maktabi, H.; Ghasemi, E.; Khorvash, M. Effects of substituting grain with forage or nonforage fiber source on growth performance, rumen fermentation, and chewing activity of dairy calves. *Anim. Feed Sci. Tech.* **2016**, *221*, 70–78. [CrossRef]

66. Imani, M.; Mirzaei, M.; Baghbanzadeh-Nobari, B.; Ghaffari, M.H. Effects of forage provision to dairy calves on growth performance and rumen fermentation: A meta-analysis and metaregression. *J. Dairy Sci.* **2017**, *100*, 1136–1150. [CrossRef] [PubMed]

67. McGavin, M.D.; Morrill, J.L. Scanning electron microscopy of ruminal papillae in calves fed various amounts and forms of roughage. *Am. J. Vet. Res.* **1976**, *37*, 497. [PubMed]

68. Porter, J.C.; Warner, R.G.; Kertz, A.F. Effect of fiber level and physical form of starter on growth and development of dairy calves fed no forage. *Prof. Anim. Sci.* **2007**, *23*, 395–400. [CrossRef]

69. Lesmeister, K.E.; Heinrichs, A.J. Effects of corn processing on growth characteristics, rumen development, and rumen parameters in neonatal dairy calves. *J. Dairy Sci.* **2004**, *87*, 3439–3450. [CrossRef]

70. Nemati, M.; Amanlou, H.; Khorvash, M.; Moshiri, B.; Mirzaei, M.; Khan, M.A.; Ghaffari, M.H. Rumen fermentation, blood metabolites, and growth performance of calves during transition from liquid to solid feed: Effects of dietary level and particle size of alfalfa hay. *J. Dairy Sci.* **2015**, *98*, 7131–7141. [CrossRef]

71. Suarez-Mena, F.X.; Heinrichs, A.J.; Jones, C.M.; Hill, T.M.; Quigley, J.D. Digestive development in neonatal dairy calves with either whole or ground oats in the calf starter. *J. Dairy Sci.* **2015**, *98*, 3417–3431. [CrossRef]

72. Muhammad, A.U.R.; Xia, C.Q.; Cao, B.H. Dietary forage concentration and particle size affect sorting, feeding behaviour, intake and growth of Chinese holstein male calves. *J. Anim. Physiol. Anim. Nutr.* **2015**, *100*, 217–223. [CrossRef]

73. Xia, C.; Muhammad, A.U.R.; Niu, W.; Shao, T.; Qiu, Q.; Su, H.; Cao, B. Effects of dietary forage to concentrate ratio and wildrye length on nutrient intake, digestibility, plasma metabolites, ruminal fermentation and fecal microflora of male Chinese Holstein calves. *J. Integr. Agric.* **2018**, *17*, 415–427. [CrossRef]

74. Miller-Cushon, E.K.; Montoro, C.; Bach, A.; DeVries, T.J. Effect of early exposure to mixed rations differing in forage particle size on feed sorting of dairy calves. *J. Dairy Sci.* **2013**, *96*, 3257–3264. [CrossRef] [PubMed]

75. Montoro, C.; Miller-Cushon, E.K.; DeVries, T.J.; Bach, A. Effect of physical form of forage on performance, feeding behavior, and digestibility of Holstein calves. *J. Dairy Sci.* **2013**, *96*, 1117–1124. [CrossRef] [PubMed]

76. Krehbiel, C.R.; Rust, S.R.; Zhang, G.; Gilliland, S.E. Bacterial direct-fed microbials in ruminant diets: Performance response and mode of action. *J. Anim. Sci.* **2003**, *81*, E120–E132.

77. Hong, H.A.; Duc, L.H.; Cutting, S.M. The use of bacterial spore formers as probiotics. *FEMS Microbiol. Rev.* **2005**, *29*, 813–835. [CrossRef] [PubMed]

78. Kumar, U.; Sareen, V.K.; Singh, S. Effect of Saccharomyces cerevisiae yeast culture supplement on ruminal metabolism in buffalo calves given a high concentrate diet. *Anim. Prod.* **1994**, *59*, 209–215. [CrossRef]

79. Xiao, J.X.; Alugongo, G.M.; Chung, R.; Dong, S.Z.; Li, S.L.; Yoon, I.; Wu, Z.H.; Cao, Z.J. Effects of Saccharomyces cerevisiae fermentation products on dairy calves: Ruminal fermentation, gastrointestinal morphology, and microbial community. *J. Dairy Sci.* **2016**, *99*, 5401–5412. [CrossRef] [PubMed]

80. Jia, P.; Cui, K.; Ma, T.; Wan, F.; Wang, W.; Yang, D.; Wang, Y.; Guo, B.; Zhao, L.; Diao, Q. Influence of dietary supplementation with Bacillus licheniformis and Saccharomyces cerevisiae as alternatives to monensin on growth performance, antioxidant, immunity, ruminal fermentation and microbial diversity of fattening lambs. *Sci. Rep.* **2018**, *8*, 16712. [CrossRef]

81. Muya, M.C.; Nherera, F.V.; Miller, K.A.; Aperce, C.C.; Moshidi, P.M.; Erasmus, L.J. Effect of Megasphaera elsdenii NCIMB 41125 dosing on rumen development, volatile fatty acid production and blood β-hydroxybutyrate in neonatal dairy calves. *J. Anim. Physiol. Anim. Nutr.* **2015**, *99*, 913–918. [CrossRef]

82. Yu, P.; Wang, J.Q.; Pu, D.F.; Liu, K.L.; Li, D.; Zhao, S.G.; Wei, H.Y.; Zhou, L.Y. Effects of bacillus subtilis natto in diets on quantities of gastrointestinal cellulytic bacteria in weaning calves. *J. China Agric. Univ.* **2009**, *14*, 111–116.

83. Zhang, R.; Dong, X.L.; Zhou, M.; Tu, Y.; Zhang, N.F.; Deng, K.D.; Diao, Q.Y. Oral administration of Lactobacillus plantarum and Bacillus subtilis on rumen fermentation and the bacterial community in calve. *Anim. Sci. J.* **2017**, *100*, 755–762. [CrossRef]

84. Agarwal, N.; Kamra, D.N.; Chaudhary, L.C.; Agarwal, I.; Sahoo, A.; Pathak, N.N. Microbial status and rumen enzyme profile of crossbred calves fed on different microbial feed additives. *Lett. Appl. Microbiol.* **2002**, *34*, 329–336. [CrossRef] [PubMed]

85. Wang, B.; Yang, C.T.; Diao, Q.Y.; Tu, Y. The influence of mulberry leaf flavonoids and Candida tropicalis on antioxidant function and gastrointestinal development of preweaning calves challenged with *Escherichia coli* O141:K99. *J. Dairy Sci.* **2018**, *101*, 6098–6108. [CrossRef] [PubMed]

86. Yohe, T.T.; O'Diam, K.M.; Daniels, K.M. Growth, ruminal measurements, and health characteristics of Holstein bull calves fed an Aspergillus oryzae fermentation extract. *J. Dairy Sci.* **2015**, *98*, 6163–6175. [CrossRef] [PubMed]

87. Guilloteau, P.; Rome, V.; Le Normand, L.; Savary, G.; Zabielski, R. Is Na-butyrate a growth factor in the preruminant calf? Preliminary results. *J. Anim. Feed Sci.* **2004**, *13* (Suppl. 1), 393–396. [CrossRef]

88. Górka, P.; Kowalski, Z.M.; Pietrzak, P.; Kotunia, A.; Kiljanczyk, R.; Flaga, J.; Holst, J.J.; Guilloteau, P.; Zabielski, R. Effect of sodium butyrate supplementation in milk replacer and starter diet on rumen development in calves. *J. Physiol. Pharmacol.* **2009**, *60*, 47–53. [PubMed]

89. Górka, P.; Kowalski, Z.M.; Pietrzak, P.; Kotunia, A.; Jagusiak, W.; Holst, J.J.; Guilloteau, P.; Zabielski, R. Effect of method of delivery of sodium butyrate on rumen development in newborn calves. *J. Dairy Sci.* **2011**, *94*, 5578–5588. [CrossRef] [PubMed]

90. Ragionieri, L.; Cacchioli, A.; Ravanetti, F.; Botti, M.; Ivanovska, A.; Panu, R.; Righi, F.; Quarantelli, A.; Gazza, F. Effect of the supplementation with a blend containing short and medium chain fatty acid monoglycerides in milk replacer on rumen papillae development in weaning calves. *Ann. Anat. Anat. Anzeiger.* **2016**, *207*, 97–108. [CrossRef] [PubMed]

91. Cummins, K.A.; Papas, A.H. Effect of isocarbon 4 and isocarbon 5 volatile fatty acids on microbial protein synthesis and dry matter digestibility in vitro. *J. Dairy Sci.* **1985**, *68*, 2588–2595. [CrossRef]

92. Liu, Q.; Wang, C.; Huang, Y.X.; Dong, K.H.; Yang, W.Z.; Wang, H. Effects of isobutyrate on rumen fermentation, urinary excretion of purine derivatives and digestibility in steers. *Arch. Anim. Nutr.* **2008**, *62*, 377–388. [CrossRef]

93. Liu, Q.; Wang, C.; Huang, Y.X.; Dong, K.H.; Yang, W.Z.; Wang, H. Effects of lanthanum on rumen fermentation, urinary excretion of purine derivatives and digestibility in steers. *Anim. Feed Sci. Tech.* **2008**, *142*, 121–132. [CrossRef]

94. Liu, Q.; Wang, C.; Zhang, Y.L.; Pei, C.X.; Zhang, S.L.; Wang, Y.X.; Zhang, Z.W.; Yang, W.Z.; Wang, H.; Guo, G.; et al. Effects of isovalerate supplementation on growth performance and ruminal fermentation in pre- and post-weaning dairy calves. *J. Agric. Sci.* **2016**, *154*, 1499–1508. [CrossRef]

95. Wang, C.; Liu, Q.; Zhang, Y.L.; Pei, C.X.; Zhang, S.L.; Wang, Y.X.; Yang, W.Z.; Bai, Y.S.; Shi, Z.G.; Liu, X.N. Effects of isobutyrate supplementation on ruminal microflora, rumen enzyme activities and methane emissions in Simmental steers. *J. Anim. Physiol. Anim. Nutr.* **2014**, *99*, 123–131. [CrossRef] [PubMed]

96. Giannenas, I.; Skoufos, J.; Giannakopoulos, C.; Wiemann, M.; Gortzi, O.; Lalas, S.; Kyriazakis, I. Effects of essential oils on milk production, milk composition, and rumen microbiota in Chios dairy ewes. *J. Dairy Sci.* **2011**, *94*, 5569–5577. [CrossRef] [PubMed]

97. Hristov, A.N.; McAllister, T.A.; Van Herk, F.H.; Cheng, K.J.; Newbold, C.J.; Cheeke, P.R. Effect of Yucca schidigera on ruminal fermentation and nutrient digestion in heifers. *J. Anim. Sci.* **1999**, *77*, 2554–2563. [CrossRef] [PubMed]

98. Calsamiglia, S.; Busquet, M.; Cardozo, P.W.; Castillejos, L.; Ferret, A. Invited review: Essential oils as modifiers of rumen microbial fermentation. *J. Dairy Sci.* **2007**, *90*, 2580–2595. [CrossRef]

99. Macheboeuf, D.; Morgavi, D.P.; Papon, Y.; Mousset, J.L.; Arturo-Schaan, M. Dose-response effects of essential oils on in vitro fermentation activity of the rumen microbial population. *Anim. Feed Sci. Tech.* **2008**, *145*, 335–350. [CrossRef]

100. Kumar, S.; Kumar, S.; Dar, A.H.; Palod, J.; Singh, S.K.; Kuma, A. Role of Aloe barbadensis supplementation on haematological parameters and rumen development of crossbred calves. *J. Entomol. Zool. Stud.* **2018**, *6*, 2261–2264.

101. Hassan, E.H.; Abdel-Raheem, S.M. Response of Growing Buffalo Calves to Dietary Supplementation of Caraway and Garlic as Natural Additives. *World Appl. Sci. J.* **2013**, *22*, 408–414.

102. Vakili, A.R.; Khorrami, B.; Danesh Mesgaran, M.; Parand, E. The Effects of Thyme and Cinnamon Essential Oils on Performance, Rumen Fermentation and Blood Metabolites in Holstein Calves Consuming High Concentrate Diet. *Asian Aust. J. Anim.* **2013**, *26*, 935–944. [CrossRef]

103. Žitňan, R.; Voigt, J.; Wegner, J.; Breves, G.; Schröder, B.; Winckler, C.; Levkut, M.; Kokardova, M.; Schonhusen, U.; Kuhla, S.; et al. Morphological and functional development of the rumen in the calf: Influence of the time of weaning. *Arch. Für Tierernaehrung.* **1999**, *52*, 351–362. [CrossRef]

104. Mao, H.; Xia, Y.; Tu, Y.; Wang, C.; Diao, Q. Effects of various weaning times on growth performance, rumen fermentation and microbial population of yellow cattle calves. *Asian Australas. J. Anim. Sci.* **2017**, *30*, 1557–1562. [CrossRef]

105. Meale, S.J.; Li, S.C.; Azevedo, P.; Derakhshani, H.; De Vries, T.J.; Plaizier, J.C.; Steele, M.A.; Khafipour, E. Weaning age influences the severity of gastrointestinal microbiome shifts in dairy calves. *Sci. Rep.* **2017**, *7*, 198. [CrossRef] [PubMed]

106. Abecia, L.; JimeÂnez, E.; MartõÂnez-Fernandez, G.; MartõÂn-GarcõÂa, A.I.; Ramos-Morales, E.; Pinloche, E.; Stuart, E.; Denman, S.E.; Newbold, C.J.; Yáñez-Ruiz, D.R. Natural and artificial feeding management before weaning promote different rumen microbial colonization but not differences in gene expression levels at the rumen epithelium of newborn goats. *PLoS ONE* **2017**, *12*, e0182235. [CrossRef] [PubMed]

107. Ji, S.; Jiang, C.; Li, R.; Diao, Q.; Tu, Y.; Zhang, N.; Si, B. Growth performance and rumen microorganism differ between segregated weaning lambs and grazing lambs. *J. Integr. Agric.* **2016**, *15*, 872–878. [CrossRef]

108. Kehoe, S.I.; Dechow, C.D.; Heinrichs, A.J. Effects of weaning age and milk feeding frequency on dairy calf growth, health and rumen parameters. *Livest. Sci.* **2007**, *110*, 267–272. [CrossRef]

109. Roth, B.A.; Keil, N.M.; Gygax, L.; Hillmann, E. Influence of weaning method on health status and rumen development in dairy calves. *J. Dairy Sci.* **2009**, *92*, 645–656. [CrossRef]

110. Meale, S.J.; Li, S.; Azevedo, P.; Derakhshani, H.; Plaizier, J.C.; Khafipour, E.; Steele, M.A. Development of Ruminal and Fecal Microbiomes Are Affected by Weaning but Not Weaning Strategy in Dairy Calves. *Front. Microbiol.* **2016**, *7*, 582. [CrossRef] [PubMed]

111. De Paula Vieira, A.; von Keyserlingk, M.A.G.; Weary, D.M. Effects of pair versus single housing on performance and behavior of dairy calves before and after weaning from milk. *J. Dairy Sci.* **2010**, *93*, 3079–3085. [CrossRef]

112. Costa, J.H.C.; Meagher, R.K.; von Keyserlingk, M.A.G.; Weary, D.M. Early pair housing increases solid feed intake and weight gains in dairy calves. *J. Dairy Sci.* **2015**, *98*, 6381–6386. [CrossRef]

113. Cowles, K.E.; White, R.A.; Whitehouse, N.L.; Erickson, P.S. Growth characteristics of calves fed an intensified milk replacer regimen with additional lactoferrin. *J. Dairy Sci.* **2006**, *89*, 4835–4845. [CrossRef]

114. Anderson, K.L.; Nagaraja, T.G.; Morrill, J.L. Rumen metabolic development in calves weaned conventionally or early. *J. Dairy Sci.* **1987**, *70*, 1000–1005. [CrossRef]

115. Schäff, C.T.; Gruse, J.; Maciej, J.; Pfuhl, R.; Zitnan, R. Effects of feeding unlimited amounts of milk replacer for the first 5 weeks of age on rumen and small intestinal growth and development in dairy calves. *J. Dairy Sci.* **2018**, *101*, 783–793. [CrossRef] [PubMed]

116. Bartlett, K.S.; McKeith, F.K.; VandeHaar, M.J.; Dahl, G.E.; Drackley, J.K. Growth and body composition of dairy calves fed milk replacers containing different amounts of protein at two feeding rates. *J. Anim. Sci.* **2006**, *84*, 1454–1467. [CrossRef] [PubMed]

117. Maccari, P.; Wiedemann, S.; Kunz, H.J.; Piechotta, M.; Sanftleben, P.; Kaske, M. Effects of two different rearing protocols for Holstein bull calves in the first 3 weeks of life on health status, metabolism and subsequent performance. *J. Anim. Physiol. Anim. Nutr.* **2015**, *99*, 737–746. [CrossRef] [PubMed]

118. Soberon, F.; Raffrenato, E.; Everett, R.W.; Van Amburgh, M.E. Preweaning milk replacer intake and effects on long-term productivity of dairy calves. *J. Dairy Sci.* **2012**, *95*, 783–793. [CrossRef] [PubMed]

119. Foldager, J.; Krohn, C.C. Heifer calves reared on very high or normal levels of whole milk from birth to 6–8 weeks of age and their subsequent milk production. *Proc. Soc. Nutr. Physiol.* **1994**, *3*, 301.

120. Terré, M.; Tejero, C.; Bach, A. Long-term effects on heifer performance of an enhanced growth feeding programme applied during the pre-weaning period. *J. Dairy Res.* **2009**, *76*, 331–339. [CrossRef]

121. Yáñez-Ruiz, D.R.; Macías, B.; Pinloche, E.; Newbold, C.J. The persistence of bacterial and methanogenic archaeal communities residing in the rumen of young lambs. *FEMS Microbiol. Ecol.* **2010**, *72*, 272–278. [CrossRef]

122. De Barbieri, I.; Hegarty, R.S.; Silveira, C.; Gulino, L.M.; Oddy, V.H.; Gilbert, R.A.; Klieve, A.V.; Ouwerkerk, D. Programming rumen bacterial communities in newborn Merino lambs. *Small Ruminant Res.* **2015**, *129*, 48–59. [CrossRef]

123. Yáñez-Ruiz, D.R.; Abecia, L.; Newbold, C.J. Manipulating rumen microbiome and fermentation through interventions during early life: A review. *Front. Microbiol.* **2015**, *6*, 1133. [CrossRef]

124. Abecia, L.; Martín-García, A.I.; Martínez, G.; Newbold, C.J.; Yáñez-Ruiz, D.R. Nutritional intervention in early life to manipulate rumen microbial colonization and methane output by kid goats postweaning. *J. Anim. Sci.* **2013**, *91*, 4832–4840. [CrossRef] [PubMed]

125. Abecia, L.; Waddams, K.E.; Martínez-Fernandez, G.; Martín-García, A.I.; Ramos-Morales, E.; Newbold, C.J.; Yáñez-Ruiz, D.R. An Antimethanogenic Nutritional Intervention in Early Life of Ruminants Modifies Ruminal Colonization by Archaea. *Archaea* **2014**, *2014*, 1–12. [CrossRef] [PubMed]

Article

Effects of Pair Versus Individual Housing on Performance, Health, and Behavior of Dairy Calves

Shuai Liu [1,†], Jiaying Ma [1,†], Jinghui Li [2], Gibson Maswayi Alugongo [1], Zhaohai Wu [3], Yajing Wang [1], Shengli Li [1] and Zhijun Cao [1,*]

[1] State Key Laboratory of Animal Nutrition, College of Animal Science and Technology, China Agricultural University, Beijing 100193, China; liushuaicau@cau.edu.cn (S.L.); majiaying@cau.edu.cn (J.M.); maswayi@yahoo.com (G.M.A.); yajingwang_cau@163.com (Y.W.); lisheng0677@163.com (S.L.)

[2] Department of Animal Science, University of California, Davis, CA 95616, USA; lgreyhui@hotmail.com

[3] State Key Laboratory of Animal Nutrition, Institute of Animal Science, Chinese Academy of Agricultural Sciences, Beijing 100193, China; wzh07128@163.com

* Correspondence: caozhijun@cau.edu.cn; Tel.: +86-010-6273-3746

† Shuai Liu and Jiaying Ma contributed to the work equally.

Received: 30 November 2019; Accepted: 22 December 2019; Published: 25 December 2019

Simple Summary: In modern dairy farming systems, calves are often housed in individual pens or hutches, which results in less social interaction with their peers during the milk-feeding period. The aim of this study was to evaluate the effects of pair versus individual housing on performance, health, and behavior of dairy calves from the milk-feeding period to the first week after mixing. Results showed that pair versus individual housing had no effects on body weight, starter intake or average daily gain during the milk-feeding period, while pair housing increased the growth performance of calves during weaning and postweaning periods, and the beneficial effects of pair housing on growth faded after calves were mixed and moved to group housing. Paired calves showed higher diarrhea frequency only in week three. The behavior of calves was altered at different periods, including increased time spent in feeding, chewing and ruminating, and decreased self-grooming time, and a drop of non-nutritive manipulation for all calves after they were mixed and moved to group housing. We also found less social contact may lead to more non-nutritive manipulation.

Abstract: The aim of this study was to evaluate the effects of pair versus individual housing on performance, health, and behavior of dairy calves. Thirty female Holstein dairy calves were assigned to individual (n = 10) or pair housing (n = 10 pairs). The results showed that both treatments had a similar starter intake and average daily gain (ADG) during the preweaning period. During weaning and postweaning periods, paired calves had a higher starter intake, and the ADG of paired calves continued to increase but calves housed individually experienced a growth check. Paired calves showed higher diarrhea frequency only in week three. The results on behavior showed that feeding, chewing and ruminating time increased, and self-grooming time decreased with age during weaning and postweaning periods, and paired calves spent less time feeding, standing and self-grooming but more time lying during this time. After mixing, feeding, and chewing and ruminating time continued to rise, and self-grooming time continued to decline for both treatments. All calves spent less time standing and non-nutritive manipulation after mixing, and previously individually housed calves tended to increase non-nutritive manipulation. These results showed that pair housing improved growth during weaning and postweaning periods and that calves altered their behavior at different phases. Less social contact may lead to more non-nutritive manipulation.

Keywords: calf; pair housing; individual housing; behavior

---

1. Introduction

Under natural conditions, calves are nursed by the dam and tend to have social interactions with their peers or other animals [1,2]. In modern dairy farming systems, however, calves are often housed in individual pens or hutches. Hence, they are less likely to interact with their peers or other animals during the milk-feeding period.

Previous work has indicated that different housing systems (group versus individual housing) affect the performance and health of dairy calves. Some studies showed that compared with individual housing, group housing increased weight gains [3], starter intake [4] and hay intake of dairy calves [5]. Conversely, other studies showed no effects [6] or even negative effects on weight gain for group-housed calves [7]. Furthermore, respiratory diseases and diarrhea were reported to occur more frequently in group-housed veal calves [8]. On the contrary, Babu et al. [9] reported that rearing calves in a group resulted in a lower disease incidence. In other cases, health outcomes were similar between different housing systems [10]. The variability among studies may be related to differences in management (e.g., the number of animals per group, milk volume provided, duration of the feeding period, weaning method, and disease diagnosis). From a behavioral standpoint, weaning from a milk-based diet to a solid diet is one typical stressor faced by dairy calves, in which case, calves vocalize more (d 37 to 55) [11,12]. After weaning, calves are mixed with unfamiliar animals and moved to a novel environment, which may cause aggressive interactions (d 91 to 126) [13]. The stress resulted from weaning and mixing can negatively affect animal welfare [14]. Social housing during the milk-feeding period may have beneficial effects on behavior and cognition ability of calves even after they were weaned and mixed with unfamiliar animals in a group. Several studies have shown that social housing improved resilience to stress (d 51 to 53) [15], as well as increased competitive behavior (d 49 to 56) [16] and interactions (d 56 to 91) [17] after weaning. Furthermore, previous research has mainly clarified the effects of social housing on lying and feeding [17–19]. However, other behavioral responses, such as standing, chewing and ruminating, self-grooming, and non-nutritive manipulation have not been well characterized when calves were weaned and moved to group housing. The primary objective of this study was to compare growth, performance and health as well as evaluate the effects of paired versus individual housing on calves' behavior when they were weaned (d 42 to 56) and moved to group housing (d 63 to 70). We hypothesized that paired-housed calves would have better performance than individually housed calves.

When calves were weaned and moved to group housing, they experience changes in the way of feeding and management, especially when they are introduced to a different diet (from a milk-based diet to a solid diet or total mixed ration) and social environment. These changes may impact the behavior of calves. Overvest et al. [18] reported the day to day changes in lying and feeding during the weaning period (d 40 to 48), and Horvath and Miller-Cushon [14] described the day to day changes in standing time of calves mixed in a group (d 60 to 74 ± 5). However, how behavior would change from one period to another was still not clear. Therefore, the secondary objective focused on calf behavioral changes from the weaning period (d 42 to 56) to when they were mixed in a group (d 63 to 70). We hypothesized that calves would exhibit less socially affiliative behavior, such as self-grooming, which may be related to greater activity and exploratory behavior when calves were initially moved to group housing [14]. We also predicted that less non-nutritive manipulation would be observed after calves were moved to group housing, as non-nutritive manipulation often occurred among individually housed calves, especially during the milk-feeding period [20].

2. Materials and Methods

2.1. Animals and Treatments

This study was conducted at China Agricultural University's Dairy Education and Research Centre (Datong, Shanxi, China) in 2016, in accordance with protocols approved by the Ethical Committee of the College of Animal Science and Technology, China Agricultural University, Beijing, China (No.

2016DR07). Thirty female Holstein dairy calves were collected from the end of March to the mid-April and were assigned to individual (n = 10) or pair housing (n = 10 pairs) based on birthdate and body weight (mean ± SEM; 43.5 ± 0.59 kg). The age difference between calves in the same pair was within 48 h. Only calves with successful passive transfer of immunity (mean ± SEM; 6.24 ± 0.09 g/dL), determined by clinical refractometer 24 h after birth, were included in the study (serum total protein ≥5.5 g/dL).

2.2. Housing and Management

2.2.1. Preweaning, Weaning, and Postweaning

All calves were born in a calving pen and separated from their dams within 1 h after birth and weighed. After that, calves were moved to a separate and clean straw-bedded nursery room adjacent to the calving facility. If the younger calf of a pair was born within 24 h of the older calf of the pair, then the two calves were moved directly to the experimental calf barn less than 200 m from the nursery room. If the younger calf of a pair was not born within 24 h of the birth of the older calf, then the older calf was kept in the nursery room until the younger calf of the pair was born, accepting an age difference within a pair of maximum 48 h. Calves were transferred to the calf pens by a cart; paired-housed calves were transferred together within 3 h to 5 h after the birth of the younger calf, whereas individually housed calves were transferred alone.

Individually reared calves were kept in individual pens (1.5 m × 2.0 m), while paired-housed calves were provided twice the area (3.0 m × 2.0 m). Calf pens were located under a 3-sided (solid, 1.1 m in height), roofed shelter with a metal gate at the front. Calves could hear calves and see calves in neighboring pens through the openings in the gate. Openings provided access to buckets (10 L for each one) placed 35 cm apart in the center of the pen for water and starter. Calves housed individually had two buckets (one for water, one for starter), while pair housing calves were provided twice the feeding facilities. All calves had free access to water and pelleted starter feed. All the feeding facilities were cleaned daily. The interior of each pen was bedded with sand and bedding was replaced weekly.

Colostrum was heated to 39 °C in a water bath. After that, colostrum was transferred to 4-L esophageal tubing bottles and fed to the calf through a tube within 2 h after the calf was born on d 1. From d 2 to 56, pasteurized waste milk (nonsaleable milk) was provided 3 times daily at 08:00, 15:00 and 20:00 and the volume of milk for each time was equal. During the preweaning period, calves were fed 6 L/d from d 2 to 7, 7.5 L/d from d 8 to 42. Weaning was carried out by reducing milk volume on d 43 and calves were fed 6 L/d from d 43 to 49 and then 3 L/d until d 56. At each milk feeding, the buckets for water were removed temporarily and milk buckets (5 L for each one) were placed in the same position of the pen. For each pair, two milk buckets were used at each milk feeding, whereas calves housed in individual pens had a single milk bucket. Milk buckets were cleaned after each feeding. After weaning, calves remained in their pens during the postweaning period (d 57 to 63 ± 1). No forage was offered before mixing.

2.2.2. Mixing Period

On d 64 ± 1, individual calves were mixed with the paired-housed calves according to the age and moved to the calf barn. There were 5 groups and each group consisted of 6 calves: 2 previously housed in individual pens and 2 pairs previously housed in pairs. The age difference between calves in the same group was within 48 h. The back wall of the group pen (5.0 m × 4.0 m) was solid with two sides made from horizontal tubular metal bars (bar diameter: 5.0 cm; distance between bars: 12 cm) and a neck rail at the front. The length of the neck rail allowed all 6 calves to eat simultaneously (83 cm per calf). Total mixed ration (TMR) was delivered twice daily at 10:00 and16:00. Each group pen was equipped with one automatic water trough (length: 120 cm, width: 40 cm, height: 70 cm, depth: 20 cm) and calves had free access to water. Sand was used as bedding material and was replenished when the group was moved on d 70. Three Digital Thermometers (Deli Electronic Commerce Co., Ltd.,

Ningbo, China) were spaced evenly and mounted above (1.0 m) the sand bedding in the calf barn to record daily temperature and humidity (maximum and minimum).

The temperature and humidity fluctuated according to weather conditions (mean ± SD; 14.7 ± 9.3 °C and 23.2 ± 10.1% relative humidity).

2.3. Sample Collection

2.3.1. Feed Sampling

Feed samples were collected weekly and immediately frozen at −20 °C until they were further analyzed. The nutritional composition (Table 1) of the dry matter, crude protein, neutral detergent fiber, acid detergent fiber, ether extract, crude ash, calcium and phosphorus were analyzed following the methods of AOAC International [21]. Throughout the study, starter intake was recorded daily based on the amount offered and refused by each calf from d 5 to 63 ± 1. TMR intake during the mixing period was not measured because of group housing.

Table 1. Nutrient compositions of milk, starter, and total mixed ration.

Nutrient Composition (%) [1]	Milk	Starter	Total Mixed Ration [2]
Dry matter (DM)	13.9	89.0	56.0
Crude protein, DM basis	3.30	22.5	13.6
Ether extract, DM basis	3.70	3.04	3.01
Crude ash, DM basis	9.0	6.07	12.8
Calcium, DM basis	0.60	0.91	-
Phosphorus, DM basis	0.60	0.52	-
Neutral detergent fiber, DM basis	-	15.0	36.0
Acid detergent fiber, DM basis	-	6.08	20.4

[1] The nutritive values are the means of the results of the analysis of samples collected each week. [2] Contained steam-flaked corn (33.5%), alfalfa hay (21.2%), oat hay (21.2%), soybean meal (19.7%), and premix compound (0.4%) on a DM basis.

2.3.2. Body Measurements and Blood Sampling

Body weights (BW) were measured weekly (d 1, 7, 14, 21, 28, 35, 42, 49, 56, 63, 70). Body length (shoulders to pins), withers height, hip height, and heart girth were also recorded at the same time points. Blood samples were collected via jugular venipuncture using vacutainer serum collection tubes containing no anticoagulant 24 h after birth. The blood samples were then centrifuged at 3500× g, 4 °C for 15 min. Serum total protein (TP) was determined by an optical refractometer (Honneur Nutritional Technology Co. Ltd., Beijing, China).

2.4. Health Check and Treatment

The health check consisted of three parts: (1) fecal scoring, (2) clinical examination of the respiratory system, (3) rectal temperature. Fecal scores were recorded daily at 10:00 each day until d 63 based on a 1 to 4 system according to the guidelines outlined by Larson et al. [22]. Scores were, 1 = firm, well-formed (not hard); 2 = soft, pudding-like; 3 = runny, pancake batter; and 4 = liquid, splatters, pulpy orange juice. Fecal score data were collected by one independent trained observer. All fecal scores were recorded by observing fecal matter on the ground of the pen or the tail and hindquarters of the calf. Fecal scoring was not conducted during the mixing period, because the fecal scores of an individual could not be accurately identified due to group housing. A diarrheic day was defined when the fecal score was >2. Weekly diarrhea frequency was calculated with the following equation: Diarrhea frequency = [(number of diarrhea calves × days of diarrhea) / (total number of calves × days of trial)] × 100%. Any calf with a fecal score >2 was treated according to the protocols established by the farm veterinarian (e.g., by administering antibiotic drugs and electrolytic solutions). Respiratory health was checked before each morning feeding through visually inspecting nasal discharge and listening to breathing difficulties with auscultation by the farm veterinarian and a member of the research team before morning milk feeding. If calves had signs of respiratory disease such as nasal discharge, cough and breathing difficulties, and a rectal temperature $\geq$39.5 °C, they were treated using an *Andrographis paniculata* injection (10 mL; Dazheng Tec-Phar. Co. Ltd., Changchun, China) for a maximum of 48 d; if respiratory disease or pyrexia was not alleviated, the calf received antibiotics treatment for a maximum of 48 d. Electrolytes were also administered intravenously to calves that had a severe respiratory disease until fully recovered. Throughout the study, one calf from individual housing was treated for 3 d during the mixing period because of nasal discharge and breathing difficulties, and no other calves had respiratory disease.

2.5. Behavioral Observations

A digital color camera (DS-7800, HIKVISION, Hangzhou, China) was placed above each selected pen (placed 2.5 m in front of the pens and 3.5 m from the pen floor), monitoring the behavior of the calves. During nighttime hours (from 17:30 to 07:30), the infrared monitoring function of the camera would turn on automatically. The recorded behaviors (Table 2) included feeding, chewing and ruminating, lying, standing, self-grooming, non-nutritive manipulation, and social contact.

Table 2. Ethogram of the recorded behaviors.

Behavior [1]	Description
Standing	Standing with all four feet on the ground either active or inactive
Lying	Lying on the sternum with head held in a raised position or down
Feeding	Head in bucket accompanied by chewing movements, including milk drinking
Chewing and ruminating	Irregular, repetitive chewing without discernible food in the mouth
Self-grooming	Movements with tongue over own body surface
Non-nutritive manipulation	Biting, sniffing, sucking or licking pen structures; may include bucket if milk is not available
Social contact	One calf's head was in contact with any part of the other calf including licking and sniffing of the other calf

[1] If one calf exhibit multiple behavior at one time point, then the multiple behavior were all recorded. Social contact was recorded for calves during the mixing period.

Six individually housed calves and 6 pairs of paired-housed calves were selected randomly for behavioral observations from d 43 to 70. Based on previous results [18,23,24], the sample sizes of behavior variables were estimated to obtain a power of 0.8 under a significance level of 0.05. During weaning and post-weaning periods, the behavioral data were recorded for 48 h on d 43, 50, and 57. For the mixing period, the behavior data were recorded for 48 h on the second day of mixing (d 65 $\pm$ 1) to avoid the effects of transition stress on calves. In order to clearly identify each selected calf from the groups during the mixing period, all selected calves were photographed from the front, back, left, right, and above. The observer could record behavior based on each calf's unique photos. For every 24 h duration (144 h in total for each calf), instantaneous scan-sampling with 5 min intervals was used to collect the lying, feeding, standing and chewing and ruminating data and continuous recording was used to collect the self-grooming, non-nutritive manipulation, and social contact data [25]. All behavioral data were recorded by one observer.

2.6. Statistical Analysis

2.6.1. Starter Intake, Growth, and Health Data

Throughout the study, data were analyzed at the pen level (based on a single calf per pen for the individual treatment and the mean of the 2 calves per pen in the pair treatment). Starter intake data were averaged by the week, except for the first week data, which were averaged across the last three days (d 5 to 7). Continuous variables with repeated measurements, including starter intake, average daily gain (ADG), BW, and structural growth, were tested for normality using the UNIVARIATE procedure of SAS (version 9.2, SAS Institute Inc., Cary, NC, USA). These data were then analyzed from week 1 to week 10 (as a whole) using the MIXED procedure of SAS. The model included the fixed effects of time, treatment, and time × treatment interaction and the random effect of pen To account for the repeated measures within-subject, the covariance structures were chosen for each repeated variable on the basis of best fit which was determined from the Bayesian information criterion. The heterogeneous first-order autoregressive structure was selected for starter intake, BW, and structural growth data, and for ADG data, the first-order autoregressive structure was selected. Data for fecal scores were summarized by the week and analyzed using the Chi-squared test.

2.6.2. Behavioral Data

Behavioral data obtained for individual calves from video were also averaged by pen (a single calf per pen for the individual treatment and the mean of the 2 calves per pen in the pair treatment) across the 48 h in each observation week (week 7 and week 8 during weaning, and week 9 during postweaning and 10 during mixing). For each 48 h behavioral observation period, the average duration of each kind of behavior per 24 h was calculated. Behavioral data were analyzed separately by two stages: (1) weaning and postweaning, and (2) mixing. The comparison of social contact between two treatments was only analyzed during the mixing period, as individually housed calves had no social interaction before mixing. For stage 1, the effect of housing on behavior was tested using the MIXED procedure of SAS. The model included the fixed effects of treatment, week, and week × treatment interaction, and the random effect of pen. To account for the repeated measures within-subject, the first-order autoregressive structure was chosen for each behavior on the basis of best fit, which was determined from the Bayesian information criterion. For stage 2, the effect of housing on behavior was tested using one-way ANOVA. Lying, standing, non-nutritive manipulation and social contact data were normally distributed. Behavioral data of feeding, and chewing and ruminating were analyzed after logarithm transformation, and self-grooming data were analyzed after square root transformation to meet the normality assumption. The transformed data were back-transformed to report.

All data were reported as least squares mean. Differences of $p < 0.05$ were considered significant and $0.05 \leq p < 0.10$ was considered a tendency.

3. Results

3.1. Starter Intake and Growth

As shown in Figure 1, starter intake showed an upward trend over time ($p < 0.001$) for both individually and pair housed calves with no difference in starter intake between treatments during the preweaning period ($p > 0.05$). During weaning and postweaning periods, starter intake tended to be higher for paired-housed calves during week seven (860.0 vs. 658.1 ± 80.1 g/d, $p = 0.09$), and than for individually housed calves during week eight (1461.4 vs. 1123.1 ± 97.0 g/d, $p = 0.02$) and week nine (2237.4 vs. 1899.5 ± 113.5 g/d, $p = 0.04$).

ADG increased over time ($p < 0.001$) for both treatments and no differences were found between treatments during the preweaning period ($p > 0.05$, Figure 2). During weaning and postweaning periods, the weight gain of paired-housed calves continued to increase, but individually housed calves experienced a growth check. The ADG for paired-housed calves tended to be higher during week seven (0.94 vs. 0.71 ± 0.07 kg/d, $p = 0.08$). Individually housed calves had higher ADG than calves housed in pairs during the mixing period (1.20 vs. 0.85 ± 0.09 kg/d, $p = 0.01$).

Throughout the study, the housing system (paired vs. individual) had no effects on BW ($p = 0.50$) and structural measurements (Table 3), including withers height ($p = 0.55$), heart girth ($p = 0.38$), abdominal girth ($p = 0.14$), and body length ($p = 0.23$).

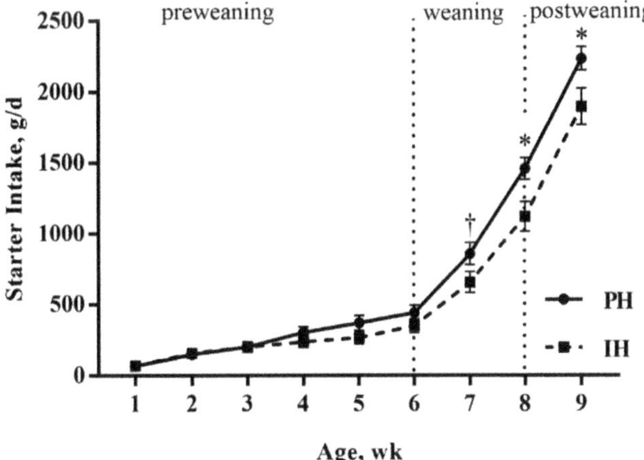

Figure 1. Starter intake (LSM ± SEM) for calves housed individually (n = 10 calves) or in pairs (n = 10 pairs) before mixing. PH = calves housed in pairs; IH = calves housed individually; wk = week. TMR intake was not measured because of group housing during the mixing period (week 10). * $p < 0.05$, † $p < 0.10$.

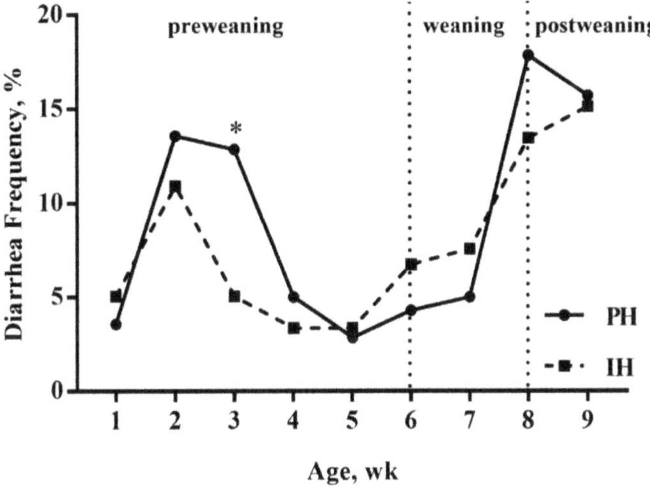

Figure 2. Average daily gain (LSM ± SEM) for calves housed individually (n = 10 calves) or in pairs (n = 10 pairs). PH = calves housed in pairs; IH = calves housed individually; wk = week. *p*-value: 0.90 (treatment), < 0.001 (week), 0.08 (treatment × week). * $p < 0.05$, † $p < 0.10$.

Table 3. Least squares mean of structural measurements and BW for calves housed individually (n = 10 calves) or in pairs (n = 10 pairs) from week 1 to week 10.

Item	Treatment [1]		SEM	*p*-Value		
	PH	IH		Treatment	Time	Treatment × Time
Body weight, kg	69.8	68.7	1.12	0.50	<0.001	0.32
Withers height, cm	86.6	86.3	0.35	0.55	<0.001	1.0
Heart girth, cm	95.7	95.3	0.47	0.38	<0.001	0.90
Abdominal girth, cm	101.4	100.2	0.73	0.14	<0.001	0.40
Body length [2], cm	78.7	78.2	0.36	0.23	<0.001	0.85

[1] PH = calves housed in pairs; IH = calves housed individually. [2] Body length was measured from shoulders to pins.

3.2. Health

Throughout the study, one calf from individual housing during the mixing period suffered from respiratory disease, and no other calves had respiratory disease. Diarrhea frequency is shown in Figure 3. Pair housing increased diarrhea frequency in comparison with individual housing of calves during week three (18.0% vs. 6.0%, $p = 0.03$), yet no differences were found between treatments in other weeks ($p > 0.05$).

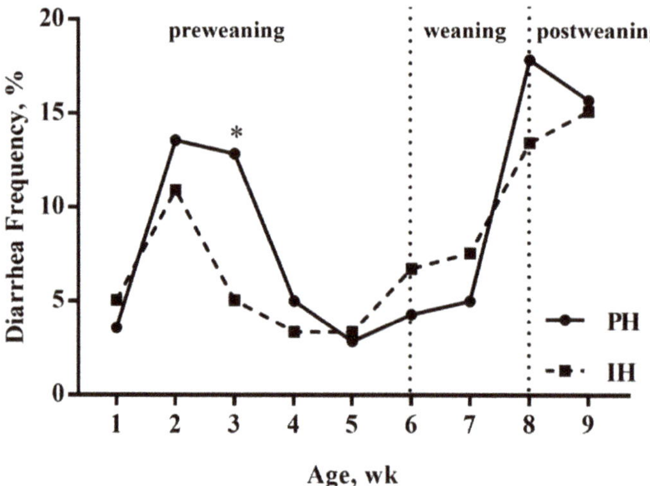

Figure 3. Effects of the housing system on diarrhea frequency before mixing for calves housed individually (n = 10 calves) or in pairs (n = 10 pairs). PH = calves housed in pairs; IH = calves housed individually; wk = week. Fecal scoring was not conducted during the mixing period (week 10), because the fecal scores of an individual could not be accurately identified due to group housing. * $p < 0.05$.

3.3. Behavior

As shown in Figure 4, during weaning (week 7–8) and postweaning (week 9) periods, feeding time increased ($p < 0.001$) for both treatments. Overall, individually housed calves spent more time feeding (83.0 vs. 53.1 ± 1.15 min/d, $p = 0.04$) compared with paired-housed calves during this period. After mixing, feeding time decreased for individually-housed calves but increased for paired-housed calves, and the previous housing system had no effects on feeding time after mixing ($p = 0.82$). Ruminating time increased over weaning and postweaning periods ($p < 0.001$) for both treatments and individually-housed calves tended to have greater ruminating time than paired-housed calves during week seven (2.56 vs. 1.79 ± 0.26 h/d, $p = 0.09$). After mixing, ruminating time continued to increase with age for all calves with no differences found between treatments ($p = 0.61$).

Lying time increased during the weaning period and decreased during the postweaning period for all calves. Standing time increased during the postweaning period for all calves. Calves housed in pairs spent more time lying (17.3 vs. 16.4 ± 0.27 h/d, $p = 0.03$) and less time standing (6.33 vs. 7.11 ± 0.18 h/d, $p = 0.01$) compared with calves housed individually during weaning and postweaning periods. After mixing, lying time remained stable and standing time decreased for all calves, and the previous housing system had no effect on lying ($p = 0.56$) and standing time ($p = 0.84$). There was a decrease in self-grooming time for both treatments over weaning and postweaning periods ($p = 0.01$), and calves housed individually exhibited more self-grooming than calves housed in pairs (40.7 vs. 20.6 ± 4.10 min/d, $p = 0.02$). After mixing, self-grooming time continued to decrease for all calves, with no differences between treatments ($p = 0.65$). In addition, non-nutritive manipulation time did not change with calf age during weaning and postweaning periods ($p = 0.62$), and treatment had no effects on non-nutritive manipulation time during this period ($p = 0.10$). After mixing, all calves decreased non-nutritive manipulation time, and the non-nutritive manipulation time tended to be longer for calves that were previously individually-housed (33.0 vs. 16.9 ± 5.62 min/d, $p = 0.07$). During the mixing period, the previous housing system had no effect on social contact (Figure 5; $p = 0.53$).

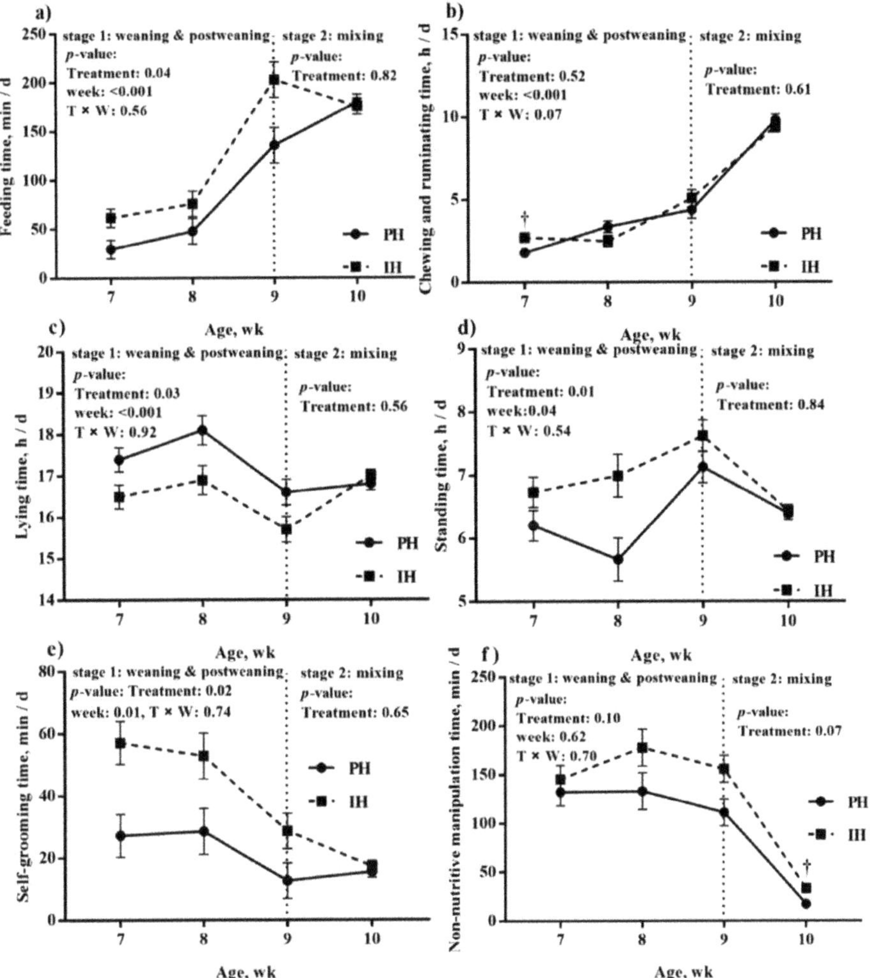

Figure 4. Effects of housing system on (**a**) feeding; (**b**) chewing and ruminating; (**c**) lying; (**d**) standing; (**e**) self-grooming; and (**f**) non-nutritive manipulation for calves housed individually (n = 6) or in pairs (n = 6). Stage 1 = from weaning to postweaning period, including weeks 7, 8, and 9. Stage 2 = mixing period, including week 10. PH = calves housed in pairs; IH = calves housed individually; wk = week. * $p < 0.05$, † $p < 0.10$.

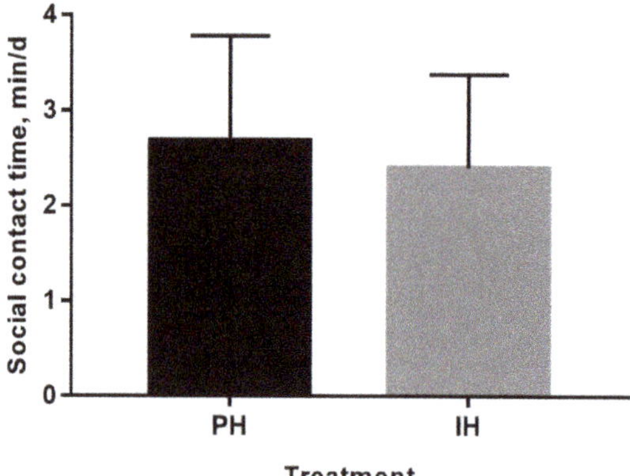

Figure 5. Effects of the housing system on social contact for calves housed individually (n = 6) or in pairs (n = 6) during the mixing period. PH = calves housed in pairs; IH = calves housed individually.

4. Discussion

4.1. Starter Intake and Growth

During the preweaning period, we did not observe any differences in starter intake, ADG or BW between treatments, yet during weaning and postweaning periods, pair housing improved growth performance. Our results were consistent with previous research that reported increased starter intake [6,26] and ADG [23] during the weaning period for paired-housed calves. Such improvements are likely due to social facilitation [12] and social learning [27], which allow calves housed in pairs to learn faster and eat more. Local enhancement is another factor affecting the feeding of calves, through which the behavior of one calf draws the attention of another in the same pair toward a particular food source [28,29]. In addition, paired calves might experience a lower level of stress during the weaning period because of social buffering [12]. The social buffering benefits of early pair housing have been discussed recently by Overvest et al. [18], who demonstrated that social housing might improve the ability to cope with the weaning stress via the positive effects on feed acceptance and behavioral flexibility. During the postweaning period, we observed greater starter intake in pair housing. Similar results were reported by Pempek et al. [20], who also attributed it to social facilitation. Besides, the competitive feeding environment among paired calves may also have resulted in more starter intake during the weaning and postweaning periods, as calves may increase the rate of feed intake in the competitive feeding environment [17]. Our results contribute to a body of evidence indicating that pair housing is particularly beneficial to solid-feed intake, growth, and supporting a smooth transition at weaning [3].

After mixing, calves were grouped together. Warnick et al. [30] and De Paula Vieira et al. [12] reported that calves previously housed in groups or pairs gained more than those previously housed in individual pens when they were mixed and placed together. Some studies attributed these results to the beneficial effects of social housing, such as reduced neophobia to new ration [31,32] and greater competitive success [16] when mixed with unfamiliar animals. On the contrary, we found that previously paired-housed calves had less ADG after mixing compared with calves housed individually, and the final BW was similar between treatments. Somewhat interestingly, Miller-Cushon and DeVries [4] reported that though paired calves had greater performance during the weaning period (d 39 to 49), previous housing (paired vs. individual) had no effect on DMI, ADG or final BW once previously individually housed calves were paired with unfamiliar calves after weaning (d 50 to 84). Similar results were reported by Overvest et al. [18], who also demonstrated that once calves previously housed individually were paired after weaning (d 49 to 56), they exhibited more feeding time and thus increased their solid feed DMI to a greater extent over time than paired-housed calves, and eventually resulted in similar DMI between treatments. These results suggested that previously individually housed calves could get the same performance (e.g., DMI, BW, and ADG) through modifying feeding behavior after they were exposed to social housing with unfamiliar calves. In this study, all calves experienced a sudden feed transition to TMR, in which case, the beneficial effects of social housing on food neophobia may be weakened by transition stress. Furthermore, we observed similar feeding time between treatments during the mixing period. Thus, we speculated that the higher ADG in calves previously housed individually may result from higher feeding rate, allowing them to consume sufficient TMR to meet or exceed their nutritional requirements and finally compensating for a previously lower starter intake. Further work to address this possibility is encouraged.

4.2. Health

In the current experiment, diarrhea frequency for calves housed in pairs was higher than that for calves housed individually in week three, yet no differences were found in other weeks. Some studies reported that housing calves in groups exhibited more health problems owing to higher levels of infectious agents and calf-calf transmission [33,34]. On the contrary, others reported a lower incidence of diarrhea for calves housed socially [9], and some found no differences in incidences of diarrhea and respiratory problems [35] between paired-housed calves and individually housed calves. The various results indicated that health problems were not consistently associated with social housing. The incidence of disease relies on many factors including calf immunity, environment management, disease diagnosis, and the ability of a calf to cope with stress [36]. These factors rather than the housing system may play a critical role in inducing health problems. Greater health problems in a group housing system may also stem from the difficulty of detecting disease in groups [23]. There is not enough evidence to support a diarrhea-increasing effect of pair housing, thus the higher incidence of diarrhea in pair housing in week three was probably due to low immunity to infection of calves aged from two to four weeks [37] and individual differences.

4.3. Behavior

Limited research has described how behavior would change at different periods from weaning to mixing, or the effects of paired or individual housing on behavior during these periods. Our results suggested that all calves experienced behavioral changes from weaning to mixing including increased feeding and ruminating time, and decreased self-grooming time.

The increase in feeding time and chewing and ruminating time over the weaning and postweaning periods aligned with the increase in feed intake. Besides, paired calves spent less time feeding but still had higher starter intake during weaning and postweaning periods, likely due to the competitive feeding environment as we discussed on starter intake and growth. Miller-Cushon et al. [17] found that calves housed in a competitive feeding environment had less time of feeding but an increased rate of feed intake compared with those housed in a noncompetitive feeding environment. Hence, paired-housed calves might increase their feeding rate rather than feeding time to consume more starter.

In the present study, lying time declined while standing time rose during the postweaning period, which could be attributed to the increase in feeding time with increasing age during this period. The previous study [18] suggested that calves may change their lying behavior to accustom themselves to feeding behavior. In addition, calves housed individually exhibited more lying time than paired calves during weaning and postweaning periods, which is contrary to previous studies [23,38]. Previous research [20] also mentioned no effects of individual vs. paired housing on lying. The variant space allowance for calves among studies may be responsible for the discrepancy in results, as space allowance was a vital factor for the expression of normal behavior [23,39]. Further research is encouraged to study the relationship between space allowance and lying. After mixing, standing time decreased for all calves. Previous studies reported that calves were more active and moved more followed by a reduction in activity after the first day of introduction to a group [14,29], and calves had diminished behavioral reactions after the first 24-h period following regrouping [40]. Thus, we speculated that calves might not be as active as the first day of introduction to a group as the behavior data were recorded for 48 h on the second day of mixing to avoid the effects of transition stress on calves in this study.

Self-grooming is expressed by calves as caring for their own body, and this behavior may be a means of satisfying socialization [27]. More self-grooming activities were observed in individually-housed calves in the present study, which was consistent with previous research [27], as the socialization was absent in these calves. In addition, self-grooming can also be an expression of stress. Taking rodent as a research model, previous studies [41–43] has reported that the relationship between stress and self-grooming can be described as an inverted U-shaped: Self-grooming typically occurs spontaneously at low stress and becomes longer during moderate stress and can be inhibited by high-stress states. Thus, the higher self-grooming of individually housed calves may respond to the higher stress (moderate stress) they faced compared with paired calves during weaning and postweaning periods.

Non-nutritive manipulation commonly occurs within artificial rearing systems [44], which can be strengthened by social deprivation [45]. In the current study, a drop in non-nutritive manipulation time for all calves after mixing was observed, which was likely due to more social interactions among calves after mixing. Bokkers and Koene [46] also indicated that less social interaction was an important factor causing dairy calves to lick objects (a nonnutritive manipulation behavior). In this study, no effects were found on non-nutritive manipulation during the weaning or postweaning period, whereas individually-housed calves tended to spend more time on non-nutritive manipulation compared with paired calves during the mixing period, which was similar to the previous study [47]. Although the effect of the previous housing system on social contact was not significant during the mixing period, calves housed in pairs previously still exhibited more social contact numerically, which may result in less non-nutritive manipulation.

5. Conclusions

Paired versus individual housing had no effects on body weight, starter intake or ADG during the preweaning period, while pair housing increased the growth performance of calves during weaning and postweaning periods, and the beneficial effects of pair housing on growth was weakened after mixing. Paired calves showed higher diarrhea frequency only in week three. Calves altered their behaviors at different periods from weaning to mixing, including increased feeding time and chewing and ruminating time, and decreased self-grooming time, and a drop of non-nutritive manipulation for all calves after mixing. Furthermore, less social contact may result in more non-nutritive manipulation.

Author Contributions: Designed the experiments: Z.C., S.L. (Shuai Liu), and J.M. Conducted the experimental work: S.L. (Shuai Liu), J.M., J.L., G.M.A. and Z.W. Conducted the data analysis: S.L. (Shuai Liu) and J.M. Wrote the paper: S.L. (Shuai Liu) and J.M. Revised the paper: S.L. (Shuai Liu), J.M., J.L., G.M.A., Z.W., Y.W., S.L. (Shengli Li), Z.C. All authors have read and agreed to the published version of the manuscript.

Funding: This research was funded by National Key Research and Development Program of China (2018YFD0501600).

Acknowledgments: We are grateful to the staff of China Agricultural University's Dairy Education and Research Centre (Datong, Shanxi, P.R. China) for their assistance with the trial.

References

1. Reinhardt, V.; Reinhardt, A. Natural sucking performance and age of weaning in zebu cattle (*Bos indicus*). *J. Agric. Sci.* **1981**, *96*, 309–312. [CrossRef]

2. Vitale, A.F.; Tenucci, M.; Papini, M.; Lovari, S. Social behaviour of the calves of semi-wild Maremma cattle, Bos primigenius taurus. *Appl. Anim. Behav. Sci.* **1986**, *16*, 217–231. [CrossRef]

3. Jensen, M.B.; Duve, L.R.; Weary, D.M. Pair housing and enhanced milk allowance increase play behavior and improve performance in dairy calves. *J. Dairy Sci.* **2015**, *98*, 2568–2575. [CrossRef]

4. Miller-Cushon, E.K.; DeVries, T.J. Effect of social housing on the development of feeding behavior and social feeding preferences of dairy calves. *J. Dairy Sci.* **2016**, *99*, 1406–1417. [CrossRef] [PubMed]

5. Hepola, H.; Hänninen, L.; Pursiainen, P.; Tuure, V.M.; Syrjälä-Qvist, L.; Pyykkönen, M.; Saloniemi, H. Feed intake and oral behaviour of dairy calves housed individually or in groups in warm or cold buildings. *Livest. Sci.* **2006**, *105*, 94–104. [CrossRef]

6. Bernal-Rigoli, J.C.; Allen, J.D.; Marchello, J.A.; Cuneo, S.P.; Garcia, S.R.; Xie, G.; Hall, L.W.; Burrows, C.D.; Duff, G.C. Effects of housing and feeding systems on performance of neonatal Holstein bull calves. *J. Anim. Sci.* **2012**, *90*, 2818–2825. [CrossRef]

7. Terré, M.; Bach, A.; Devant, M. Performance and behaviour of calves reared in groups or individually following an enhanced-growth feeding programme. *J. Dairy Res.* **2006**, *73*, 480–486. [CrossRef]

8. Svensson, C.; Liberg, P. The effect of group size on health and growth rate of Swedish dairy calves housed in pens with automatic milk-feeders. *Prev. Vet. Med.* **2006**, *73*, 43–53. [CrossRef]

9. Babu, L.K.; Pandey, H.; Patra, R.C.; Sahoo, A. Hemato-biochemical changes, disease incidence and live weight gain in individual versus group reared calves fed on different levels of milk and skim milk. *Anim. Sci. J.* **2009**, *80*, 149–156. [CrossRef]

10. Cobb, C.J.; Obeidat, B.S.; Sellers, M.D.; Pepper-Yowell, A.R.; Ballou, M.A. Group housing of Holstein calves in a poor indoor environment increases respiratory disease but does not influence performance or leukocyte responses. *J. Dairy Sci.* **2014**, *97*, 3099–3109. [CrossRef]

11. Watts, J.M.; Stookey, J.M. Vocal behaviour in cattle: The animal's commentary on its biological processes and welfare. *Appl. Anim. Behav. Sci.* **2000**, *67*, 15–33. [CrossRef]

12. De Paula Vieira, A.; von Keyserlingk, M.A.G.; Weary, D.M. Effects of pair versus single housing on performance and behavior of dairy calves before and after weaning from milk. *J. Dairy Sci.* **2010**, *93*, 3079–3085. [CrossRef] [PubMed]

13. Veissier, I.; Gesmier, V.; Le Neindre, P.; Gautier, J.Y.; Bertrand, G. The effects of rearing in individual crates on subsequent social behaviour of veal calves. *Appl. Anim. Behav. Sci.* **1994**, *41*, 199–210. [CrossRef]

14. Horvath, K.C.; Miller-Cushon, E.K. Characterizing social behavior, activity, and associations between cognition and behavior upon social grouping of weaned dairy calves. *J. Dairy Sci.* **2018**, *101*, 7287–7296. [CrossRef]

15. Bolt, S.L.; Boyland, N.K.; Mlynski, D.T.; James, R.; Croft, D.P. Pair housing of dairy calves and age at pairing: Effects on weaning stress, health, production and social networks. *PLoS ONE* **2017**, *12*, e0166926. [CrossRef]

16. Duve, L.R.; Weary, D.M.; Halekoh, U.; Jensen, M.B. The effects of social contact and milk allowance on responses to handling, play, and social behavior in young dairy calves. *J. Dairy Sci.* **2012**, *95*, 6571–6581. [CrossRef]

17. Miller-Cushon, E.K.; Bergeron, R.; Leslie, K.E.; Mason, G.J.; DeVries, T.J. Competition during the milk-feeding stage influences the development of feeding behavior of pair-housed dairy calves. *J. Dairy Sci.* **2014**, *97*, 6450–6462. [CrossRef]

18. Overvest, M.A.; Crossley, R.E.; Miller-Cushon, E.K.; DeVries, T.J. Social housing influences the behavior and feed intake of dairy calves during weaning. *J. Dairy Sci.* **2018**, *101*, 8123–8134. [CrossRef]

19. DeVries, T.J.; Von keyserlingk, M.A.G. Competition for feed affects the feeding behavior of growing dairy heifers. *J. Dairy Sci.* **2009**, *92*, 3922–3929. [CrossRef]

20. Pempek, J.A.; Eastridge, M.L.; Swartzwelder, S.S.; Daniels, K.M.; Yohe, T.T. Housing system may affect behavior and growth performance of Jersey heifer calves. *J. Dairy Sci.* **2016**, *99*, 569–578. [CrossRef]

21. AOAC. *Official Methods of Analysis of AOAC INTERNATIONAL*; Horwitz, W., Lalimer, G.W., Eds.; AOCA International: Richmond, VA, USA, 2006; ISBN 935584-77-3A.

22. Larson, L.L.; Owen, F.G.; Albright, J.L.; Appleman, R.D.; Lamb, R.C.; Muller, L.D. Guidelines Toward More Uniformity in Measuring and Reporting Calf Experimental Data. *J. Dairy Sci.* **1977**, *60*, 989–991. [CrossRef]

23. Chua, B.; Coenen, E.; van Delen, J.; Weary, D.M. Effects of Pair Versus Individual Housing on the Behavior and Performance of Dairy Calves. *J. Dairy Sci.* **2002**, *85*, 360–364. [CrossRef]

24. Abdelfattah, E.M.; Schutz, M.M.; Lay, D.; Marchant-Forde, J.; Eicher, S.D. Effect of group size on behavior, health, production, and welfare of veal calves. *J. Anim. Sci.* **2013**, *91*, 5455–5465. [CrossRef] [PubMed]

25. Martin, P.; Bateson, P. Measuring Behaviour: An Introductory Guide. In *Measuring Behaviour: An Introductory Guide*; Cambridge University Press: Cambridge, UK, 2007; p. 120. ISBN 9781565763005.

26. Costa, J.H.C.; Meagher, R.K.; von Keyserlingk, M.A.G.; Weary, D.M. Early pair housing increases solid feed intake and weight gains in dairy calves. *J. Dairy Sci.* **2015**, *98*, 6381–6386. [CrossRef]
27. Babu, L.K.; Pandey, H.N.; Sahoo, A. Effect of individual versus group rearing on ethological and physiological responses of crossbred calves. *Appl. Anim. Behav. Sci.* **2004**, *87*, 177–191. [CrossRef]
28. Nicol, C.J. The social transmission of information and behaviour. *Appl. Anim. Behav. Sci.* **1995**, *44*, 79–98. [CrossRef]
29. De Paula Vieira, A.; de Passillé, A.M.; Weary, D.M. Effects of the early social environment on behavioral responses of dairy calves to novel events. *J. Dairy Sci.* **2012**, *95*, 5149–5155. [CrossRef]
30. Warnick, V.D.; Arave, C.W.; Mickelsen, C.H. Effects of Group, Individual, and Isolated Rearing of Calves on Weight Gain and Behavior. *J. Dairy Sci.* **1977**, *60*, 947–953. [CrossRef]
31. Costa, J.H.C.; Daros, R.R.; von Keyserlingk, M.A.G.; Weary, D.M. Complex social housing reduces food neophobia in dairy calves. *J. Dairy Sci.* **2014**, *97*, 7804–7810. [CrossRef]
32. Whalin, L.; Weary, D.M.; von Keyserlingk, M.A.G. Short communication: Pair housing dairy calves in modified calf hutches. *J. Dairy Sci.* **2018**, *101*, 5428–5433. [CrossRef]
33. Maatje, K.; Verhoeff, J.; Kremer, W.D.; Cruijsen, A.L.; van den Ingh, T.S. Automated feeding of milk replacer and health control of group-housed veal calves. *Vet. Rec.* **1993**, *133*, 266–270. [CrossRef] [PubMed]
34. Gulliksen, S.M.; Lie, K.I.; Løken, T.; Østerås, O. Calf mortality in Norwegian dairy herds. *J. Dairy Sci.* **2009**, *92*, 2782–2795. [CrossRef] [PubMed]
35. Jensen, M.B.; Larsen, L.E. Effects of level of social contact on dairy calf behavior and health. *J. Dairy Sci.* **2014**, *97*, 5035–5044. [CrossRef] [PubMed]
36. Costa, J.H.C.; von Keyserlingk, M.A.G.; Weary, D.M. Invited review: Effects of group housing of dairy calves on behavior, cognition, performance, and health. *J. Dairy Sci.* **2016**, *99*, 2453–2467. [CrossRef] [PubMed]
37. Chase, C.C.L.; Hurley, D.J.; Reber, A.J. Neonatal Immune Development in the Calf and Its Impact on Vaccine Response. *Vet. Clin. N. Am. Food Anim. Pract.* **2008**, *24*, 87–104. [CrossRef]
38. Tapki, İ. Effects of individual or combined housing systems on behavioural and growth responses of dairy calves. *Acta Agric. Scand. Sect. A Anim. Sci.* **2007**, *57*, 55–60. [CrossRef]
39. Jensen, M.B.; Vestergaard, K.S.; Krohn, C.C. Play behaviour in dairy calves kept in pens: The effect of social contact and space allowance. *Appl. Anim. Behav. Sci.* **1998**, *56*, 97–108. [CrossRef]
40. Veissier, I.; Boissy, A.; Depassillé, A.M.; Rushen, J.; Van Reenen, C.G.; Roussel, S.; Andanson, S.; Pradel, P. Calves' responses to repeated social regrouping and relocation. *J. Anim. Sci.* **2001**, *79*, 2580–2593. [CrossRef]
41. Fernández-Teruel, A.; Estanislau, C. Meanings of self-grooming depend on an inverted U-shaped function with aversiveness. *Nat. Rev. Neurosci.* **2016**, *17*, 591. [CrossRef]
42. Song, C.; Berridge, K.C.; Kalueff, A.V. "Stressing" rodent self-grooming for neuroscience research. *Nat. Rev. Neurosci.* **2016**, *17*, 591. [CrossRef]
43. Kalueff, A.V.; Stewart, A.M.; Song, C.; Berridge, K.C.; Graybiel, A.M.; Fentress, J.C. Neurobiology of rodent self-grooming and its value for translational neuroscience. *Nat. Rev. Neurosci.* **2016**, *17*, 45–59. [CrossRef] [PubMed]
44. Jensen, M.B. The effects of feeding method, milk allowance and social factors on milk feeding behaviour and cross-sucking in group housed dairy calves. *Appl. Anim. Behav. Sci.* **2003**, *80*, 191–206. [CrossRef]
45. Veissier, I.; Chazal, P.; Pradel, P.; Le Neindre, P. Providing Social Contacts and Objects for Nibbling Moderates Reactivity and Oral Behaviors in Veal Calves. *J. Anim. Sci.* **1997**, *75*, 356–365. [CrossRef] [PubMed]
46. Bokkers, E.A.M.; Koene, P. Activity, oral behaviour and slaughter data as welfare indicators in veal calves: A comparison of three housing systems. *Appl. Anim. Behav. Sci.* **2001**, *75*, 1–15. [CrossRef]
47. Veissier, I.; Ramirez De La Fe, A.R.; Pradel, P. Nonnutritive oral activities and stress responses of veal calves in relation to feeding and housing conditions. *Appl. Anim. Behav. Sci.* **1998**, *57*, 35–49. [CrossRef]

Communication

Effects of Dietary Rumen-Protected Betaine Supplementation on Performance of Postpartum Dairy Cows and Immunity of Newborn Calves

Beibei Wang [1], Chong Wang [2], Ruowei Guan [1], Kai Shi [1], Zihai Wei [1], Jianxin Liu [1] and Hongyun Liu [1,*]

[1] College of Animal Sciences, Zhejiang University, Hangzhou 310058, China; 21617083@zju.edu.cn (B.W.); 21617025@zju.edu.cn (R.G.); shik_1992@163.com (K.S.); weizihai365@163.com (Z.W.); liujx@zju.edu.cn (J.L.)

[2] College of Animal Sciences and Technology, Zhejiang A & F University, Hangzhou 311300, China; wangcong992@163.com

* Correspondence: hyliu@zju.edu.cn; Tel.: +86-571-88982965

Received: 8 March 2019; Accepted: 12 April 2019; Published: 15 April 2019

Simple Summary: Betaine plays an important role in growth, lactation, protein synthesis, and fat metabolism in animals, but there are few studies on transition dairy cows and newborn calves. The aim of the current study was to evaluate the effects of rumen-protected betaine supplementation from four weeks before expected calving to six weeks postpartum regarding the lactation performance and blood metabolites of dairy cows and immunity of newborn calves. The results suggested that betaine supplementation tended to increase fat mobilization of postpartum dairy cows. Furthermore, compared to the control calves, the betaine calves had greater plasma total protein and globulin concentrations, which indicates that the immunity of the betaine calves might have improved.

Abstract: The objective of this study was to evaluate the effects of rumen-protected betaine supplementation on performance of postpartum dairy cows and immunity of newborn calves. Twenty-four multiparous Holstein dairy cows were randomly divided into the control (CON, $n = 12$) and rumen-protected betaine (BET, $n = 12$) groups after blocking by parity and milk yield during the previous lactation cycle. The cows were fed a basal total mixed ration diet without BET (CON) or with BET at 20 g/d per cow (BET) from four weeks before expected calving to six weeks postpartum. The results showed that betaine supplementation had no effect on dry matter intake and milk yield of the cows. The BET cows tended to increase feed efficiency (energy-corrected milk/dry matter intake) and body weight loss postpartum compared to the CON cows. The plasma β-hydroxybutyrate concentrations of the BET cows were greater at d seven after calving than those of the CON cows. Moreover, compared to the CON calves, the BET calves had greater plasma total protein and globulin concentrations. The plasma glucose concentrations of the BET calves tended to decrease relative to CON cows. In conclusion, rumen-protected betaine supplementation from four weeks before expected calving tended to increase fat mobilization of postpartum dairy cows, and might improve the immunity of newborn calves.

Keywords: betaine; dairy cows; newborn calves; fat mobilization; immunity

---

1. Introduction

During the transition period, dairy cows are in a state of great metabolic stress because of the increased demand for nutrients to maintain fetal growth and milk synthesis. Transition dairy cows tend to have negative energy and amino acid balance after calving, which leads to an increase in fat and protein mobilization in tissues [1,2]. A negative methyl donor balance also likely occurs in transition cows because milk is high in methylated compounds [3]. Moreover, the last two months of gestation, where 60% of the body weight gain before birth occurs [4], is critical for bovine fetal development.

Betaine functions as a methyl donor and an organic osmolyte [2,5], which plays an important role in growth, lactation, protein synthesis, and fat metabolism in animals [6]. Betaine supplementation in the diets of steers increased body weight gain and fat deposition [7]. In lactating dairy cows, feeding betaine improved the milk yield and milk protein [8,9]. Supplementing betaine reduced plasma concentrations of non-esterified fatty acids (NEFA) and β-hydroxybutyrate (BHB) of lactating dairy cows [10], but elevated the concentrations of NEFA and BHB of transition dairy cows to change lipid metabolism [11]. Furthermore, betaine is vital for fetal development [12], and is related to the offspring's weight and immunity [13]. However, due to the fast rumen degradation (approximately 45%/h) of betaine in vivo [14], unprotected betaine cannot be absorbed efficiently. Our previous study showed that dietary rumen-protected betaine supplementation in lactating dairy cows improved lactation performance and fat metabolism [15]. Whether it improves the performance of postpartum dairy cows and the immunity of newborn calves remains unexplored. Therefore, the objectives of the current study were to evaluate the effects of rumen-protected betaine supplementation from four weeks before expected calving to six weeks postpartum on the lactation performance and blood metabolites of dairy cows and immunity of newborn calves.

2. Materials and Methods

2.1. Animals and Treatments

All the experimental protocols used in this study were approved by the Animal Care Committee of Zhejiang University (Hangzhou, China) (Approval Number: ZJU20160379). Twenty-four multiparous prepartum Holstein dairy cows were selected and divided randomly into the control (CON, n = 12) and rumen-protected betaine (BET, n = 12) groups after blocking by parity (2.27, SD = 1.4) and milk yield during the previous lactation cycle (24.9 kg/d, SD = 6.0). The cows were fed a basal total mixed ration (TMR) diet (Table 1) without BET (CON) or with BET at 20 g/d per cow (BET), according to Zhang et al. (2014) [9] from four weeks before expected calving to six weeks postpartum. The basal diets were formulated based on the NRC (2001) [16]. The cows were fed three times daily at approximately 06:30, 13:30, and 19:30 h, and BET (BET with 30% purity, Hangzhou King Technology Feed Co., Ltd, Hangzhou, China) was supplemented twice daily in the morning and evening by top-dressing the TMR during feeding. All cows were housed in tie-stall barns and given access to fresh water ad libitum. After parturition, the cows were milked three times daily at approximately 07:00, 14:00, and 20:00 h. Fourteen calves randomly selected (CON calves: n = 7, BET calves: n = 7) were studied from birth to 24 h. All calves were weighed with a digital scale immediately after birth and were fed fresh colostrum from their dams within 2 h of birth. The calves were individually housed in hutches, and water was offered ad libitum.

Table 1. Ingredient and chemical composition of the diets fed during the prepartum and postpartum periods.

Item	Prepartum	Postpartum
Ingredient, % of DM [1]		
Corn flour	12.41	13.39
Steam-flaked corn	7.15	11.93
Soybean meal	8.61	13.68
Bran	5.30	-
Sodium bicarbonate	0.40	0.72
Calcium hydrophosphate	0.40	0.48
Limestone	0.60	0.66
Fatty acid calcium salts	-	0.78
Salt	0.39	0.46

Table 1. *Cont.*

Item	Prepartum	Postpartum
Ingredient, % of DM [1]		
Premix [2]	0.37	0.44
Mycotoxin binder	0.05	0.07
Active yeast	-	0.07
Brewer's grains	7.31	4.55
Beet pulp	6.70	9.26
Corn silage	25.39	21.05
Alfalfa hay	6.78	16.86
Oat grass	18.08	5.62
Chemical composition, % of DM		
Crude protein	10.99	17.49
Ether extract	3.27	4.31
Crude ash	7.96	7.84
Neutral detergent fiber	48.48	37.08
Acid detergent fiber	27.57	20.78
NEL, Mcal/kg of DM	-	1.63
Lys: Met	2.76:1	3.13:1

[1] DM = dry matter. [2] Formulated to contain (per kilogram of premix) 220 to 400 KIU of vitamin A, 50 to 100 KIU of vitamin D3, ≥2250 IU of vitamin E, ≥40 mg of D-Biotin, ≥380 mg of niacinamide, ≥40 mg of Beta-carotene, 0.2 to 0.7 g of Cu, 1.0 to 3.8 g of Zn, 0.8 to 3.0 g of Mn, 12.5 to 100 mg of I, 8.0 to 25 mg of Se, 2.5 to 50 of mg Co, 10.0% to 30.0% of Ca, 10.0% to 30.0% of NaCl, and ≥1.5% of total phosphorus.

2.2. Sample Preparation

The amounts of feed offered and refused were recorded according to Gu et al. (2018) [17] to determine dry matter intake (DMI). The TMR samples were collected weekly for dry matter (DM, 105 °C for 5 h), crude ash, ether extract (EE), crude protein (CP), and acid detergent fiber (ADF), according to AOAC procedures (method 942.05, 920.39, 988.05, and 973.18, respectively), and neutral detergent fiber (NDF) with sodium sulfite and amylase was analyzed [18]. Body weight (BW) was measured on d 0 and 42 after calving according to Wang et al. (2017) [19].

Milk yield was recorded for two consecutive days each week and milk samples from three consecutive milking were taken each week in the amounts proportional to the yield (4:3:3, composite from each daily milking). The samples were stored at 4 °C with bronopol tablets (D & F Control System Inc., San Ramon CA, USA) for later determination of protein, fat, lactose, total solids, and milk urea nitrogen (MUN) using a Combi Foss FT+ instrument (Foss Electric, Hillerød, Denmark). The 3.5% FCM (Fat-corrected milk) and ECM (Energy-corrected milk) were calculated by the formula [20]: 3.5% FCM = (milk yield, kg/d × 0.4324) + (milk fat, kg/d × 16.216), ECM = (milk yield, kg/d × 0.327) + (milk fat, kg/d × 12.95) + (milk protein, kg/d × 7.20).

Blood samples from the cows were collected from the coccygeal vein in sodium-heparinized tubes at approximately 4 h after the morning feeding on −21, −10, 0, 7, 14, 28, and 42 d relative to calving. Blood samples from the calves were collected via the jugular vein using a sodium-heparinized tube shortly after birth before colostrum feeding and at approximately 24 h after birth. The samples were centrifuged for 10 min at 3000 g at 4 °C to harvest plasma, which was stored at -20 °C until analysis. The plasma samples were analyzed using an Auto Analyzer 7020 instrument (Hitachi High-Technologies Corp., Tokyo, Japan) with colorimetric commercial kits (Ningbo Medical System Biotechnology Co., Ltd., Ningbo, China) for total protein (TP), albumin (ALB), alanine aminotransferase (ALT), aspartate aminotransferase (AST), alkaline phosphatase (ALP), total bilirubin (TBIL), triglyceride (TG), cholesterol (CHOL), glucose (GLU), superoxide dismutase (SOD), NEFA, and BHB. The concentrations of globulin (GLOB) were calculated by the formula [21]: GLOB (g/L) = TP (g/L) − ALB (g/L).

2.3. Data Analysis

A randomized block design with repeated measures was used. The DMI, lactation performance, feed efficiency, and blood metabolites of the cows were analyzed with PROC MIXED of SAS 9.2 (SAS Institute Inc., Cary, NC, USA). Treatment, week, treatment × week, and block were included as the fixed effects in the model, and cow within treatment was used as a random effect. The blood metabolites of the calves were analyzed using the same procedure in SAS 9.2, except sampling hour instead of week was used as the repeated measure. The BW change of the cows, colostrum composition, and calves birth weight were analyzed using PROC MIXED of SAS 9.2 without the repeated statement. All associated interactions were removed from the model. The results are presented as least squares means. Statistical significance was determined at $p \leq 0.05$ and tendencies at $0.05 < p \leq 0.10$.

3. Results and Discussion

Betaine supplementation had no effect on DMI, milk yield, and composition ($p > 0.1$, Table 2). Monteiro et al. (2017) [11] found that cows supplemented with betaine-containing molasses from 60 d before expected calving had higher milk yield, whereas no differences were observed in milk yield of cows supplemented with betaine-containing molasses from 24 d before expected calving, which is consistent with our results. The addition of betaine during the transition period increased the milk yield in a time-dependent manner, which might be related to the functions of betaine as an organic osmolyte to maintain the cell function by stabilizing cellular proteins and promoting proper protein folding [5,22]. The dry period is critical for the renewal and growth of mammary cells [23]. Hence, betaine addition during the far-off period has a positive effect on prepartum mammary growth, which increases the subsequent milk yield. The Lys: Met ratio in the postpartum diets was estimated to be 3.13:1 in our study, which had met the ideal Lys: Met ratio of 3.0:1 for an optimal milk protein content and yield [16,24]. Methyl donors (choline) additional supplementation had no detectable effect on cow performance when the Lys: Met ratio in diets had reached 3.0:1 [25]. This might also be a reason for BET additional supplementation, which has no effect on milk yield and composition.

Compared to CON cows, BET cows tended to increase fat-corrected milk/dry matter intake (FCM/DMI, $p = 0.09$), energy-corrected milk/dry matter intake (ECM/DMI, $p = 0.08$), and BW loss postpartum ($p = 0.10$) (Table 2). The plasma BHB concentrations of the BET cows were greater at d seven after calving than those of the CON cows (treatment × time: $p = 0.07$, Table 3). The greater number of animals in the study might have increased the statistical significance. The BET cows tended to have greater feed efficiency and BW loss postpartum in our study, coupled with greater concentrations of BHB at d seven after calving, which indicates that the BET cows might have an enhanced fat mobilization in early lactation due to higher milk yield numerically (milk yield was approximately 2.53 kg/d higher in BET cows than in CON cows) [11,26].

Table 2. Effects of supplementing cows without rumen-protected betaine (CON) or with rumen-protected betaine (BET) on dry matter intake, lactation performance, and body weight change during the first six weeks of lactation.

	Treatment		SEM	*p*-Value		
Items	CON	BET		Treat	Week	Treat × Week
DMI, kg/d	20.33	20.21	0.76	0.92	<0.01	0.10
Milk yield, kg/d	30.44	32.97	1.68	0.31	<0.01	0.61
Milk composition						
Fat, %	4.30	4.18	0.09	0.35	<0.01	0.44
Protein, %	3.23	3.13	0.06	0.30	<0.01	0.71
Lactose, %	5.00	4.94	0.04	0.28	0.00	0.50
Total solids, %	12.93	12.77	0.13	0.42	<0.01	0.93
MUN, mgN/dL	10.74	10.64	0.63	0.91	0.00	0.23
3.5% FCM [1], kg/d	34.36	36.35	1.83	0.46	0.05	0.49

Table 2. *Cont.*

Items	Treatment		SEM	*p*-Value		
	CON	BET		Treat	Week	Treat × Week
ECM [2], kg/d	34.24	35.70	1.79	0.58	0.33	0.38
FE (FCM/DMI)	1.80	2.00	0.08	0.09	<0.01	0.22
FE (ECM/DMI)	1.77	1.96	0.07	0.08	<0.01	0.26
BW change, kg/d	−1.18	−1.51	0.13	0.10	-	-

[1] 3.5% FCM (Fat-corrected milk) = (milk yield, kg/d × 0.4324) + (milk fat, kg/d × 16.216) [20]. [2] ECM = (milk yield, kg/d × 0.327) + (milk fat, kg/d × 12.95) + (milk protein, kg/d × 7.20) [20].

Table 3. Effects of supplementing cows without rumen-protected betaine (CON) or with rumen-protected betaine (BET) on blood metabolites from four weeks before expected calving to six weeks postpartum.

Items [1]	Treatment		SEM	*p*-Value		
	CON	BET		Treat	Week	Treat × Week
TP, g/L	78.78	78.95	1.61	0.94	<0.01	0.36
ALB, g/L	25.77	25.42	0.35	0.50	<0.01	0.28
GLOB, g/L	53.01	53.53	2.02	0.86	<0.01	0.56
A/G	0.50	0.49	0.02	0.76	<0.01	0.76
ALT, U/L	14.41	13.90	0.94	0.71	<0.01	0.81
AST, U/L	71.86	73.67	4.15	0.76	<0.01	0.40
ALP, U/L	33.29	36.34	1.81	0.26	<0.01	0.91
TBIL, μmol/L	2.61	2.75	0.25	0.70	<0.01	0.99
TG, mmol/L	0.08	0.08	0.01	0.32	<0.01	0.39
CHOl, mmol/L	2.52	2.37	0.09	0.26	<0.01	0.42
GLU, mmol/L	3.34	3.30	0.07	0.72	<0.01	0.83
NEFA, μmol/L	246.57	243.90	20.74	0.93	<0.01	0.95
BHB, μmol/L	802.72	812.65	92.55	0.94	<0.01	0.07

[1] TP = total protein. ALB = albumin. GLOB = globulin. A/G = albumin/globulin. ALT = alanine aminotransferase. AST = aspartate aminotransferase. ALP = alkaline phosphatase. TBIL = total bilirubin. TG = triglyceride. CHOL = cholesterol. GLU = glucose. NEFA = non-esterified fatty acids. BHB = β-hydroxybutyrate.

The plasma TP and GLOB concentrations of the BET calves were greater than those of the CON calves ($p = 0.04$, $p = 0.05$, respectively, Table 4), although no differences in calves birth weight were found between treatments (37.80 ± 1.68 kg vs. 36.03 ± 1.68 kg). The plasma TP and GLOB concentrations of calves increased significantly with maternal betaine supplementation, which indicates that it might improve the immunity of newborn calves because of maternal methyl donors supplementation [13,27]. Maternal dietary supplementation with methyl donors could program the health of offspring through the epigenetic regulation of the DNA molecule and cell signaling [27,28], which might improve the capacity for GLOB absorption of the intestine to improve the immunity of newborn calves. Furthermore, the lactose content of the BET colostrum tended to increase compared to the CON colostrum in our study (3.50% vs. 3.07%, $p = 0.07$), which might also have contributed to the results. Lactose plays a key role in the energy supply, absorption of minerals, and gastrointestinal functions of calves [29,30].

The plasma GLU concentrations of the BET calves tended to decrease compared with those of the CON calves ($p = 0.09$, Table 4). The plasma SOD concentrations of the BET calves were greater at 2 h after birth than those of the CON cows (treatment × time: $p = 0.01$, Table 4). A positive correlation between neonatal glucose and cortisol concentrations proved that the lower concentrations of glucose in the BET calves were likely to have less stress around calving [31]. In turn, the degree of stress might influence newborn calves' energetic mobilization [31,32]. The greater plasma SOD concentrations 2 h after birth in our study suggested that the BET calves were in a state of less stress [33]. The plasma glucose concentrations in newborn calves might be related to the uteroplacental transport of glucose via mTOR signaling [34] and hepatic gluconeogenic gene expression via epigenetic mechanisms [35], which deserve further study.

Table 4. The blood metabolites during the 24 h after birth of calves born to dams supplemented without rumen-protected betaine (CON) or with rumen-protected betaine (BET) during the peripartal period.

Items [1]	Treatment		SEM	*p*-Value		
	CON Calves	BET Calves		Treat	Hour	Treat × Hour
TP, g/L	54.71	59.63	1.48	0.04	<0.01	0.15
ALB, g/L	18.48	18.59	0.42	0.85	0.00	0.03
GLOB, g/L	36.23	41.03	1.57	0.05	<0.01	0.06
A/G	0.58	0.56	0.02	0.54	<0.01	0.04
ALT, U/L	7.67	8.24	0.37	0.30	<0.01	0.21
AST, U/L	53.67	55.31	4.98	0.82	<0.01	0.87
ALP, U/L	265.38	257.78	35.73	0.88	<0.01	0.86
TBIL, μmol/L	6.21	6.60	0.87	0.76	0.05	0.97
GLU, mmol/L	5.62	4.82	0.30	0.09	0.00	0.36
SOD, U/ml	66.54	70.52	2.91	0.35	0.00	0.01

[1] TP = total protein. ALB = albumin. GLOB = globulin. A/G = albumin/globulin. ALT = alanine aminotransferase. AST = aspartate aminotransferase. ALP = alkaline phosphatase. TBIL = total bilirubin. GLU = glucose. SOD = superoxide dismutase.

4. Conclusions

Dietary rumen-protected betaine supplementation from four weeks before expected calving had no detectable effect on dry matter intake and milk yield, but tended to increase fat mobilization of postpartum dairy cows. Furthermore, the BET calves had greater plasma total protein and globulin concentrations, which indicates that the immunity of the BET calves might improve.

Author Contributions: Conceived and designed the experiments: B.W., C.W., J.L. and H.L. Conducted the experimental work: B.W., R.G. and K.S. Conducted the data analysis: B.W. and Z.W. Wrote the paper: B.W. Revised the paper: B.W., C.W. and H.L.

Funding: Grants from the National Key Research and Development Program of China (2018YFD0501600), the China Agriculture Research System (No. CARS-37), and the Zhejiang Provincial Natural Science Foundation (LY18C170002) supported this research.

Acknowledgments: The authors thank Hangjiang Dairy Farm (Hangzhou, China) for providing and caring for the cows.

Conflicts of Interest: The authors declare no conflict of interest.

References

1. Coonen, J.M.; Maroney, M.J.; Crump, P.M.; Grummer, R.R. Short communication: Effect of a stable pen management strategy for precalving cows on dry matter intake, plasma nonesterified fatty acid levels, and milk production. *J. Dairy Sci.* **2011**, *94*, 2413–2417. [CrossRef] [PubMed]
2. Zhou, Z.; Vailati-Riboni, M.; Luchini, D.N.; Loor, J.J. Methionine and choline supply during the periparturient period alter plasma amino acid and one-carbon metabolism profiles to various extents: Potential role in hepatic metabolism and antioxidant status. *Nutrients* **2017**, *9*, 10. [CrossRef] [PubMed]
3. Pinotti, L.; Baldi, A.; Dell'Orto, V. Comparative mammalian choline metabolism with emphasis on the high-yielding dairy cow. *Nutr. Res. Rev.* **2002**, *15*, 315–332. [CrossRef]
4. Bauman, D.E.; Currie, W.B. Partitioning of nutrients during pregnancy and lactation: A review of mechanisms involving homeostasis and homeorhesis. *J. Dairy Sci.* **1980**, *63*, 1514–1529. [CrossRef]
5. Engin, F.; Hotamisligil, G.S. Restoring endoplasmic reticulum function by chemical chaperones: An emerging therapeutic approach for metabolic diseases. *Diabetes Obes. Metab.* **2010**, *12*, 108–115. [CrossRef] [PubMed]
6. Eklund, M.; Bauer, E.; Wamatu, J.; Mosenthin, R. Potential nutritional and physiological functions of betaine in livestock. *Nutr. Res. Rev.* **2005**, *18*, 31–48. [CrossRef]
7. Pas, B.J.B.; Pas, J.R.B.; Harmoney, K.R.; Pas, S.R.G. Influence of betaine on pasture, finishing, and carcass performance in steers. *Prof. Anim. Sci.* **2004**, *20*, 53–57.

8. Peterson, S.E.; Rezamand, P.; Williams, J.E.; Price, W.; Chahine, M.; Mcguire, M.A. Effects of dietary betaine on milk yield and milk composition of mid-lactation Holstein dairy cows. *J. Dairy Sci.* **2012**, *95*, 6557–6562. [CrossRef] [PubMed]

9. Zhang, L.; Ying, S.J.; An, W.J.; Lian, H.; Zhou, G.B.; Han, Z.Y. Effects of dietary betaine supplementation subjected to heat stress on milk performances and physiology indices in dairy cow. *Genet. Mol. Res.* **2014**, *13*, 7577–7586. [CrossRef]

10. Wang, C.; Liu, Q.; Yang, W.Z.; Wu, J.; Zhang, W.W.; Zhang, P.; Dong, K.H.; Huang, Y.X. Effects of betaine supplementation on rumen fermentation, lactation performance, feed digestibilities and plasma characteristics in dairy cows. *J. Agric. Sci.* **2010**, *148*, 487–495. [CrossRef]

11. Monteiro, A.P.A.; Bernard, J.K.; Guo, J.R.; Weng, X.S.; Emanuele, S.; Davis, R.; Dahl, G.E.; Tao, S. Effects of feeding betaine-containing liquid supplement to transition dairy cows. *J. Dairy Sci.* **2017**, *100*, 1063–1071. [CrossRef]

12. Lever, M.; Slow, S. The clinical significance of betaine, an osmolyte with a key role in methyl group metabolism. *Clin. Biochem.* **2010**, *43*, 732–744. [CrossRef] [PubMed]

13. Van, L.L.; Tint, M.T.; Aris, I.M.; Quah, P.L.; Fortier, M.V.; Lee, Y.S.; Yap, F.K.; Saw, S.M.; Godfrey, K.M.; Gluckman, P.D. Prospective associations of maternal betaine status with offspring weight and body composition at birth: The GUSTO (Growing Up in Singapore Toward healthy Outcomes) cohort study. *Am. J. Clin. Nutr.* **2016**, *104*, 1327–1333. [CrossRef]

14. Mitchell, A.D.; Chappell, A.; Knox, K.L. Metabolism of betaine in the ruminant. *J. Anim. Sci.* **1979**, *49*, 764–774. [CrossRef]

15. Liu, H. Effects of Dietary Supplementation of Rumen-Protected Betaine on Lactation Performance and Serum Metabolites of Dairy Cows. Master's Thesis, Zhejiang University, Hangzhou, China, 2017.

16. NRC. *Nutrient Requirements of Dairy Cattle*, 7th ed.; National Academy Press: Washington, DC, USA, 2001.

17. Gu, F.F.; Liang, S.L.; Wei, Z.H.; Wang, C.P.; Liu, H.Y.; Liu, J.X.; Wang, D.M. Short communication: Effects of dietary addition of N-carbamylglutamate on milk composition in mid-lactating dairy cows. *J. Dairy Sci.* **2018**, *101*, 10985–10990. [CrossRef] [PubMed]

18. Van Soest, P.J.; Robertson, J.B.; Lewis, B.A. Methods for dietary fiber, neutral detergent fiber, and nonstarch polysaccharides in relation to animal nutrition. *J. Dairy Sci.* **1991**, *74*, 3583–3597. [CrossRef]

19. Wang, D.M.; Zhang, B.X.; Wang, J.K.; Liu, H.Y.; Liu, J.X. Short communication: Effects of dietary 5,6-dimethylbenzimidazole supplementation on vitamin B 12 supply, lactation performance, and energy balance in dairy cows during the transition period and early lactation. *J. Dairy Sci.* **2017**, *101*, 1–4.

20. Tyrrell, H.F.; Reid, J.T. Prediction of the energy value of cow's milk. *J. Dairy Sci.* **1965**, *48*, 1215–1223. [CrossRef]

21. Fouché, N.; Graubner, C.; Howard, J. Correlation between serum total globulins and gamma globulins and their use to diagnose failure of passive transfer in foals. *Vet. J.* **2014**, *202*, 384–386. [CrossRef]

22. Ratriyanto, A.; Mosenthin, R.; Bauer, E.; Eklund, M. Metabolic, osmoregulatory and nutritional functions of betaine in monogastric animals. *Asian-Australas. J. Anim. Sci.* **2009**, *22*, 1461–1476. [CrossRef]

23. Bernier-Dodier, P.; Girard, C.L.; Talbot, B.G.; Lacasse, P. Effect of dry period management on mammary gland function and its endocrine regulation in dairy cows. *J. Dairy Sci.* **2011**, *94*, 4922–4936. [CrossRef] [PubMed]

24. Wang, C.; Liu, H.Y.; Wang, Y.M.; Yang, Z.Q.; Liu, J.X.; Wu, Y.M.; Yan, T.; Ye, H.W. Effects of dietary supplementation of methionine and lysine on milk production and nitrogen utilization in dairy cows. *J. Dairy Sci.* **2010**, *93*, 3661–3670. [CrossRef] [PubMed]

25. Zhou, Z.; Vailati-Riboni, M.; Trevisi, E.; Drackley, J.K.; Luchini, D.N.; Loor, J.J. Better postpartal performance in dairy cows supplemented with rumen-protected methionine compared with choline during the peripartal period. *J. Dairy Sci.* **2016**, *99*, 8716–8732. [CrossRef] [PubMed]

26. Dann, H.M.; Litherland, N.B.; Underwood, J.P.; Bionaz, M.; D'Angelo, A.; Mcfadden, J.W.; Drackley, J.K. Diets during far-off and close-up dry periods affect periparturient metabolism and lactation in multiparous cows. *J. Dairy Sci.* **2006**, *89*, 3563–3577. [CrossRef]

27. Liu, J.; Yao, Y.; Yu, B.; Mao, X.; Huang, Z.; Chen, D. Effect of maternal folic acid supplementation on hepatic proteome in newborn piglets. *Nutrition* **2013**, *29*, 230–234. [CrossRef] [PubMed]

28. Ji, Y.; Wu, Z.; Dai, Z.; Sun, K.; Wang, J.; Wu, G. Nutritional epigenetics with a focus on amino acids: Implications for the development and treatment of metabolic syndrome. *J. Nutr. Biochem.* **2016**, *27*, 1–8. [CrossRef] [PubMed]

29. Frizzo, L.S.; Soto, L.P.; Zbrun, M.V.; Signorini, M.L. Effect of lactic acid bacteria and lactose on growth performance and intestinal microbial balance of artificially reared calves. *Livest. Sci.* **2011**, *140*, 246–252. [CrossRef]

30. Liu, J.; Chang, J.; Yao, A.; Hu, Y.; Yuan, Y.; Yu, F.; Ma, Z.; Wang, G.; Zhao, X. Diagnosis and clinical observation of lactose-free milk powder on treatment of neonatal diarrhea. *Pak. J. Pharm. Sci.* **2016**, *29*, 309–314. [PubMed]

31. Vannucchi, C.I.; Rodrigues, J.A.; Silva, L.C.G.; Lúcio, C.F.; Veiga, G.A.L.; Furtado, P.V.; Oliveira, C.A.; Nichi, M. Association between birth conditions and glucose and cortisol profiles of periparturient dairy cows and neonatal calves. *Vet. Rec.* **2015**, *176*, 358. [CrossRef]

32. Jacometo, C.B.; Zhou, Z.; Luchini, D.; Trevisi, E.; Corrêa, M.N.; Loor, J.J. Maternal rumen-protected methionine supplementation and its effect on blood and liver biomarkers of energy metabolism, inflammation, and oxidative stress in neonatal Holstein calves. *J. Dairy Sci.* **2016**, *99*, 6753–6763. [CrossRef] [PubMed]

33. Hassanpour, A.; Sabegh, Y.G.; Sadeghi-nasab, A. Assessment of serum antioxidant enzymes activity in cattle suffering from Theileriosis. *Eur. J. Exp. Biol.* **2013**, *3*, 493–496.

34. Batistel, F.; Alharthi, A.S.; Wang, L.; Parys, C.; Pan, Y.X.; Cardoso, F.C.; Loor, J.J. Placentome nutrient transporters and mammalian target of rapamycin signaling proteins are altered by the methionine supply during late gestation in dairy cows and are associated with newborn birth weight. *J. Nutr.* **2017**, *147*, 1640–1647. [CrossRef]

35. Cai, D.; Jia, Y.; Song, H.; Sui, S.; Lu, J.; Jiang, Z.; Zhao, R. Betaine supplementation in maternal diet modulates the epigenetic regulation of hepatic gluconeogenic genes in neonatal piglets. *PLoS ONE* **2013**, *9*, e105504. [CrossRef]

Communication

The Effect of Intravenous Infusions of Glutamine on Duodenal Cell Autophagy and Apoptosis in Early-Weaned Calves

Xusheng Dong [1,†], Ruina Zhai [2,†], Zhaolin Liu [1], Xueyan Lin [1], Zhonghua Wang [1] and Zhiyong Hu [1,*]

1   Ruminant Nutrition and Physiology Laboratory, College of Animal Science and Technology, Shandong Agricultural University, Taian 271018, China
2   College of Animal Science, Xinjiang Agricultural University, Urumqi 830052, China
*   Correspondence: hzy20040111@126.com; Tel.: +86-1856-236-0982
†   These authors contributed equally to this work.

Received: 14 March 2019; Accepted: 24 June 2019; Published: 1 July 2019

Simple Summary: The objective of this study was to determine the effects of intravenous infusions of L-glutamine (Gln) on the autophagy and apoptosis of duodenum cells in weaned calves. The results showed that the autophagy level of duodenal cells was increased with an increasing Gln infusion dose (0 to 20 g/d) and dropped when Gln was further increased to 40 g/d. We also found that the level of apoptosis was decreased with an increasing Gln infusion dose from 0 to 20 g/d, and then rose as the dose increased to 40 g/d. This knowledge will provide a reference for weaned calf health management.

Abstract: The objectives of this study were to determine the effects of intravenous infusions of L-glutamine (Gln) on the autophagy and apoptosis of duodenum cells in early-weaned calves. Holstein male calves were weaned at day 35 (20 male calves, birth weight 43 ± 1.8 kg; 35 ± 3 d of age) and randomly allocated to four treatments (5 calves/treatment). The treatments were: (1) infusion of NaCl, representing the control group (C); (2) infusion of 10 g/d of Gln solution (L); (3) infusion of 20 g/d of Gln solution (M); and (4) infusion of 40 g/d of Gln solution (H). The solutions were infused for 2 h daily for 3 consecutive days after weaning. All calves were killed on the third day post-weaning. The results showed that the autophagy level of the duodenal cells was increased as the Gln infusions increased from 0 to 20 g/d and dropped with a further increase in dose (40 g/d). We also found that the level of apoptosis was decreased with Gln infusion from 0 to 20 g/d and rose as the dose increased to 40 g/d. This knowledge provides a reference for weaned calf health management.

Keywords: calf; glutamine; autophagy; apoptosis

---

1. Introduction

Glutamine (Gln) is the most abundant amino acid in vivo and is a major respiratory fuel and metabolic precursor for many cell types [1]. Glutamine, which once was regarded as a nonessential amino acid, has recently been termed conditionally essential during injury or oxidative stress [2,3]. A previous study suggested that Gln could protect the small intestine from various harmful injuries in rats [4]. Kallweit et al. [5] showed that Gln protects intestinal cells from both heat and oxidant stress. Recently, researchers attempted to evaluate the impact of Gln on autophagy and apoptosis [6–8]. Sakiyama et al. [7] suggested that Gln could protect intestinal epithelial cells by enhancing autophagy.

Autophagy is a specific protein degradation process that functions in the bulk degradation of cellular components and has been recognized as an important mechanism for cell survival under conditions of stress [7,9]. When cells lack nutrients, autophagy is activated to supply amino acids in order to maintain cell survival [10]. In vivo apoptosis and autophagy are two forms of physiological and conserved programmed cell death [11]. Apoptosis is characterized by a series of morphological changes, including plasma membrane blebbing, nuclear condensation, and fragmentation, which lead to the formation of apoptotic bodies [12]. When cells are under stress, autophagy and apoptosis are activated [12]. In general, autophagy is activated first and maintains cell homeostasis [13]. When stress is prolonged or exceeds a threshold, apoptosis is activated [12,14]. The protein microtubule-associated protein 1 light chain 3-II(LC3-II), which is a useful marker of autophagic membranes, is essential for the expansion of the early autophagosome in the context of cellular house-keeping and autophagic cell death [15,16]. Caspase-3 is an executioner caspase, which is activated by apoptosis [12]. Furthermore, PI3K/Akt/mTOR signaling pathways, which inhibit autophagy, have been found to be essential for the regulation of autophagy [17]. A previous study found that caffeine could induce autophagy by abolishing AKT phosphorylation [17]. The kinase mammalian target of rapamycin (mTOR) is a downstream target of the PI3K/AKT pathway [17]. Deactivation of mTOR signaling induces autophagy [18]. Amino acids, which are provided by autophagy, can restore mTOR complex 1 (mTORC1) activity during amino acid starvation [8]. The restoration of mTORC1 in turn inhibits autophagy, which completes the feedback loop [8]. The feedback loop could protect the cells by mitigating damage from stress and starvation.

Weaning is the transition from the ingestion of milk to solid feed for calves with dramatic gastrointestinal transformations [19]. Weaning is a particularly vulnerable period for mammals, with an increased risk of malnutrition, intestinal infections, and poor growth [20,21]. The function and morphology of the small intestine are severely disturbed after weaning, such as villous shortening in pigs [22,23]. Our previous study found that an exogenous supply of Gln increased the autophagy level of liver cells and increased growth rates, villus height, and crypt depth of the duodenum in early-weaned calves [20,24]. Whether Gln induces autophagy and apoptosis of the duodenum cells in early-weaned calves remains unknown. The purpose of this study was to evaluate the effect of Gln on the autophagy and apoptosis of duodenum cells in early-weaned calves. We hypothesized that intravenous infusions of Gln would increase the level of autophagy and reduce the level of apoptosis. This would provide a reference for weaned calf health management.

2. Materials and Methods

Animal care and use were approved and conducted under established standards of the Ethics Committee on animals of Shandong Agricultural University (SDAUA-2018-012). The study was conducted during November of 2018 at the Shandong high-speed modern dairy farm in Ji Nan, Shandong, China. The animals were individually housed in a pen with free access to water and fresh calf starter. The ingredient and nutrient composition of the calf starter is given in Table 1.

All calves received 4 L of colostrum in the 2-hour period after birth and were then fed 6 L of whole milk 3 times daily until weaning. Fresh calf starter was offered ad libitum beginning at 3 d of age. Water was offered daily ad libitum. Holstein calves were weaned at day 35 (20 male calves, birth weight 43 ± 1.8 kg; 35 ± 3 d of age) and randomly allocated to four treatments (5 calves/treatment). Starting from day 35, the calves were given the following treatments for 3 consecutive days. The treatments were: (1) infusion of 1.5 L of 0.85% NaCl, representing the control group (C); (2) infusion of 10 g/d of Gln mixed with 1.5 L of 0.85% NaCl solution (L); (3) infusion of 20 g/d of Gln mixed with 1.5 L of 0.85% NaCl solution (M); and (4) infusion of 40 g/d of Gln mixed with 1.5 L of 0.85% NaCl solution (H). The dose of intravenous infusion Gln referred to that in a previous study [20]. At the beginning of the experiment, all calves had milk removed from their daily diet. The solutions were infused for 2 h daily for 3 consecutive days after weaning. Starter intake for each calf was measured daily during the infusion period.

Table 1. Ingredient and nutrient composition of the experimental starter of calves.

Items	Content (% of DM)
Ingredients	
Corn grain	48
Wheat bran	12.6
Soybean meal	18.8
Extruded soybean	7
Corn gluten meal	9
Salt	0.55
Calcium carbonate	2
Dicalcium phosphate	1.15
Vitamin and trace mineral premix [1]	0.9
Nutrients, % of DM	
DM, %	89.3
CP, %	22.13
Crude fat, %	4.32
NDF, %	17.14
ADF, %	6.62
Ca, %	1.07
P, %	0.56
ME, Mcal/kg	2.83

DM: dry matter; CP: crude protein; NDF: neutral detergent fiber; ADF: acid detergent fiber; ME: metabolizable energy. [1] Premix contained (mg/kg): vitamin A, 4035; vitamin D, 1740; vitamin E, 39; Fe, 18; Zn, 37; Cu, 10.6; Mn, 15.3; Co, 0.12; I, 0.47; and Se, 0.35.

All calves were euthanized following captive bolt gun stunning on the 3 d post-weaning day for measuring the autophagy and apoptosis of duodenum cells. After opening the body cavity, the samples of duodenum (entire wall from 6 cm distal to the pylorus) were immediately frozen in liquid nitrogen and stored at −80 °C until western blotting was performed.

Briefly, the tissue sample blocks (entire duodenum from 6 cm distal to the pylorus) were washed with phosphate buffer saline (PBS, Solarbio, P1020-500 mL, Beijing, China), cut into small pieces, homogenized in PBS at 4 °C using a Servicebio KZ-II homogenizer, kept on ice for 0.5 h, oscillated to ensure complete tissue cracking every 5 min, and then centrifuged (3000× g, 10 min, 4 °C). Protein concentration was determined in the supernatant (BCA Protein Assay Kit, G2026, Servicebio, Wuhan, China). The sample was then diluted with an equal volume of Laemmli sample buffer (Bio-Rad, 1610737, Shanghai, China) and boiled for 5 min. Sodium dodecyl sulfate-PAGE, electro-transfer of proteins, and immunoblotting were performed as previously described [25,26]. Antibodies used for immunoblotting were anti-LC3 (Sigma-Aldrich, L8918, Shanghai, China), anti-Caspase-3 (Sigma-Aldrich, C8487, Shanghai, China), anti-mTOR (Sigma-Aldrich, SAB2701843, Shanghai, China), anti-phospho-mTOR (Sigma-Aldrich, SAB4301526, Shanghai, China), anti-β-actin (Sigma-Aldrich, A2066, Shanghai, China), and appropriate secondary antibodies (Servicebio, GB23303, Wuhan, China). The chemiluminescence of bands of interest were detected with a digital G: Box imager (Syngene, Frederick, MD, USA). The band density was quantified with ImageJ software (National Institutes of Health, Bethesda, MD, USA).

The data were analyzed as a completely randomized design using one-way ANOVA of SAS 8.2 (SAS Institute Inc., Cary, NC). The individual calf was considered as the experimental unit. The analysis used the following model: $y_{ij} = \mu + \alpha_i + \varepsilon_{ij}$ (y = western blot data, μ = mean, i = dose of infusions, and ε = residuals). The means were compared using Duncan's multiple range test. Significance was declared at $p < 0.05$.

3. Results

The starter intake of group C, L, M, and H were 1.12 kg/d, 1.15 kg/d, 1.22 kg/d, and 1.19 kg/d, respectively. Starter intake was not different between treatments. The results reported in this research showed that the autophagy level of the duodenal cells was increased with an increasing Gln infusion dose (0 to 20 g/d) and dropped when Gln was further increased to 40 g/d (Figure 1). We also found that the level of apoptosis was decreased with an increasing Gln infusion dose from 0 to 20 g/d, and then rose with an increasing dose of Gln to 40 g/d (Figure 1). In group M, the level of autophagy reached the highest level; in contrast, the level of apoptosis reached a lowest point. The expression of mTOR was significantly decreased after Gln infusion ($p < 0.05$). The expression of p-mTOR in group M was lower than that in other groups ($p < 0.05$).

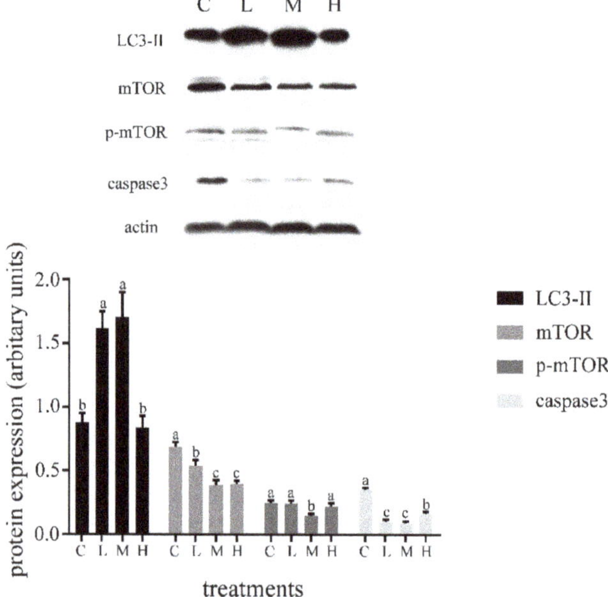

Figure 1. Effects of glutamine (Gln) infusions on the microtubule-associated protein 1 light chain 3-II (LC3-II), mTOR, p-mTOR, and caspase3 expression of duodenum in weaned calves. Treatment was as follows: (1) C: infusion of 1.5 L of 0.85% NaCl; (2) L: infusion of 10 g/d of Gln mixed with 1.5 L of 0.85% NaCl; (3) M: infusion of 20 g/d of Gln mixed with 1.5 L of 0.85% NaCl; (4) H: infusion of 40 g/d of Gln mixed with 1.5 L of 0.85% NaCl. Insets depict representative blots. Values represent means ± SD. Response from statistical result, $p < 0.05$. β-Actin was used to normalize the expression of target proteins. The letters below the bar graph indicate different treatments. Different letters above the bar indicate differences between different groups ($p < 0.05$).

4. Discussion

In this study, we demonstrated that low dose infusion of Gln could induce autophagy and retard apoptosis. We further increased the infusion of Gln to 40 g/d and found that the effect of Gln on calves was reduced. We concluded that Gln-induced autophagy is mainly dependent on the inhibition of mTOR phosphorylation. Gln is involved in stress protection by way of the stimulation of autophagy in intestinal cells [7,27]. A previous study found that Gln infusion increased growth rates, villus height, and crypt depth in the duodenum of early-weaned calves [24]. In this study, when the concentration of

Gln was increased from 0 to 20 g/d, the autophagy levels increased as the Gln infusion dose increased. The results suggested that Gln can promote autophagy in the duodenum. This finding is consistent with that of Sakiyama et al. [7], who confirmed that Gln is essential for maintaining autophagy and mounting an autophagic response under stress in intestinal cells. It has been suggested that Gln can induce autophagy in intestinal epithelial cells through restraining mTOR and p38 MAP kinase pathways [7]. The expression of mTOR and p-mTOR in our study is also consistent with this study. The expression of p-mTOR was significantly decreased after Gln infusion. To further investigate the effects of high dose Gln, we further increased the infusion of Gln to 40 g/d. Interestingly, we found that the autophagy level was decreased as the Gln infusion dose increased from 20 to 40 g/d. A possible explanation for this might be that the activity of Gln synthetase in the body gradually decreases when the blood concentration of Gln was excessive. In *Escherichia coli*, Gln synthetase activity is subject to inhibition by different end products of Gln metabolism [28]. A previous report suggested that over expression of Gln synthetase inhibited mTOR activity and activated autophagy [29]. Thus, the level of autophagy induced by Gln synthetase was decreased as the activity of Gln synthetase decreased.

Normally, autophagy restrains the activity of apoptosis, and apoptosis-associated caspase activation shuts off the autophagic process [12]. In our study, we also found that the level of apoptosis fell to a low point with the intravenous infusion dose of 20 g/d. The tendency of apoptosis was opposite to that of autophagy, which suggested that autophagy may inhibit the activity of apoptosis. Kallweit et al. [5] found that Gln protects intestinal cells from both heat and oxidant injury, which are key mechanisms in the prevention of apoptosis. Han et al. [30] demonstrated that the combination of Gln has the ability to maintain the integrity of the intestinal mucosal barrier by inhibiting the apoptosis of intestinal epithelial cells. These results are similar to our study. Our result suggests that the effect of Gln on apoptosis is contrary to that on autophagy. AKT is a kinase with dual autophagy–apoptosis regulatory potential, which can phosphorylate Beclin 1 and B-cell lymphoma-2 antagonists of cell death (BAD) to inhibit autophagic and apoptotic functions, respectively [12]. The activation of AKT could inhibit autophagy by inducing mTOR [12]. Thus, considerably more work needs to be done to determine the effect of Gln on Akt/mTOR signaling pathways in early-weaned calf.

In commercial dairy farms, dairy calves are weaned early to reduce milk costs. However, the gastrointestinal tract of the calf is not ready for early weaning [19]. In our research, intravenous infusions of low dose Gln could increases autophagy, which probably relieved weaning stress. A further study could assess the effects of diet supplement Gln on the early-weaned calf. This knowledge will provide a reference for Gln supplementation for weaned calf health management.

5. Conclusions

In conclusion, Gln could induce autophagy and decrease the level of apoptosis in the duodenum of early-weaned calves. The intravenous infusion moderate dose (20 g/d) of Gln is most effective. This knowledge will provide a reference for weaned calf health management.

Author Contributions: Z.H. and Z.W. conceived and designed the experiments; X.D. and R.Z. performed the experiments; X.D. and X.L. analyzed the data; X.D. and Z.L. wrote the paper.

Funding: This research was funded by the National Natural Science Foundation of China, grant number 31772624, and the National Key Research and Development Program of China, grant number 2018YFD0501600.

Conflicts of Interest: The authors declare no conflict of interest.

References

1. Newsholme, P. Why is l-glutamine metabolism important to cells of the immune system in health, postinjury, surgery or infection? *J. Nutr.* **2001**, *131*, 2515S–2522S. [CrossRef] [PubMed]
2. Peng, Z.Y.; Hamiel, C.R.; Banerjee, A.; Wischmeyer, P.E.; Friese, R.S.; Wischmeyer, P. Glutamine attenuation of cell death and inducible nitric oxide synthase expression following inflammatory cytokine-induced injury is dependent on heat shock factor-1 expression. *J. Parenter. Enter. Nutr.* **2006**, *30*, 400–406. [CrossRef]

3. Wischmeyer, P.E. Glutamine and heat shock protein expression. *Nutrition* **2002**, *18*, 225–228. [CrossRef]

4. Papaconstantinou, H.T.; Hwang, K.O.; Rajaraman, S.; Hellmich, M.R.; Townsend, C.M., Jr.; Ko, T.C. Glutamine deprivation induces apoptosis in intestinal epithelial cells. *Surgery* **1998**, *124*, 152–159. [CrossRef]

5. Kallweit, A.R.; Baird, C.H.; Stutzman, D.K.; Wischmeyer, P.E. Glutamine prevents apoptosis in intestinal epithelial cells and induces differential protective pathways in heat and oxidant injury models. *J. Parenter. Enter. Nutr.* **2012**, *36*, 551–555. [CrossRef] [PubMed]

6. Chen, L.; Cui, H. Targeting glutamine induces apoptosis: A cancer therapy approach. *Int. J. Mol. Sci.* **2015**, *16*, 22830–22855. [CrossRef]

7. Sakiyama, T.; Musch, M.W.; Ropeleski, M.J.; Tsubouchi, H.; Chang, E.B. Glutamine increases autophagy under Basal and stressed conditions in intestinal epithelial cells. *Gastroenterology* **2009**, *136*, 924–932. [CrossRef]

8. Tan, H.W.S.; Sim, A.Y.L.; Long, Y.C. Glutamine metabolism regulates autophagy-dependent mTORC1 reactivation during amino acid starvation. *Nat. Commun.* **2017**, *8*, 338. [CrossRef]

9. Huang, H.; Li, X.; Zhuang, Y.; Li, N.; Zhu, X.; Hu, J.; Ben, J.; Yang, Q.; Bai, H.; Chen, Q. Class A scavenger receptor activation inhibits endoplasmic reticulum stress-induced autophagy in macrophage. *J. Biomed. Res.* **2014**, *28*, 213–221. [CrossRef]

10. Zhu, Y.; Lin, G.; Dai, Z.; Zhou, T.; Li, T.; Yuan, T.; Wu, Z.; Wu, G.; Wang, J. L-glutamine deprivation induces autophagy and alters the mTOR and MAPK signaling pathways in porcine intestinal epithelial cells. *Amino Acids* **2015**, *47*, 2185–2197. [CrossRef]

11. Yi, H.Y.; Yang, W.Y.; Wu, W.M.; Li, X.X.; Deng, X.J.; Li, Q.R.; Cao, Y.; Zhong, Y.J.; Huang, Y.D. BmCalpains are involved in autophagy and apoptosis during metamorphosis and after starvation in *Bombyx mori*. *Insect Sci.* **2018**, *25*, 379–388. [CrossRef] [PubMed]

12. Marino, G.; Niso-Santano, M.; Baehrecke, E.H.; Kroemer, G. Self-consumption: The interplay of autophagy and apoptosis. *Nat. Rev. Mol. Cell Biol.* **2014**, *15*, 81–94. [CrossRef] [PubMed]

13. Liu, G.; Pei, F.; Yang, F.; Li, L.; Amin, A.D.; Liu, S.; Buchan, J.R.; Cho, W.C. Role of autophagy and apoptosis in non-small-cell lung cancer. *Int. J. Mol. Sci.* **2017**, *18*, 367. [CrossRef] [PubMed]

14. Kroemer, G.; Marino, G.; Levine, B. Autophagy and the integrated stress response. *Mol. Cell* **2010**, *40*, 280–293. [CrossRef] [PubMed]

15. Abeliovich, H. Guidelines for the use and interpretation of assays for monitoring autophagy. *Autophagy* **2012**, *8*, 445–544. [CrossRef]

16. Tanida, I.; Ueno, T.; Kominami, E. LC3 conjugation system in mammalian autophagy. *Int. J. Biochem. Cell Biol.* **2004**, *36*, 2503–2518. [CrossRef] [PubMed]

17. Saiki, S.; Sasazawa, Y.; Imamichi, Y.; Kawajiri, S.; Fujimaki, T.; Tanida, I.; Kobayashi, H.; Sato, F.; Sato, S.; Ishikawa, K.; et al. Caffeine induces apoptosis by enhancement of autophagy via PI3K/Akt/mTOR/p70S6K inhibition. *Autophagy* **2011**, *7*, 176–187. [CrossRef]

18. Guan, H.; Piao, H.; Qian, Z.; Zhou, X.; Sun, Y.; Gao, C.; Li, S.; Piao, F. 2,5-Hexanedione induces autophagic death of VSC4.1 cells via a PI3K/Akt/mTOR pathway. *Mol. Biosyst.* **2017**, *13*, 1993–2005. [CrossRef]

19. Steele, M.A.; Penner, G.B.; Chaucheyras-Durand, F.; Guan, L.L. Development and physiology of the rumen and the lower gut: Targets for improving gut health. *J. Dairy Sci.* **2016**, *99*, 4955–4966. [CrossRef]

20. Hu, Z.Y.; Li, S.L.; Cao, Z.J. Short communication: Glutamine increases autophagy of liver cells in weaned calves. *J. Dairy Sci.* **2012**, *95*, 7336–7339. [CrossRef]

21. Spreeuwenberg, M.A.; Verdonk, J.M.; Gaskins, H.R.; Verstegen, M.W. Small intestine epithelial barrier function is compromised in pigs with low feed intake at weaning. *J. Nutr.* **2001**, *131*, 1520–1527. [CrossRef] [PubMed]

22. Wijtten, P.J.; van der Meulen, J.; Verstegen, M.W. Intestinal barrier function and absorption in pigs after weaning: A review. *Br. J. Nutr.* **2011**, *105*, 967–981. [CrossRef] [PubMed]

23. Tsukahara, T.; Inoue, R.; Yamada, K.; Yajima, T. A mouse model study for the villous atrophy of the early weaning piglets. *J. Vet. Med. Sci.* **2010**, *72*, 241–244. [CrossRef] [PubMed]

24. Hu, Z.Y.; Su, H.W.; Li, S.L.; Cao, Z.J. Effect of parenteral administration of glutamine on autophagy of liver cell and immune responses in weaned calves. *J. Anim. Physio. Anim. Nutr.* **2013**, *97*, 1007–1014. [CrossRef] [PubMed]

25. Nestal de Moraes, G.; Carvalho, E.; Maia, R.C.; Sternberg, C. Immunodetection of caspase-3 by western blot using glutaraldehyde. *Anal. Biochem.* **2011**, *415*, 203–205. [CrossRef]

26. Wohlgemuth, S.E.; Seo, A.Y.; Marzetti, E.; Lees, H.A.; Leeuwenburgh, C. Skeletal muscle autophagy and apoptosis during aging: Effects of calorie restriction and life-long exercise. *Exp. Gerontol.* **2010**, *45*, 138–148. [CrossRef]
27. Nakagawa, I.; Amano, A.; Mizushima, N.; Yamamoto, A.; Yamaguchi, H.; Kamimoto, T.; Nara, A.; Funao, J.; Nakata, M.; Tsuda, K.; et al. Autophagy defends cells against invading group A *Streptococcus*. *Science* **2004**, *306*, 1037–1040. [CrossRef]
28. Stadtman, E.R. Regulation of glutamine synthetase activity. *EcoSal Plus* **2004**. [CrossRef]
29. Van der Vos, K.E.; Eliasson, P.; Proikas-Cezanne, T.; Vervoort, S.J.; van Boxtel, R.; Putker, M.; van Zutphen, I.J.; Mauthe, M.; Zellmer, S.; Pals, C.; et al. Modulation of Gln metabolism by the PI(3)K-PKB-FOXO network regulates autophagy. *Nat. Cell Biol.* **2012**, *14*, 829–837. [CrossRef]
30. Han, T.; Li, X.L.; Cai, D.L.; Zhong, Y.; Geng, S.S. Effects of glutamine-supplemented enteral or parenteral nutrition on apoptosis of intestinal mucosal cells in rats with severe acute pancreatitis. *Eur. Rev. Med. Pharmacol. Sci.* **2013**, *17*, 1529–1535.

Article

Changed Caecal Microbiota and Fermentation Contribute to the Beneficial Effects of Early Weaning with Alfalfa Hay, Starter Feed, and Milk Replacer on the Growth and Organ Development of Yak Calves

Shengru Wu [1,*,†], Zhanhong Cui [1,2,†], Xiaodong Chen [1], Peiyue Wang [1] and Junhu Yao [1,*]

[1] College of Animal Science and Technology, Northwest A&F University, Xianyang 712100, China; cuizhanhong27@126.com (Z.C.); xiaodongchen2017@nwafu.edu.cn (X.C.); wpy991120@163.com (P.W.)

[2] Qinghai Academy of Animal Husbandry and Veterinary Sciences, Qinghai University, Xining 810016, China

* Correspondence: wushengru2013@nwafu.edu.cn (S.W.); yaojunhu2008@nwsuaf.edu.cn (J.Y.)

† These authors contribute equally to this study.

Received: 27 September 2019; Accepted: 1 November 2019; Published: 5 November 2019

Simple Summary: Yak calves during the pre-weaning period are mainly fed by maternal grazing and nursing, which is beneficial to the oestrus and mating of female yaks or the survival and growth of calves. Barn feeding and early weaning with mixed rations of available roughage and grains was presented as an alternative to maternal grazing and was supposed to be beneficial to the tremendous ruminal and intestinal development and growth of yak calves. The caecum is also the primary site of microbial fermentation, but the limited research has focused on the role of caecal microbiota in regulating the growth of yaks. The findings of the current study indicated that early weaning by supplying calves with milk replacer, alfalfa hay, and starter feed improves yak calf growth performance compared with maternal grazing and nursing, in part through alterations of caecal microbiota and caecal volatile fatty acid (VFA) production induced by supplementation with alfalfa hay and starter feed.

Abstract: This study aimed to investigate the effect of early weaning by supplying calves with alfalfa hay, starter feed, and milk replacer on caecal bacterial communities and on the growth of pre-weaned yak calves. Ten 30-day-old male yak calves were randomly assigned to 2 groups. The maternal grazing (MG) group was maternally nursed and grazed, and the early weaning (EW) group was supplied milk replacer, starter feed, and alfalfa hay twice per day. Compared with the yak calves in the MG group, the yak calves in the EW group showed significantly increased body weight, body height, body length, and chest girth. When suffering to the potential mechanism of improved growth of yak calves, except for the enhanced ruminal fermentation, the significantly increased total volatile fatty acids, propionate, butyrate, isobutyrate, and valerate in the caecum in the EW group could also serve to promote the growth of calves. By using 16S rDNA sequencing, some significantly increased caecal phylum and genera, which were all related to the enhanced caecal fermentation by utilizing both the fibrous and non-fibrous carbohydrates, were identified in the EW group. In conclusion, early weaning of yak calves by supplying them with alfalfa hay, starter feed, and milk replacer is more beneficial to the growth of yak calves when compared with maternal grazing and nursing, in part due to alterations in caecal microbiota and fermentation.

Keywords: yak calf; early weaning; caecal microbiota; 16S rRNA gene sequencing; growth performance

---

1. Introduction

Yak calves during the pre-weaning period are mainly fed by maternal grazing and nursing, which are not beneficial to the oestrus and mating of female yaks or the survival and growth of calves [1]. However, the pre-weaning period is a critical period for the developmental plasticity and, subsequently, biological function changes of young ruminants [2,3]. Adequate nutrition during early life is beneficial to gastrointestinal microbiota establishment, development, and the subsequent functional transition from metabolizing the glucose from milk to the volatile fatty acids (VFAs) from a solid diet [4,5]. Barn feeding and early weaning with mixed rations of available roughage and grains was presented as an alternative to maternal grazing and was supposed to be beneficial to the tremendous gastrointestinal ramifications and growth of yak calves and other juvenile ruminants [5,6]. In previous studies, the significantly enhanced rumen fermentation and changed rumen microbiota condition were implicated as the main reasons for the observed improved growth performance of cattle and lamb by supplying them with alfalfa hay, starter feed, and milk replacer in barn feeding and early weaning groups [4–6], which were rarely studied in the yak calves.

In addition to rumen fermentation, hindgut fermentation, which includes caecal fermentation, is also an important factor that affects growth performance and healthy conditions [4,7]. The caecum is also the major site of fermentation and absorption in the large intestine of ruminants, and approximately 17% of digested cellulose is broken down there [8]. The VFAs produced in the caecum account for 12% of total VFA production in sheep [9]. However, compared with the extensive studies focusing on the rumen microbiota and fermentation, caecal microbiota and fermentation is also an important factor that affects growth performance, which was comparatively limited in studies but worth further studying of the roles of microbiota from different segments in utilizing the nutrients and promote the growth of yak calves. In the present study, the effect of early weaning with alfalfa hay, starter feed, and milk replacer versus maternal grazing and nursing on the caecal microbiota and fermentation of yak calves was evaluated and compared, with the aim of further adding knowledge of changed caecal microbiota in regulating the growth of yak calves. Moreover, we further compared the differences between ruminal and caecal microbiota and fermentation to justify the contribution of caecal microbiota and fermentation on the growth of yak calves.

2. Materials and Methods

2.1. Ethics Approval Statement

This study was carried out in accordance with the recommendations of the Administration of Affairs Concerning Experimental Animals (Ministry of Science and Technology, China, revised 2004). The protocol was approved by the Institutional Animal Care and Use Committee of the Northwest A&F University (protocol number NWAFAC1118).

2.2. Animals, Experimental Design, and Sample Collection

Before the commencement of the trial, all yak calves were only fed with the milk by maternal nursing in Datong Yak Breeding Farm of Qinghai Province. A total of ten 30-day-old male yak calves (34.86 ± 2.06 kg) with similar body conditions were randomly assigned to 2 groups with 5 calves per group. The maternal grazing (MG) group was maternally nursed and grazed, and the early weaning (EW) group was supplied with milk replacer, starter feed, and alfalfa hay. The yak calves in the maternal nursing group had access to fresh grass and yak milk. Briefly, the MG yak calves were allowed to graze a rangeland for a period of 8 h. Water was offered ad libitum twice a day at 08:00 and 16:00 h. Specifically, the experiment was performed from July to October and lasted for 90 d, allowing for the sufficient grazing of fresh grass. Moreover, at the last day of the feeding experiment, the fresh grass and the yak milk were collected and provided to the their yak calves, and the dairy intake were recorded and used to calculate the dry matter intake (DMI). The yak calves in the early weaning group were housed in a barn and kept in individual pens (7 × 4 m). The pens included a sawdust-bedded pack area and a feed lane equipped with an automatic cable scraping system. In addition to free access to starter feed and alfalfa hay, all yak calves in the early weaning group were supplied with milk replacer reconstituted from 100–350 g milk replacer powder (the supplementation of milk replacer were increased along with the increasing body weight) dissolved in 1 L 60 °C water twice per day at 08:00 and 16:30. Water was supplied ad libitum to the yak calves during the experimental period. Feed (include the alfalfa and starter feed) offered was adjusted daily to ensure at least 10% orts. Feed offered and refused by each calf was weighed and recorded on the last day of the feeding experiment. Meanwhile, the daily intake was calculated for further analysis of DMI (overall consideration of the dry matter intake of milk replacer, alfalfa, and starter feed). After the feeding experiment, the yak calves were weighed, and their body size indexes, including the body height, body length, and chest girth, were measured and recorded. Then all animals were euthanized by exsanguination after anaesthesia and immediately dissected, and the liver, thymus, spleen, and pancreas were collected and weighed immediately. At last, the ruminal fluid and caecal contents were collected and stored in −80 °C for further analyses. Specifically, rumen fluid was strained through 4 layers of sterile cheesecloth and collected for VFA and NH3-N analyses and 16S rRNA gene sequencing.

Composites of the fresh grass, starter feed, alfalfa hay, and milk replacer were measured (AOAC International, 2000) for DM (oven method 930.15), ash (oven method 942.05), CP (Kjeldahl method 988.05), fat (alkaline treatment with Röse–Gottlieb method 932.06 for MR; diethyl ether extraction method 2003.05 for starters and hay), Ca and P (dry ashing, acid digestion, and analysis by inductively coupled plasma, method 985.01), NDF with ash without sodium sulfite or α-amylase, ADF with ash, starch (α-amylase method), and sugar (colorimetric method), and the details of the nutrient composition are given in Tables S1 and S2.

2.3. Determination Of VFA and NH₃-N in Ruminal Fluid and Caecal Contents

For the VFA and NH3-N measurements, the rumen fluid and caecal contents dissolved in the buffer were centrifuged at 13,000× g for 10 min. The VFAs were analysed on an Agilent 6850 gas chromatograph (Agilent Technologies Inc., Santa Clara, CA, USA) equipped with a polar capillary column (HP-FFAP, 30 m × 0.25 mm × 0.25 µm) and a flame ionization detector (FID), as previously described [10]. The NH3-N in the supernatant was quantified using a continuous-flow analyser (SKALAR San, Skalar Co., Breda, The Netherlands).

2.4. Microbial DNA Extraction and 16S rRNA Gene Sequencing

The ruminal fluid and caecal content samples from yak calves were subjected to DNA extraction using the QIAamp DNA Stool Mini Kit (Qiagen, Hilden, Germany). The quantity and quality of those DNA samples were further assessed by a Nanodrop ND-1000 spectrophotometer (Thermo Scientific, Waltham, MA, USA). The 16S rRNA gene amplicons of 8 DNA samples (4 samples from EW group and 4 samples from MG group) with high quality were used to determine the diversity and compare the community structures of the bacterial species in each of these samples using Illumina HiSeq sequencing at Novogene Bioinformatics Technology Co., Ltd., Beijing, China. The preparation of the amplicon library was performed by polymerase chain reaction amplification of the V3–V4 region of the 16S rRNA gene using the primer set 341F 5′-CCTAYGGGRBGCASCAG-3′ and 806R 5′-GGACTACNNGGGTATCTAAT-3′ with barcode. The identified sequences were deposited in the NCBI sequence archive (SRA) under the accession no. PRJNA552771.

Sequencing data splicing and quality filtering of the raw tags were performed using Trimmomatic (V0.36) and Usearch (V9.2.64) [11,12]. All sequences shorter than 200 bp and those with quality scores lower than 15 in the raw reads were removed, and high-quality clean tags were obtained. These sequences were classified into operational taxonomic units (OTUs) at an identity threshold of 97% similarity using UPARSE software [12]. For each OTU, by, a representative sequence was screened and used to assign taxonomic composition by comparison with the RDP 16S Training set (v16) and the core set using the SINTA (Usearch V9.2.64) and PyNAST (QIIME) programmed algorithms [13,14].

Subsequent analysis of alpha and beta diversity was performed based on the output of this normalized data. The taxon abundance for each sample was determined according to phylum, class, order, family, and genus. The *t*-test was performed to estimate the differential microbiota between the treatments. The threshold was set at p value < 0.05.

2.5. Statistical Analysis

Analysis was performed using Student's *t* test with SPSS 21.0 software with replicates as experiment units, and differences were considered statistically significant at $p < 0.05$.

3. Results and Discussion

3.1. Early Weaning of Yak Calves with Alfalfa Hay, Starter Feed, and Milk Replacer Significantly Promoted Growth and Organ Development

Compared with the yaks in the maternal grazing group, the yaks in the early weaning group showed significantly increased body weight, withers height, body length, and chest girth (Table 1). Additionally, the significantly increased weight of the liver, spleen, and thymus, as well as the significantly increased indexes of spleen and thymus (g/kg body weight) were also identified in the early weaning group (Table 1). Meanwhile, the ruminal fermentation characteristics of yak calves under the grazing and barn feeding conditions are presented in Table 2. The PH and NH3-N showed no differences between the different feeding groups. The total VFA concentration was significantly higher in the early weaning group than in the grazing group; of these, the propionate, butyrate, isobutyrate, and valerate were also significantly increased in the early weaning group (Table 2). Furthermore, the ratio of acetate/propionate and acetate/total VFA were significantly decreased in the early weaning group, while the ratio of butyrate/total VFA, isobutyrate/total VFA, and valerate/total VFA were all significantly increased in the early weaning group. Moreover, those significantly increased growth performance and ruminal fermentation were mostly resulted from the significant differences between the treatments in the daily DMI of yak calves, where the increased intake was found for calves on early weaning group, especially the increased intakes of concentrate supplement (Table 1).

Table 1. Effect of early-weaning feeding and maternal grazing feeding on body weight, body size indexes, and organ weight of yak calves.

Items	Treatments		SEM	*p*-Value
	Early Weaning	**Maternal Grazing**		
Body weight (kg)	87.90 [a]	64.50 [b]	4.347	<0.001
Chest girth (cm)	116.20 [a]	107.25 [b]	1.956	0.009
Withers height (cm)	95.60 [a]	77.25 [b]	31.563	0.001
Body length (cm)	109.20 [a]	86.00 [b]	4.215	<0.001
Liver (g)	1391.50 [a]	1058.38 [b]	61.067	<0.001
Spleen (g)	237.84 [a]	150.60 [b]	17.554	0.002
Thymus (g)	252.96 [a]	105.25 [b]	26.176	<0.001
Pancreas (g)	50.56	50.85	1.694	0.939
Liver index (g/kg body weight)	1.584	1.605	0.023	0.677
Spleen index (g/kg body weight)	0.272 [a]	0.228 [b]	0.011	0.050
Thymus index (g/kg body weight)	0.288 [a]	0.160 [b]	0.023	<0.001
Pancreas index (g/kg body weight)	0.057 [b]	0.077 [a]	0.039	0.002
DMI (g)	1774.60	1147.52	21.303	<0.001

[a,b] within a row with different superscripts means significantly difference.

Table 2. Effect of early-weaning feeding and maternal grazing feeding on caecal fermentation of yak calves.

	Items	Treatments		SEM	*p*-Value
		Early Weaning	**Maternal Grazing**		
Rumen	pH	6.88	7.17	0.128	0.298
	Ammonia nitrogen, NH_3N (mg/dL)	6.91	6.55	0.234	0.486
	Total VFA (mmol/L)	66.82 [a]	58.00 [b]	1.767	0.002
	Acetate (mmol/L)	42.09	40.95	0.758	0.493
	Propionate (mmol/L)	11.53 [a]	9.85 [b]	0.346	0.004
	Butyrate (mmol/L)	8.60 [a]	4.08 [b]	0.801	<0.001
	Isobutyrate (mmol/L)	1.56 [a]	1.03 [b]	0.117	0.009
	Valerate (mmol/L)	1.18 [a]	0.55 [b]	0.118	<0.001
	Isovalerate (mmol/L)	1.87	1.54	0.086	0.058
	Acetate/Propionate	3.66 [b]	4.16 [a]	0.111	0.011
	Acetate/Total VFA	0.630 [b]	0.706 [a]	0.014	<0.001
	Propionate/Total VFA	0.173	0.170	0.002	0.580
	Butyrate/Total VFA	0.129 [a]	0.071 [b]	0.010	<0.001
	Isobutyrate/Total VFA	0.023 [a]	0.018 [b]	0.001	0.035
	Valerate/Total VFA	0.018 [a]	0.010 [b]	0.002	<0.001
	Isovalerate/Total VFA	0.0279	0.0265	0.001	0.482
cecum	pH	6.92	6.95	0.033	0.736
	Ammonia nitrogen, NH_3N(mg/dL)	6.86	6.80	0.177	0.874
	Total VFA (mmol/L)	63.53 [a]	56.37 [b]	1.601	0.012
	Acetate (mmol/L)	41.45	40.52	0.861	0.624
	Propionate (mmol/L)	11.93 [a]	9.60 [b]	0.427	<0.001
	Butyrate (mmol/L)	5.51 [a]	3.08 [b]	0.465	<0.001
	Other (mmol/L)	4.64 [a]	3.17 [b]	0.306	0.004
	Acetate/Propionate	3.48 [b]	4.22 [a]	0.148	0.002
	Acetate/Total VFA	0.652 [b]	0.719 [a]	0.012	0.001
	Propionate/Total VFA	0.188 [a]	0.171 [b]	0.004	0.024
	Butyrate/Total VFA	0.087 [a]	0.055 [b]	0.006	0.002

[a,b] within a row with different superscripts means significantly difference.

Our results indicated that the early weaning yak calves provided with milk replacer, starter feed, and alfalfa hay during early life showed improved growth and development, in accordance with the results of previous studies on lambs during early life [5,6]. Supplementation of the diets of ruminants with carbohydrates such as alfalfa hay and starter feed during the pre-weaning period has a crucial long-term impact on ruminal fermentation in other ruminants that has been shown to be beneficial to their growth performance [15]. In accordance with the previous studies, DMI and ruminal VFA production were both significantly increased, which contributed to the significantly promoted growth performance of yak calves in early weaning group [5,6,10,15]. However, except for the VFAs from ruminal fermentation, the caecum VFAs produced accounted for 12% of total VFA production in sheep [9,16], while limited research focused on caecal fermentation in response to the early weaning with starter feed and alfalfa hay in yaks [7].

3.2. Significantly Enhanced Caecel Fermentation Was Identified in the Yak Calves in the Early Weaning Group

The caecal fermentation characteristics of yak calves under the grazing and early weaning conditions were further measured (Table 2). The pH and NH_3-N also showed no differences between the two groups. The total VFA concentration was significantly higher in the early weaning group than in the grazing group (p = 0.026). Significantly higher concentrations of propionate, butyrate, and other VFAs were also identified in the early weaning group (p < 0.01), whereas concentration of acetate was not significantly altered by treatment in the present study. Meanwhile, the ratio of acetate/propionate and acetate/total VFA in caecum were also significantly decreased in the early weaning group, while the ratio of butyrate/total VFA and propionate/total VFA were both significantly increased in the early weaning group. Moreover, according to our results, we found that the concentration and production of VFAs in the caecum could not be ignored when compared with the ruminal VFA concentration. In ruminants including the yak calves, the VFAs are absorbed by the rumen and caecal epithelium, and then metabolized into glucose, triglycerides, and amino acids and further provide energy and nutrients resources for the growth of yak calves. Our results indicated that changed caecal VFAs in early weaning, especially the increased acetate and propionate induced by supplementation of the diet with alfalfa hay and starter feed, can also be absorbed by the caecal epithelium [5,17], and then metabolized into glucose, triglycerides, and amino acids and further provide energy and nutrients resources for the growth of yak calves [7]. Moreover, the increasing ratio of butyrate/total VFA and propionate/total VFA further represented the improved energy utilization efficiency when compared with the acetate type fermentation and the promoted caecal development process.

3.3. Different Responses of Caecal Microbiota to Early Weaning or Maternal Grazing Feeding Contribute to Enhanced Caecal Fermentation of Yak Calves

Considering the significantly increased ruminal and caecal VFAs, the ruminal and caecal microbiota were both further analysed. The beta diversity analyses revealed that the compositions of the gastrointestinal prokaryotic community of the yak calves in two different feeding groups were significantly different (Figure 1A,B, and Figure 2A,B). Moreover, Chao1 indexes indicated that early weaning with the starter feed and alfalfa hay was beneficial to the diversity of ruminal and caecal microbiota (Table S3). Recently, several studies have focused on the effect of different feeding paradigms on the gastrointestinal microbiota of several animals, including lizards, cheetahs, yaks, lambs, deer mice, and seals [18–21]. These studies all identified that the diversity and abundance of the gastrointestinal microbiota were increased in animals from the wild environment than in captive animals. However, our study identified that early weaning and barn feeding significantly increased the diversity of ruminal and caecel microbiota. Different from the previous studies which focused on adult animals which obtained more varied nutrients in wild feeding paradigms, the early weaning in the barn feeding paradigm in the present study, which supplied the calves with milk replacer, starter feed, and alfalfa hay, provided enough carbohydrate, protein, and lipid for the growth and proliferation of microbes, which indicated that more microbial species could survive in the gastrointestinal tract due to the abundant sources of carbon and nitrogen [7,17]. In contrast, the yak calves from the maternal grazing group obtained limited nutrients from maternal milk and fresh grass, which resulted in less diversity and abundance of the ruminal and caecal microbiota. Moreover, the similar results of the increased diversity of ruminal and caecal microbiota also indicated that the dietary changes can simultaneously altered the ruminal and caecal microbiota, which also indicated that caecal microbiota could also have potential response to dietary supplementation and further influence the growth of yak calves. Overall, our results indicated that the nutrition supplementation is beneficial to the richness of caecal microbiota such as the alteration of the diversity of ruminal microbiota, which could even changeover the beneficial effect of a wild environment on the diversity and richness of gut microbiota. Our results also indicated that nutrition was the main effect among environmental indices which influence the gastrointestinal microbiota [22].

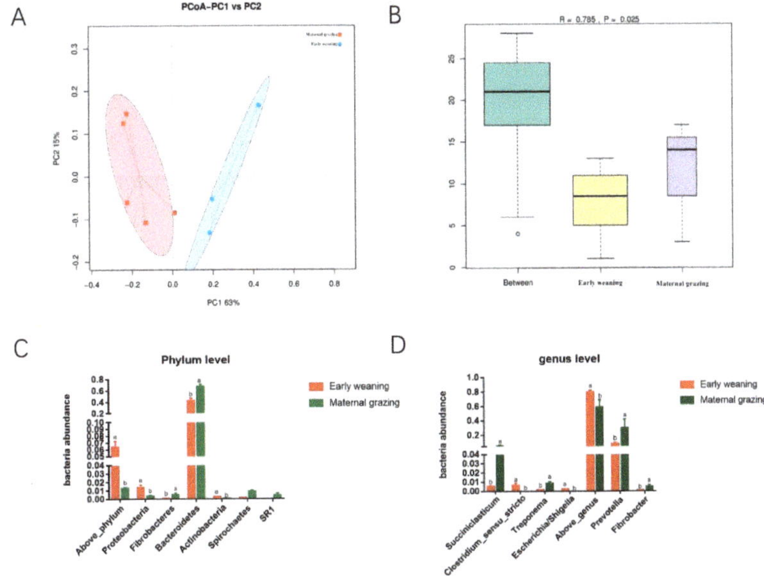

Figure 1. Ruminal microbial community difference between the different feeding paradigm groups (n = 4). (**A**) PCoA analysis. (**B**) Anosium analysis. (**C**) Differential ruminal microbes at phylum level based on *t*-test analysis. (**D**) Differential ruminal genera based on *t*-test analysis.

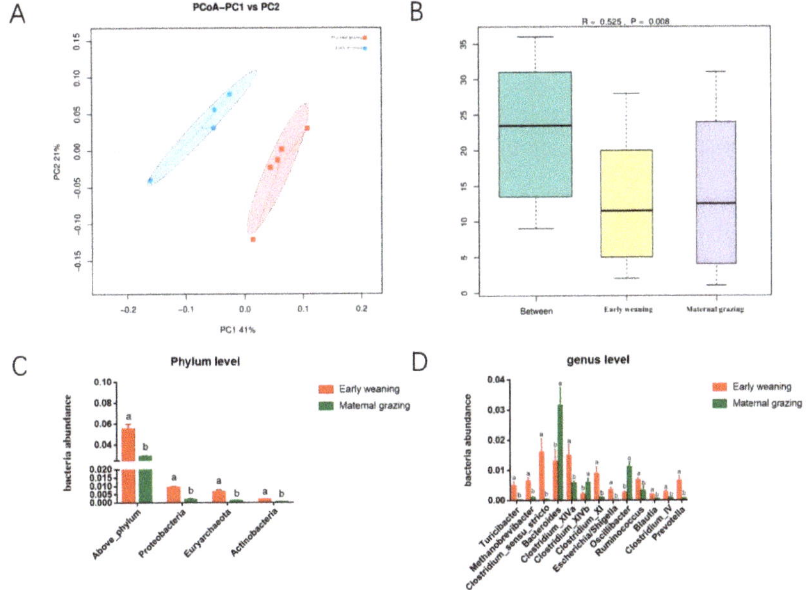

Figure 2. Caecal microbial community difference between the different feeding paradigm groups (n = 4). (**A**) PCoA analysis. (**B**) Anosium analysis. (**C**) Differential caecal microbes at phylum level based on *t*-test analysis. (**D**) Differential caecal genera based on *t*-test analysis.

The differential microbiota were further identified based on the counts of different microbes by using *t*-test analyses. In rumen, the significantly increased phylum of Proteobacteria, Fibrobacteres,

Bacteroidetes, Actinobacteria, Spirochaetes, and SR1 (Figure 1C), as well as the genus of Succiniclasticum, Clostridium_sensu_stricto, Treponema, Escherichia/Shigella, Prevotella, and Fibrobacter (Figure 1D) were identified in the early weaning group. In caecum, the significantly increased Proteobacteria, Euryarchaeota, and Actinobacteria in the phylum level (Figure 2C), as well as significantly increased genera of Turicibacter, Methanobrevibacter, Clostridium_sensu_stricto, Bacteroides, Clostridium_XIVb, Clostridium_XI, Escherichia/Shigella, Oscillibacter, Ruminococcus, Blautia, Clostridium_IV, and Prevotella (Figure 2D) were identified in the early weaning group. Accordingly, the phylum of Proteobacteria, Euryarchaeota, and Actinobacteria and the genera of Clostridium_sensu_stricto, Escherichia/Shigella, and Prevotella were co-influenced by the early weaning with alfalfa hay, starter feed, and milk replacer, which were all involved in the utilization of fibrous and non-fibrous carbohydrates and the production of propionate and butyrate. Meanwhile, these results again proved that dietary alteration could have a similar effect on the ruminal and caecal microbiota. Moreover, the main finding of our study lies in the fact that supplemental feeding with alfalfa hay and starter feed exceeded maternal grazing and nursing in shaping hindgut functional achievement. The significantly increased caecal genera of yak calves identified in the early weaning groups, including the *Prevotella*, *Clostridium_XIVb*, *Turicibacter*, *Clostridium_IV*, *Clostridium_XI*, *Clostridium_sensu_stricto*, *Bacteroides*, *Oscillibacter* and, *Ruminococcus* were mainly involved in the utilization of fibrous and non-fibrous carbohydrates and the production of acetate, propionate, and butyrate [17,23–27]. The primary determinant for this could be that the early-weaning calves consumed a greater amount of concentrate and alfalfa hay, and dietary fiber and starch were the suitable fermentation substrate when they reached the hindgut in significant quantities. Considering the identified significantly increased VFAs in the present study, the effect of early weaning with alfalfa hay and starter feeds on the identified variational microbiota and the roles of these changed microbiota were again proved. According to previous studies, in goats during the early life, caecal propionate, butyrate, and isobutyrate concentrations also significantly increased in response to a grain-rich diet [7,17,28]. In accordance with these previous studies, significantly higher concentrations of propionate, butyrate and total VFAs were also identified in the early weaning group of the present study, which were produced by our identified differential microbes by using starch or fibrous carbohydrates. Moreover, the effect of differential supplementing of carbohydrates during early life, induced by early weaning with alfalfa hay and starter feed, on the subsequent gastrointestinal microbiota and the related caecal fermentation, could further increase the absorbed VFAs from the caecal epithelium, and further provide more energy for the growth of yak calves [29,30]. Overall, except for ruminal fermentation, caecal fermentation could also be enhanced by providing enough fermentable carbohydrates in the EW group, which was induced by the increased abundance of microbes involved in the utilization of fibrous and non-fibrous carbohydrates and subsequently increased; and then the increased caecal VFAs could contribute to promoting the growth of yak calves.

4. Conclusions

Early weaning and barn feeding with milk replacer, alfalfa hay, and starter feed is recommended during pre-weaning to improve yak calf growth performance. Except for their beneficial roles in ruminal microbiota construction and ruminal VFAs production, the facilitating caecal starch-using and fibre-using microbial colonization and the subsequently improved caecal fermentation can also contribute to the growth of yak calves, which may play similar roles to the changed ruminal microbiota.

Supplementary Materials: The following are available online at http://www.mdpi.com/2076-2615/9/11/921/s1, Table S1: Nutrient composition of the alfalfa, starter feed, and milk replacement used in the present study; Table S2. Nutrient content of fresh grass and yak milk for yak calves from maternal grazing group; Table S3 Effect of early-weaning feeding and maternal grazing feeding on caecal microbial alpha diversity index of yak calves.

Author Contributions: S.W., Z.C., and J.Y. conceived and designed the experiments; S.W., Z.C., and X.C. mainly performed the experiments; S.W. analyzed the data; J.Y., S.W., and Z.C. contributed reagents/materials/analysis tools; S.W. wrote the manuscript. J.Y. and S.W. had primary responsibility for final content. All authors (including P.W.) read and approved the final manuscript.

Funding: This work was funded by the National Key R&D Program of China (2016YFC0501805), the China Postdoctoral Science Foundation (2019M653774), the Thousand-person Plan of Qinghai High-end Innovative Talents (top-notch talents Train), and the Program for Qinghai Science & Technology (3-4), and the National natural science foundation of China (31902184).

Conflicts of Interest: The authors declare that the research was conducted in the absence of any commercial or financial relationships that could be construed as a potential conflict of interest.

References

1. Zi, X.D. Reproduction in female yaks (Bos grunniens) and opportunities for improvement. *Theriogenology* **2003**, *59*, 1303–1312. [CrossRef]
2. Baldwin, R.L.; McLeod, K.R.; Klotz, J.L.; Heitmann, R.N. Rumen development, intestinal growth and hepatic metabolism in the pre-and postweaning ruminant. *J. Dairy Sci.* **2004**, *87*, 55–65. [CrossRef]
3. Soberon, F.; Van Amburgh, M.E. Lactation Biology Symposium: The effect of nutrient intake from milk or milk replacer of pre-weaned dairy calves on lactation milk yield as adults: A meta-analysis of current data. *J. Anim. Sci.* **2013**, *91*, 706–712. [CrossRef]
4. Liu, J.; Bian, G.; Sun, D.; Zhu, W.; Mao, S. Starter feeding supplementation alters colonic mucosal bacterial communities and modulates mucosal immune homeostasis in newborn lambs. *Front. Microbiol.* **2017**, *8*, 429. [CrossRef]
5. Yang, B.; Le, J.; Wu, P.; Liu, J.; Guan, L.L.; Wang, J. Alfalfa intervention alters rumen microbial community development in hu lambs during early life. *Front. Microbiol.* **2018**, *9*, 574. [CrossRef]
6. Laarman, A.H.; Ruiz-Sanchez, A.L.; Sugino, T.; Guan, L.L.; Oba, M. Effects of feeding a calf starter on molecular adaptations in the ruminal epithelium and liver of Holstein dairy calves. *J. Dairy Sci.* **2012**, *95*, 2585–2594. [CrossRef]
7. Sun, D.; Mao, S.; Zhu, W.; Liu, J. Effects of starter feeding on caecal mucosal bacterial composition and expression of genes involved in immune and tight junctions in pre-weaned twin lambs. *Anaerobe* **2019**, *59*, 167–175. [CrossRef]
8. Gray, F.V. The digestion of cellulose by sheep. The extent of cellulose digestion at successive levels of the alimentary tract. *J. Exp. Biol.* **1947**, *24*, 15–19.
9. Faichney, G.J. Volatile fatty acids in the caecum of the sheep. *Aust. J. Biol. Sci.* **1968**, *21*, 177–180. [CrossRef]
10. Xue, D.; Chen, H.; Zhao, X.; Xu, S.; Hu, L.; Xu, T.; Jiang, L.; Zhan, W. Rumen prokaryotic communities of ruminants under different feeding paradigms on the Qinghai-Tibetan Plateau. *Syst. Appl. Microbiol.* **2017**, *40*, 227–236. [CrossRef]
11. Bolger, A.M.; Lohse, M.; Usadel, B. Trimmomatic: A flexible trimmer for illumina sequence data. *Bioinformatics* **2014**, *30*, 2114–2120. [CrossRef]
12. Edgar, R.C. UPARSE: Highly accurate OTU sequences from microbial amplicon reads. *Nat. Methods* **2013**, *10*, 996–998. [CrossRef] [PubMed]
13. Caporaso, J.G.; Bittinger, K.; Bushman, F.D.; DeSantis, T.Z.; Andersen, G.L.; Knight, R. PyNAST: A flexible tool for aligning sequences to a template alignment. *Bioinformatics* **2010**, *26*, 266–267. [CrossRef] [PubMed]
14. Edgar, R.C. SINTAX: A simple non-Bayesian taxonomy classifier for 16S and ITS sequences. *bioRxiv* **2016**. [CrossRef]
15. Saro, C.; Hohenester, U.M.; Bernard, M.; Lagrée, M.; Martin, C.; Doreau, M.; Boudra, H.; Popova, M.; Morgavi, D.P. Effectiveness of interventions to modulate the rumen microbiota composition and function in pre-ruminant and ruminant lambs. *Front. Microbiol.* **2018**, *9*, 1273. [CrossRef] [PubMed]
16. Dixon, R.M.; Nolan, J.V. Studies of the large intestine of sheep. 1. Fermentation and absorption in sections of the large intestine. *Br. J. Nutr.* **1982**, *47*, 289–300. [CrossRef]
17. Jiao, J.Z.; W, Z.; Guan, L.L.; Tan, Z.L.; Han, X.F.; Tang, S.X.; Zhou, C.S. Postnatal bacterial succession and functional establishment of hindgut in supplemental feeding and grazing goats. *J. Anim. Sci.* **2015**, *93*, 3528–3538. [CrossRef]
18. Nelson, T.M.; Rogers, T.L.; Carlini, A.R.; Brown, M.V. Diet and phylogeny shape the gut microbiota of Antarctic seals: A comparison of wild and captive animals. *Environ. Microbiol.* **2013**, *15*, 1132–1145. [CrossRef]

19. Kohl, K.D.; Brun, A.; Magallanes, M.; Brinkerhoff, J.; Laspiur, A.; Acosta, J.C.; Caviedes-Vidal, E.; Bordenstein, S.R. Gut microbial ecology of lizards: Insights into diversity in the wild, effects of captivity, variation across gut regions and transmission. *Mol. Ecol.* **2017**, *26*, 1175–1189. [CrossRef]

20. Wasimuddin Menke, S.; Melzheimer, J.; Thalwitzer, S.; Heinrich, S.; Wachter, B.; Sommer, S. Gut microbiomes of free-ranging and captive Namibian cheetahs: Diversity, putative functions and occurrence of potential pathogens. *Mol. Ecol.* **2017**, *26*, 5515–5527. [CrossRef]

21. Schmidt, E.; Mykytczuk, N.; Schulte-Hostedde, A.I. Effects of the captive and wild environment on diversity of the gut microbiome of deer mice (Peromyscus maniculatus). *ISME J.* **2019**, *13*, 1293–1305. [CrossRef] [PubMed]

22. Rothschild, D.; Weissbrod, O.; Barkan, E.; Kurilshikov, A.; Korem, T.; Zeevi, D.; Costea, P.I.; Godneva, A.; Kalka, I.N.; Bar, N.; et al. Environment dominates over host genetics in shaping human gut microbiota. *Nature* **2018**, *555*, 210–215. [CrossRef] [PubMed]

23. Mansfield, H.R.; Endres, M.I.; Stern, M.D. Influence of non-fibrous carbohydrate and degradable intake protein on fermentation by ruminal microorganisms in continuous culture. *J. Anim. Sci.* **1994**, *72*, 2464–2474. [CrossRef]

24. Lu, C.D.; Kawas, J.R.; Mahgoub, O.G. Fibre digestion and utilization in goats. *Small Rumin. Res.* **2005**, *60*, 45–52. [CrossRef]

25. Cai, S.; Li, J.; Hu, F.Z.; Zhang, K.; Luo, Y.; Janto, B.; Boissy, R.; Ehrlich, G.; Dong, X. Cellulosilyticum ruminicola, a newly described rumen bacterium that possesses redundant fibrolytic-protein-encoding genes and degrades lignocellulose with multiple carbohydrate-borne fibrolytic enzymes. *Appl. Environ. Microbiol.* **2010**, *76*, 3818–3824. [CrossRef]

26. Edwards, J.E.; Forster, R.J.; Callaghan, T.M.; Dollhofer, V.; Dagar, S.S.; Cheng, Y.; Chang, J.; Kittelmann, S.; Fliegerova, K.; Puniya, A.K.; et al. PCR and Omics Based Techniques to Study the Diversity, Ecology and Biology of Anaerobic Fungi: Insights, Challenges and Opportunities. *Front. Microbiol.* **2017**, *8*, 1657. [CrossRef]

27. Zhang, J.; Shi, H.; Wang, Y.; Li, S.; Cao, Z.; Ji, S.; He, Y.; Zhang, H. Effect of dietary forage to concentrate ratios on dynamic profile changes and interactions of ruminal microbiota and metabolites in holstein heifers. *Front. Microbiol.* **2017**, *8*, 2206. [CrossRef]

28. Wang, Y.; Xu, L.; Liu, J.; Zhu, W.; Mao, S. A high grain diet dynamically shifted the composition of mucosa-Associated microbiota and induced mucosal injuries in the colon of sheep. *Front. Microbiol.* **2017**, *8*, 2080. [CrossRef]

29. Khan, M.A.; Bach, A.; Weary, D.M.; von Keyserlingk, M.A.G. Invited review: Transitioning from milk to solid feed in dairy heifers. *J. Dairy Sci.* **2016**, *99*, 885–902. [CrossRef]

30. Meale, S.J.; Li, S.; Paula, A.; Hooman, D.; Plaizier, J.C.; Khafipour, E.; Steele, M.A. Development of ruminal and fecal microbiomes are affected by weaning but not weaning strategy in dairy calves. *Front. Microbiol.* **2016**, *7*, 582. [CrossRef]

Article

Effects of Body Condition Score Changes During Peripartum on the Postpartum Health and Production Performance of Primiparous Dairy Cows

Yujie Wang, Pengju Huo, Yukun Sun and Yonggen Zhang *

College of Animal Science and Technology, Northeast Agricultural University, Harbin 150030, China; wangyjanet@163.com (Y.W.); huopengju@163.com (P.H.); sun_yukun@126.comg (Y.S.)
* Correspondence: zhangyonggen@sina.com

Received: 8 November 2019; Accepted: 9 December 2019; Published: 17 December 2019

Simple Summary: This study systematically describes the effects of body condition score (BCS) changes in primiparous cows during the peripartum period on hormone indexes, health, and production. The BCS and its changes indirectly measure the degree of fat mobilization and is a good predictor of the risk of postpartum disease. In production practice, confounding the management of primiparous and multiparous cow risks neglecting the postpartum characteristics of primiparous cows. A prospective observational study observed that primiparous cows that have a lower BCS have higher non-esterified fatty acids (NEFA) and β-hydroxybutyrate (BHBA) concentrations and more dramatic hormonal changes. Prepartum BCS changes were inconsistent and small, while after calving, there was a drastic decline in the BCS, suggesting that even a slight drop in the prepartum BCS may be a warning of a postpartum risk for primiparous cows. It is suggested that operators attach importance to the primiparous cow prepartum BCS and keep it stable through prepartum management adjustments, since an ideal BCS at calving reduces the incidence of postpartum disease.

Abstract: This is a prospective observational study that evaluates the effects of body condition score (BCS) changes in primiparous Holstein cows during peripartum on their NEFA and BHBA concentrations, hormone levels, postpartum health, and production performance. The cows under study (n = 213) were assessed to determine their BCS (5-point scale; 0.25-point increment) once a week during the whole peripartum by the same researchers; backfat was used for corrections. Blood samples were collected 21 and 7 days before calving and 7, 21, and 35 days after calving, and were assayed for NEFA, BHBA, growth hormone (GH), insulin, leptin, and adiponectin concentrations. The incidence of disease and milk yield were recorded until 84 days after calving. Cows were classified according to their BCS changes during peripartum as follows: Those that gained BCS (G; ΔBCS ≥ 0.25), maintained BCS (M; ΔBCS = 0–0.25), or lost BCS (L; ΔBCS ≥ 0.5). The BCS at −21 days and at 7, 14, and 21 days were different (p < 0.01), but trended toward uniformity in all groups at calving. The L group had higher NEFA and BHBA concentrations and hormone levels (p < 0.01) than the M and G groups at 21 and 35 days after calving, and had a higher incidence of uterine and metabolic diseases; however, there were no differences in production performance between the various groups. In conclusion, a lower BCS in primiparous cows during peripartum influences the NEFA and BHBA concentrations, hormone levels, and occurrence of health problems postpartum. The postpartum effects of BCS changes appear prior to calving.

Keywords: body condition score; peripartum; fat mobilization; primiparous dairy cow

---

1. Introduction

For a cow approaching calving, the periparturient period, from three weeks prepartum to three weeks postpartum, is an important stage that determines whether milk yield and dry matter intake (DMI) will rapidly increase postpartum; recovery of postpartum DMI is an important measure for avoiding a negative energy balance (NEB) [1,2]. Numerous metabolic and hormonal changes, together with a series of stress reactions, such as calving, lactating, and ration changes, involving feeding management during this period, have a direct effect on the health, reproduction, and lactation performance of cows [3,4]. Parity is a well-known risk factor for disease: Multiparous cows are more likely to develop ketosis and hypocalcemia [5,6]; the evolution of metabolic profiles in healthy and sick cows during the periparturient period varies according to parity [7]. Studies have found that primiparous cows have higher concentrations of insulin-like growth factor-I, lower concentrations of BHBA throughout periparturient, and higher concentrations of leptin; these differences are associated with significantly lower milk production and body condition scores [8,9]. These results suggest that the management of primiparous cows during the periparturient period should be different.

Body condition score (BCS) is strongly correlated with energy reserves, directly reflecting the fat reserves of individual dairy cows. Changes in the BCS rather than a single BCS measurement, which is frequently used to monitor energy balance as a practical tool for dairy farm management, are widely used and easy to determine [10]. The peripartum BCS and a series of changes, including the BCS at calving and the rate and degree of BCS reduction after calving, may indicate the increase of non-esterified fatty acids, possibility of postpartum diseases and differences in production performance. Studies have shown that the BCS of multiparous cows can be regarded as a prediction tool due to the strong association between the BCS and metabolic diseases, including hepatic lipidosis, ketosis, and abomasum displacement [11–13]. The main reason for this strong association is that weight loss over 50 kg due to improper prepartum feeding significantly inhibits DMI and milk production; meanwhile, high milk production and the consequent synthesis of milk fat result in a high degree of fat mobilization, causing cows to go through NEB [14]. Health and performance at the primiparous stage have a profound impact on later stage incidence of disease and production potential; thus, reasonable management of cows' body condition is particularly important. Little is known about the characteristics and the reference range of the BCS in primiparous cows on disease prediction, hormonal levels, and lactation performance. Considering that primiparous cows represent a high proportion of cows in production, it is particularly necessary to further study the prepartum BCS of primiparous dairy cows and to use a hormonal index to specifically investigate the influence of BCS changes on fat metabolic, health, and lactation.

Adipose tissue reserves are predominantly controlled by the energy balance and abundance of insulin, with the expression and tissue responsiveness of key hormones being altered to maintain physiological equilibrium at the beginning of chronic energy deficiency [15]. Growth hormone (GH) directly regulates ruminant adipose stores. Insulin is an antagonist of the lipolytic actions of GH and lowers mobilization of the tissue reserves. Adiponectin is recognized to play an important role in metabolic syndrome. Leptin serves as an intake satiety signal by predominantly acting on the brain [10]. The somatotropic axis, primarily consisting of growth hormone (GH; somatotropin) and insulin-like growth factor-I (IGF-I; somatomedin), is essential for the regulation of intrahepatic lipid metabolism [16]. Association between peripartum BCS, fat mobilization, and postpartum serum insulin concentration has been demonstrated in several studies [17–20]; besides, an increase in serum concentrations of NEFA and BHBA, as well as decreased serum concentrations of insulin and glucose, are indicators of NEB [21]. Studies convincingly demonstrate that exogenous bovine somatotropin (BST) results in an increase in milk yield in treated animals and results in a series of coordinated adaptations in their body tissues to support the increased use of nutrients for milk synthesis [22]. A recent study exploring the association between postpartum plasma insulin and NEFA and BHBA concentrations demonstrated that cows with low plasma insulin had significantly higher concentrations of circulating NEFA; moreover, cows with low plasma insulin during early postpartum produced more milk and had higher FCM (fat corrected milk) or ECM (energy corrected milk) compared with cows with high plasma insulin [23]. The serum adiponectin concentration was positively associated with the insulin responsiveness of glucose and NEFA metabolism [24]. Most of the studies that investigated hormone concentration in the context of fat and glucose metabolism used multiparous cows in their experiments; however, differences and changes in fat metabolic hormone levels among primiparous cows, especially the direct effects mechanism, are not fully understood and deserve further investigation.

This study highlights the need for the preferential treatment of primiparous cows to ensure that their BCS trajectory is sufficient for calving, and the need to adopt a management strategy for adjusting the prepartum BCS to maximize the prevention of postpartum disease and production potential during later stages. Furthermore, it was hypothesized that testing the levels of the key regulatory hormones related to fat mobilization would result in a better understanding of the factors that influence BCS mobilization and replenishment. The objectives of the current study were to evaluate the association between BCS changes and hormone levels during peripartum and postpartum with the health and performance characteristics of primiparous lactating Holstein cows, and to examine the herd- and cow-level factors that influence the BCS profile, thus promoting animal management aimed at improving farm productivity, profit, and animal welfare.

2. Materials and Methods

2.1. Animals and Management

This prospective observational experiment was conducted on a commercial farm in Harbin City, Heilongjiang Province, China from August 2018 to January 2019. Two hundred and thirteen primiparous cows from a total of 692 lactation cows met the enrollment criteria and were used in the current study. All cows were synchronized using a Double-Ovsynch protocol for first TAI (Timed Artificial Insemination) with a progesterone implant during the Ovsynch. Protocol of synchronization started when cows had 60–65 days in milk. Primiparous cows entered the peripartum period during August and September. The calving date was synchronized to be in fall, during September and October; thus, there was no calving month effect. Milk yield data were collected from 5 to 84 days after calving. Disease records were kept to 84 days postpartum. The collection of postpartum data was completed in January. The living environment of the experimental primiparous Holstein cows ($n =$ 213) during peripartum was consistent: Cows were housed in free-stall barns that included mattresses composed of rice hulls, and were equipped with self-locking head gates at the feed line and with a cross-ventilation system with fans and spray devices. One to 14 days after calving, cows were relocated to the fresh-cows cowshed for postpartum care. Cows were milked three times a day at 7:00 a.m., 2:30 p.m., and 10:30 p.m. The milk yield of individual cows was recorded and stored in the software of the automatic milking system. Cows were fed total mixed ration (TMR) formulated according to NRC (2001) to meet the nutritional requirements of each period (pre- and postpartum). Prepartum cows were fed beginning at 4 p.m., fresh cows were fed at 7:35 a.m., with ad libitum access to food and water. All cows were synchronized using a Double-Ovsynch protocol for the first TAI with a progesterone implant during Ovsynch. The synchronization protocol was started when cows were in milk for 60–65 days.

2.2. Collection, Treatment, and Assessment of Blood Samples

Blood samples were collected throughout the experimental period relative to parturition at −21, −7, 7, 21, and 35 days from coccygeal vessels using a sterile syringe prior to the morning feeding. The samples were collected in a 5 mL evacuated centrifuge tube containing heparin anticoagulant and centrifuged at 2000× *g* for 15 min. Supernatant was collected, aliquoted into 1.5 mL centrifuge tubes, and stored at −20 °C for the measurement of blood biochemical and hormonal indexes, including NEFA, BHBA, GH, insulin, adiponectin, and leptin. The analyses were performed using commercial kits (Jiang Lai biotechnology co. LTD, Shanghai, China) by the enzymatic colorimetric endpoint method. The values were recorded using a Switzerland Tecan multifunctional microplate reader.

2.3. Assessment of the BCS and Classification

The BCS were recorded at the beginning of peripartum (21 d before calving) and scored weekly throughout peripartum at 21, 14, and 7 days before calving; the day of calving; and 7, 14, and 21 days after calving. The BCS assessment was completed by a doctor and a master trained by the College of Animal Science and Technology, Northeast Agricultural University. These personnel had also completed four-level course training, had practical operation experience in a commercial farm, and used the visual and tactile technique to determine the BCS based on the US 5-point system with a 0.25 increment, with 1 being too thin and 5 being too obese. Each cow was evaluated on each occasion by both researchers. Before each morning feeding, cows were kept in a normal standing posture [25]. To guarantee the accuracy and objectivity of the final BCS, a portable ultrasound backfat instrument was used to measure the fat thickness of the rump at each BCS assessment, and developed a linear regression model of BCS according to US 5-point system rules to determine BCS. The maximum penetration depth of the backfat instrument probe was 10 cm. The probe was placed vertically at 1/4 to 1/5 of the line connecting the ischial tuberosity and the hip tuberosity, and all values were measured on the right side. The final BCS of each cow was calculated using the average of three data points. Cows were classified according to BCS changes during peripartum (BCS at 21 days minus BCS at −21 days). The classifications included: Gained BCS (Gained; ΔBCS > 0), maintained BCS (Maintained; $-0.25 \leq \Delta$BCS ≤ 0), and lost BCS (Lost; ΔBCS < −0.25).

2.4. Disease Definition

Data on health status (mastitis, metabolic and digestive disorders, and metritis) were collected from the day of calving to 84 days in milk. Early warning of disease was based on a system that was fitted to the cows: A neck-mounted electronic rumination and activity monitoring tag (HR Tags, SCR Dairy, Netanya, Israel). At all times, cows that the tag identified were examined to establish a preliminary diagnosis by the same personnel. Information on the treatment and collected data were stored on the on-farm software. Fresh cows were observed daily from calving to 14 days postpartum until relocation to a cowshed for healthy cows. Cows with health disorders were subjected to required milk withdrawal and placed in a separate pen, and their milk was discarded until it became saleable. The clinical examination included a direct observation (general appearance and attitude, muscle strength, presence of fetal membranes outside the vulva, evaluation of vaginal discharge, foot health, udder health, and manure consistency), rectal temperature, urinary ketones, and rumen auscultation.

Cows were evaluated for metritis postpartum by palpation. Metritis was characterized by an enlarged uterus with a fetid watery red–brown discharge within 21 days postpartum. The rectal temperature was measured for cows with metritis, and those with a temperature of 39.5 °C were diagnosed with puerperal metritis. Abortion was defined as failure to deliver a normal calf. Retained fetal membrane was defined as failure to detach fetal membranes within 24 h postpartum. At every milking, all cows were examined for signs of clinical mastitis by the herd personnel immediately before milking: Clinical mastitis was characterized by the presence of abnormal milk or by signs of inflammation in one or more quarters. The herd personnel milked three handfuls and checked whether characterized by the presence of abnormal milk. A case of milk fever was defined as a prostrated cow with minimal rumen contractions that responded to an intravenous calcium treatment within 30 min. Cows with a decreased appetite and altered patterns of milk production had their urine tested for ketone bodies (Keto-Stix, Bayer Diagnostics, Tarrytown, NY, USA), and those that tested at or above moderate were diagnosed with ketosis. Cows with a metallic (ping) sound at percussion auscultation of the left or right abdomen (between the 4th and 13th ribs) were diagnosed with a displacement of the abomasum. Cows with scant manure, lack of appetite, and rumen stasis were diagnosed with indigestion. Respiratory disease was characterized by panting, rectal temperatures >39.5 °C, crackling, rales, or percussion dullness when auscultating the lungs. Cows with traumatic events (cesarean section, udder/teat cuts, and broken limbs) were excluded from the study. Occurrences of retained fetal membranes, abortion, and metritis were grouped into one variable: Uterine disease. Milk fever, ketosis, and displacement of the abomasum were grouped into one variable: Metabolic disease. Cows with a diagnosis of respiratory disease and cows with undefined sickness were grouped into the category: Other diseases.

2.5. Statistical Analysis

This study was a prospective observational study. The BCS at various times and the NEFA, BHBA, GH, insulin, leptin, and adiponectin concentrations were analyzed by GLM using MIXED PROC using SPSS software (version 22.0, IBM SPSS Statistics, Chicago, state of Illinois, USA). The model included the fixed effects of the experimental group (G, M, or L), the fixed effect of week in lactation, and interaction of the groups by week in lactation. A BCS of −21 days was used as covariance. To verify significant differences between the groups, data were analyzed using ANOVA in SPSS. Milk yield data were processed using ANOVA in SPSS. Significance was declared at $p < 0.05$ unless otherwise indicated. The distribution of the BCS in cows between various groups at day −21 was processed using Microsoft Office Excel. Health event statistics were recorded using Microsoft Office Excel (version MSO 16.0, Microsoft, Redmond, state of Washington, USA).

3. Results

Primiparous cows were divided into three groups based on changes in the prepartum BCS: Gained BCS (Gained, G), maintained BCS (Maintained, M), and lost BCS (Lost, L); the proportions of each group were 15.96% (34/213), 30.99% (66/213), and 53.05% (113/213), respectively. The BCS at −21 days were different; the mean BCS (± SEM) at −21 d were 3.09 ± 0.06, 3.39 ± 0.03, and 3.45 ± 0.02 for the G, M, and L groups, respectively. The L group had the highest BCS (3.45), followed by the M group (3.39). The G group (3.09) had a mean BCS lower than the other groups ($p < 0.01$; Table 1). Cows had similar BCS on days −14, −7, and 0. However, the postpartum BCS were different from the prepartum BCS: The L group had the lowest BCS versus the BCS of the other two groups ($p < 0.01$) at 7 days. The BCS at 14 and 21 days had very significant differences between groups.

Table 1. Comparison of body condition score (BCS; least squares means ± SEM) on days 21, 14, 7 before calving, in relation to calving, and 7, 14, and 21 after calving for primiparous cows in different groups.

Item	Groups			*p*-Value
	G	M	L	
N	34	66	113	–
−21 dBCS	3.09 ± 0.06 [a]	3.39 ± 0.03 [b]	3.45 ± 0.02 [b]	<0.01
−14 dBCS	3.25 ± 0.05	3.35 ± 0.03	3.38 ± 0.02	0.19
−7 dBCS	3.30 ± 0.06	3.33 ± 0.04	3.30 ± 0.03	0.33
0 dBCS	3.34 ± 0.07	3.33 ± 0.04	3.28 ± 0.02	0.42
7 dBCS	3.35 ± 0.08 [a]	3.23 ± 0.05 [a]	3.00 ± 0.04 [b]	<0.01
14 dBCS	3.34 ± 0.08 [a]	3.23 ± 0.05 [b]	2.89 ± 0.03 [c]	<0.01
21 dBCS	3.35 ± 0.07 [a]	3.25 ± 0.05 [b]	2.8 ± 0.03 [c]	<0.01

[a–c] Values within a row with different superscript letters differ at $p < 0.05$. Cows had their BCS evaluated during the transition period (−21 to 21) using a 5-point scale with 0.25 increments. G, gained BCS; M, maintained BSC; L, lost BCS.

The G group experienced a slow rise in the BCS throughout the prepartum period, increasing significantly from −21 days to calving and remaining essentially constant postpartum. Cows that gained BCS had a lower BCS at −21 days. In contrast, the L group had a declining trend over the entire prepartum period: Cows that lost BCS had the highest BCS at the beginning, followed by a slow downward trend and a more dramatic decline after calving. High BCS cows entering the prepartum period had a higher likelihood of losing BCS. This slight decrease was apparent even before calving (Figure 1a). Moreover, when entering the prepartum period, the average BCS of the G group cows was below 3.25. The BCS of the other groups were higher than 3.25. All the BCS were higher than 3.25 at −21 days, except that of the G group. The L group had a higher percentage of cows with BCS greater than or equal to 3.25 ($p < 0.01$; Figure 1b) than the other groups, suggesting that cows with a lower prepartum BCS at −21 days are more likely to have an increased BCS during the prepartum period.

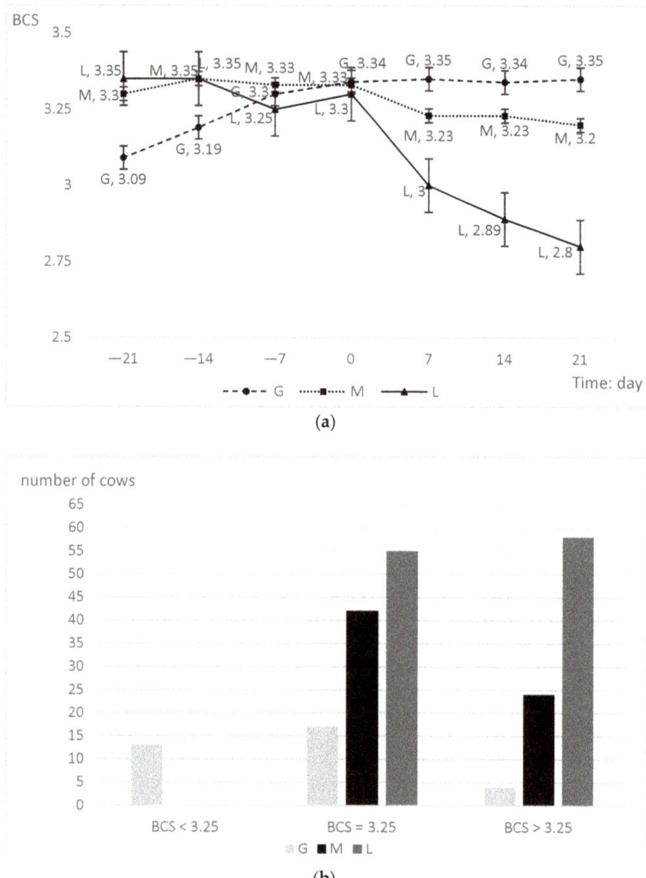

Figure 1. (**a**) Comparison of variation trend of body condition score (BCS) on days 21, 14, 7 before calving, calving day, and 7, 14 and 21 after calving for primiparous cows in different groups. (**b**) Distribution of cows that G ($n = 34$), M ($n = 66$), and L ($n = 113$) groups on −21 days BCS.

The NEFA and BHBA concentrations in the three experimental groups are presented in Table 2. The NEFA and BHBA concentrations differed ($p < 0.01$) between groups at postpartum 21 and 35 days. The L group had higher concentrations compared with those in cows that gained or maintained BCS. NEFA and BHBA did not change in a time-dependent manner. There were group-time interaction effects on NEFA ($p < 0.01$) and BHBA ($p = 0.02$). The changes during the prepartum period are shown in Figure 2. The variation trends of the three indicators in the L group had a common feature: The prepartum concentration was slightly higher than that in the other two groups and reached its lowest value on 7 days; then, the concentration sharply increased to significantly higher levels than those of the other two groups, reaching a maximum on 21 days. In the G group, the fluctuations in the prepartum changes were higher than those during postpartum. Cows that maintained BCS showed a smaller change with only slight fluctuations.

Table 2. Comparison of NEFA and BHBA contents in different groups of primiparous cattle before and after delivery.

Item	Time	Groups			SEM	p-Value		
		G	M	L		Time	Group	G × T
NEFA mmol/L	−21 days	0.99	0.82	1	0.061	0.22	<0.01	<0.01
	−7 days	1.05	0.7	1.24				
	7 days	0.76	0.72	0.88				
	21 days	0.76 [a]	0.72 [a]	1.96 [b]				
	35 days	0.67 [a]	0.66 [a]	1.76 [b]				
BHBA mmol/L	−21 days	4.37	3.01	4.51	0.257	0.21	<0.01	0.02
	−7 days	3.79	4.12	5.86				
	7 days	3.95	3.43	4.22				
	21 days	3.48 [a]	3.63 [a]	8.75 [b]				
	35 days	3.31 [a]	3.39 [a]	7.45 [b]				

[a,b] Values within a row with different superscript letters differ at $p < 0.05$.

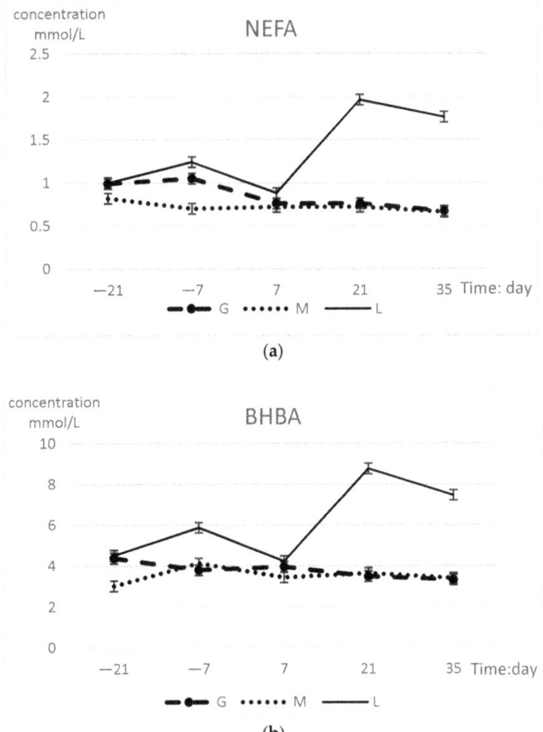

Figure 2. (a) Serum NEFA (upper panel) and (b) BHBA (lower panel) concentrations (least squares means ± SEM) in different groups during prepartum period.

The changes in hormone indexes during peripartum are shown in Table 3. The hormone levels of the groups were highly significantly different ($p < 0.01$), but time had no effect. The GH and adiponectin concentrations exhibited significant difference in group-time interaction ($p = 0.04$). At

21 and 35 days, the L group values were significantly higher than those of the other two groups with regard to GH, leptin, and adiponectin, while the insulin concentrations differed only at 35 days. Each hormone index in the L group showed the most dramatic changes during the prepartum period, with the lowest concentrations observed at 7 days and the highest concentrations at 21 days. The insulin concentrations continued to show an upward trend after 21 days. The GH and insulin indexes of the G group dramatically changed, while the hormone indexes in the M group remained essentially unchanged (Figure 3). The concentrations in all three groups tended to be consistent at 7 days.

Table 3. Comparison of growth hormone (GH), insulin, leptin, and adiponectin contents in different groups of primiparous cattle before and after delivery.

Item	Time	Groups			SEM	p-Value		
		G	M	L		Time	Group	G × T
GH ng/mL	−21 days	17.88	11.25	18.55	2.904	0.16	<0.01	0.04
	−7 days	19.69	17.14	23.62				
	7 days	12	12.73	15.51				
	21 days	14.25 [a]	13.98 [a]	33.17 [b]				
	35 days	10.76 [a]	15.12 [a]	26.69 [b]				
Insulin mIU/L	−21 days	32.23	26.95	32.04	1.028	0.09	<0.01	0.07
	−7 days	25.36	26.18	42.45				
	7 days	27.76	29.22	29.61				
	21 days	35.04	26.83	42.58				
	35 days	23.82 [a]	24.92 [a]	52.78 [b]				
Leptin ng/mL	−21 days	15.6	12.39	19.95	1.666	0.77	<0.01	0.32
	−7 days	15.32	15.94	22.31				
	7 days	13.67	13.7	17.26				
	21 days	15.7 [a]	15.07 [a]	35.51 [b]				
	35 days	12.96 [a]	13.95 [a]	31.75 [b]				
Adiponectin ug/mL	−21 days	58.09	50.89	55.9	1.047	0.07	<0.01	0.04
	−7 days	56.99	47.86	74.52				
	7 days	46.15	44.28	54.48				
	21 days	48.91 [a]	44.97 [a]	103.41 [b]				
	35 days	45.65 [a]	44.45 [a]	87.59 [b]				

[a,b] Values within a row with different superscript letters differ at $p < 0.05$.

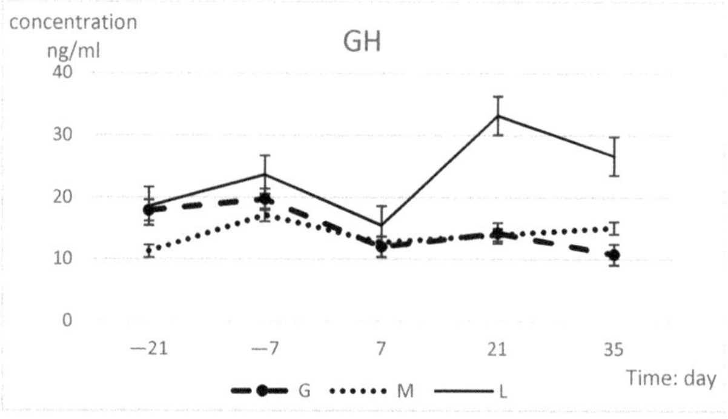

Figure 3. *Cont.*

(a)

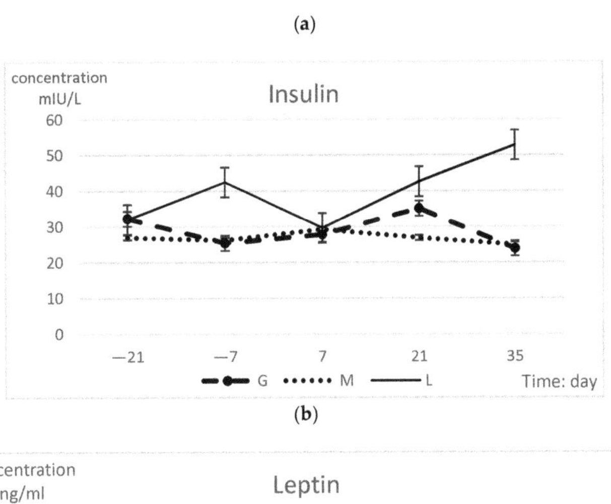

(b)

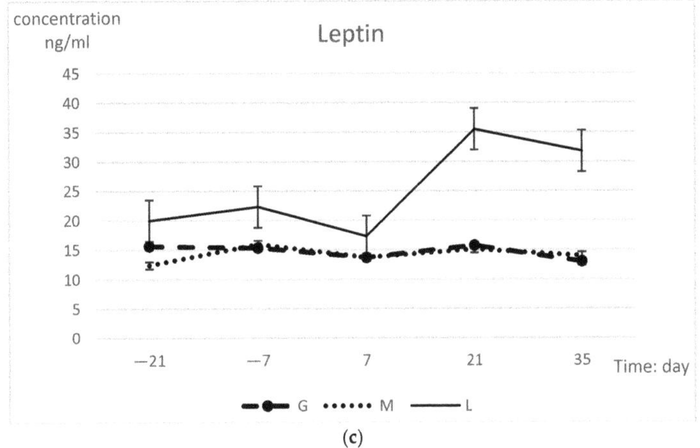

(c)

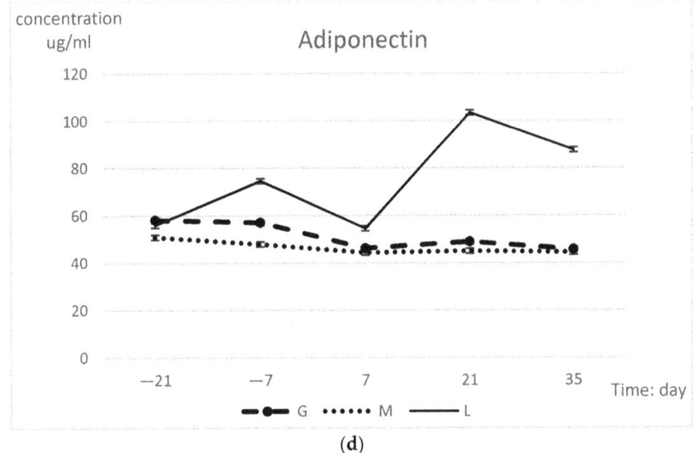

(d)

Figure 3. Serum GH (**a**), insulin (**b**), leptin (**c**), and adiponectin (**d**) concentration (least squares means ± SEM) in different groups during prepartum period.

The postpartum incidence of disease in cows that gained, maintained, and lost BCS is presented in Table 4. Cows that gained and maintained BCS had fewer health problems than cows that lost BCS. Moreover, when we evaluated cows with one or more health problems, cows that gained and maintained BCS had fewer health events than cows that lost BCS. Milk yield was similar among the experimental groups, and there was no group–time interaction. As reported in Table 5, the cows averaged 24.89 kg/d; however, the time of lactation influenced milk production ($p < 0.01$).

Table 4. Postpartum incidence (%) of health problems for primiparous cow in different groups.

Item	Groups		
	G	M	L
N	34	66	113
Abortion	11.76 (4/34)	4.55 (3/66)	12.39 (14/113)
Metritis		7.58 (5/66)	8.85 (10/113)
Obstetric canal strain			9.73 (11/113)
Retaine fetal membrane	17.65 (6/34)		12.39 (14/113)
Lameness			2.65 (3/113)
Milk fever			11.50 (13/113)
Mastitis		9.09 (6/66)	11.50 (13/113)
Health problems >1			10.62 (12/113)

Table 5. Milk yield (least squares means ± SEM) for primiparous cow in different groups.

Item	Groups			p-Value		
	G	M	L	Time	Group	G × T
N	34	66	113			
Milk yield	24.63 ± 1.65	24.92 ± 1.27	25.12 ± 12.09	<0.01	0.67	0.43

4. Discussion

The BCS and its changes are used as indirect indicators to measure fat mobilization and energy balance in individual cows and as a good predictor of the risk of disease. To our knowledge, most studies focus on high-yield multiparous cows, and little is known about the health status, milk yield, hormonal levels, and interrelationships with BCS changes in primiparous cows. A novel contribution of this study is its summary of BCS changes during peripartum in primiparous cows and its emphasis on the effects of varying management of BCS changes on fat mobilization, milk yield, and health between primiparous and multiparous cow. Blood biochemical and hormone indices were analyzed, focusing on primiparous cow management before calving. It is commonly accepted that milk production gradually increases after calving and reaches a peak at approximately 4 weeks postpartum. By contrast, changes in the BCS are inversely related to the lactation curve; however, the use of BCS values as a management tool can be enhanced when the prepartum period is added to the analysis [10,26,27]. Interestingly, it was found that the trend of changes in the prepartum BCS was different from that in the early postpartum lactation period, showing irregular variability. In actual production, it is generally recommended that modern high productivity dairy cows have a moderate BCS (≥3.25 and <3.5) at the beginning of the peripartum period, consequently resulting in lower mobilization of body reserves [28]. Similarly, our results show that primiparous cows with a higher BCS (3.40) at the outset had higher fat mobilization that was specifically reflected in higher NEFA and BHBA concentrations and hormone levels after calving. Surprisingly, the BCS of these cows tended to be consistent at calving, despite the initial variability, implying that the BCS is likely to reach the ideal body condition for calving through prepartum management adjustments, ensuring a suitable BCS prior to calving to minimize fat mobilization.

Ricardo C. et al. (2017) [29] found that BCS changes are strongly related to the BCS at dry off. Dechow et al. (2002) [30] evaluated correlations between the BCS and loss of BCS and found that a higher BCS at calving was phenotypically associated with a higher loss of BCS during early lactation. Our results show that BCS changes during early lactation are largely dependent on the BCS at −21 days, that is, the same as previous finding stated; but have no association to calving BCS because the BCS values at calving were similar between groups. The most important is that the similar BCS for calving comes from changes during prepartum, indicating that change in prepartum BCS plays an important role in the change of early lactation BCS, which determines the change of postpartum. Cows that lost BCS had the highest BCS at −21 days; in contrast, the BCS of cows that gained BCS as lower. This effect is similar to multiparous cows and it is more likely to happen in primiparous cows: Roche et al. (2009) [10] suggested that it is easier for primiparous cows to reach the optimal BCS and have better results in their metabolic and hormonal profiles than multiparous cows. Due to different nutrient partitioning, these cows have a high requirement to maintain good appetite and DMI for continued body growth and development [9,31], leading to a uniform BCS during calving. Moreover, these cows do not go through the drying milk stage and avoid stress caused by changes in feed structure.

Previous studies have confirmed that avoiding BCS loss and maintaining energy balance have positive effects on the fertility, health, and performance of lactating dairy cows. R.V. Barletta reported that a BCS loss was initiated even prior to calving. Our results are clearly consistent with this observation and indicate that BCS changes have an identifiable trend during the prepartum period [32]. The general conclusion is that cows with a relatively consistent BCS exhibit approximately similar changes. The postpartum changes were generally consistent with prepartum changes. The prepartum changes exhibited only a slight change, while the postpartum changes were especially evident in cows that lost BCS. However, there were no significant differences in NEFA, BHBA, or hormone indexes during the prepartum period, indicating that a loss of the prepartum BCS did not involve substantial fat mobilization. Higher circulating concentrations of NEFA, BHBA, and hormones postpartum and a BCS loss happened at the same time, suggesting that primiparous cows are in a parturition stress environment that causes a sharp decrease in DMI, as suggested by Proudfoot et al. (2009) [33]. The majority of studies have reported that DMI is a major driver of variations in the BCS during the early postpartum period. In contrast, insulin and GH are associated with growth [34] and a higher BCS [10], which may explain, in part, the higher concentrations of these metabolites in cows that lost BCS in our study. Additionally, insulin plays a role in energy metabolism, and its concentration is positively correlated with energy intake; however, this relationship has been observed in multiparous cows but not in primiparous cows [8], as confirmed by our experiments.

Usually, circulating NEFA concentrations and DMI have an inverse relationship and are associated with negative effects on health and production. Hayirli et al. (2002) [35] demonstrated that cows classified as obese 21 days before the expected calving date had lower DMI (as a percent of BW, body weight) from 21 days before the expected calving date to calving compared with thinner cows. The reduction in DMI from 21 days before the expected calving date to calving was 40, 29, and 28% for obese, moderate, and thin cows, respectively. In our study, cows that lost BCS had higher levels of NEFA and BHBA than those in the other groups. The effects remained significant after calving, thus confirming that obese cows have low DMI, significant reductions in BCS, and underwent higher fat mobilization after calving. Parity is a well-known pivotal factor for differences in feed intake. Studies have shown that cows with different parity have different characteristics in terms of milk yield, incidence of metabolic disease postpartum, and reproductive performance, with significant differences in feeding habits and behavior during the prepartum period. Primiparous cows have a lower dry matter intake, eat more slowly, are replaced at the feeder more frequently and are typically smaller than multiparous cows. Proudfoot et al. (2009) [33] showed that primiparous cows ate less than multiparous cows during weeks 1–2 after calving. Similarly, H. W. Neave found that even after controlling for BW and milk production, primiparous cows ate less than multiparous cows, with the differences increasing over the postpartum period [36]. Primiparous cows have higher pregnancy rate, lower milk yield, and higher glucose receptor sensitivity than multiparous cows, as well as their physiological functions are vigorous, hormone levels are adequate, there is almost no disorders of glycometabolism and metabolic disease, which is also proven in the study that no ketosis among all the experimental cows. Additionally, cows that lost BCS had higher adiponectin, implying fat mobilization. C. Urh (2018) [37] confirmed the involvement of adiponectin in the regulation of energy partitioning in primiparous cows, with adiponectin concentrations higher than those in multiparous cows. Therefore, nutritional management should be differentiated from multiparous cows that high nutrient and energy concentration to avoid excessive fat in body condition and low DMI postpartum. As for BCS management, it is better to keep a moderate BCS, about 3.25, to enter the peripartum period; higher BCS cows show poor health and performance postpartum, and even a slight drop in the prepartum BCS can be a warning of a postpartum risk of low DIM in primiparous cows. A slight decrease in the prepartum body condition is correlated with a BCS loss after calving and with the magnitude of postpartum NEB.

The increased energy requirements due to lactogenesis and reduced dry matter intake mean that all dairy cows undergo a state of negative energy balance (NEB) from late gestation to early lactation. Additionally, feeding peaks appear after lactation peaks, forcing cows to mobilize their fat reserves or proteins to meet their needs. These problems are increasingly understood to be rooted in DMI 2–3 weeks before calving, arguing for the importance of nutritional management in the prepartum period [38]. Although DMI was not included in this study's statistics, cows that lost a condition had an increase in leptin, demonstrating that a decrease in DMI leads to lower energy storage levels in the group that lost BCS. At the same time, an increase in the GH concentration meets the higher requirement for body growth and development and fat metabolism. The data indicate that the concentration of insulin increases in cows that lost BCS; however, it is possible that a negative energy balance causes the cows to become insulin resistant, an early warning sign of metabolic disease. One of the purposes of this study was to identify hormone indicators that can efficiently predict changes in the body condition score before calving and can be used as an effective tool for nutritional management; however, these results do not show unified and regular changes in the hormones as a clear means of guiding management.

Our study showed that cows with higher losses of BCS during the peripartum period had higher incidences of abortion, metritis, obstetric canal strain, retention of the fetal membrane, and mastitis during early lactation compared with cows that maintained or gained BCS. Meanwhile, cows that lost BCS had higher risks of lameness and milk fever and were more likely to have more than one health event. In general, cows that maintained BCS had better health than cows that gained or lost BCS. Health status is also associated with elevated concentrations of NEFA and BHBA. An increase in the circulating concentration of NEFA 7 to 10 days prepartum and of BHBA in the postpartum period can indicate metabolic problems, indirectly related to multiple common diseases of energy metabolism [39,40]. Excessive fat mobilization and a long period of NEB can attenuate the function of the immune system [41]. Ospina et al. (2010) [40] demonstrated that an increase in prepartum and postpartum NEFA levels is associated with an increased risk of retained fetal membranes, metritis, clinical ketosis, and displacement of the abomasum. It can be speculated that cows that lost BCS had a negative energy balance and, consequently, an impaired immune response during the periparturient period, possibly predisposing them to metabolic disorders. It should be noted that cows that lost BCS postpartum had higher insulin levels, which is a key adaptive mechanism that raises a concern for the development of insulin resistance and changes in insulin responsiveness [37,42]. Furthermore, primiparous cows experience a suite of stressful events that they have not experienced previously, including regrouping, diet changes, parturition, and the onset of lactation, resulting in reduced food intake and excessive NEB [36].

Intensive genetic selection to increase milk production increases the demands for dietary nutrients and body tissue reserves, resulting in poor health and infertility [43]. The changes in nutrient metabolism required to support lactation in high producing dairy cows are controlled by hormones that coordinate a variety of processes. If the nutritional environment is adequate, cows can meet their energy demands from DMI, and consequently, tissue mobilization will be minimized [10]. Insulin is an especially powerful mediator of a number of various physiological effects, most of which serve to acutely maintain metabolic equilibrium in the face of short-term variations in nutrient supply and demand. Insulin levels were higher after calving in cows that lost BCS: Fewer nutrients are directed to body fat reserves and other non-mammary tissues because of the altered response to insulin, and more nutrients are taken up by the mammary gland consistent with an increase in milk synthesis [44]. However, there were no significant differences in milk production between cows that gained, maintained, or lost BCS. Cows that lost BCS did not have a better production performance.

5. Conclusions

In conclusion, it may be easier for primiparous cows to attain an ideal BCS at calving through sufficient prepartum management adjustments, even though a slight drop in the prepartum BCS may constitute a warning of a postpartum risk of large changes in the BCS, greater health problems, and a poor productive performance. Meanwhile, simultaneous increases in hormone levels in cows that lost BCS allows the prevention of excessive lipolysis and metabolic disease, suggesting that their vigorous physiological function and adequate hormone secretion is enough to sustain postpartum demand. In production, it is recommended a moderate BCS of approximately 3.25 for primiparous cows entering peripartum, keeping BCS stable during the prepartum period. Attention should be paid to primiparous cows in the prepartum stage through efficient management, avoiding excessive nutrition, leading to a higher BCS, possibly ensuring higher postpartum DMI and enabling the maximization of potential productivity.

Author Contributions: Conceptualization, Y.W. and Y.S.; methodology, Y.W.; investigation, P.H.; writing—original draft preparation, Y.W.; writing—review and editing, Y.S. and Y.Z.; supervision, Y.Z.; project administration, P.H.; funding acquisition, Y.Z.

Funding: This research was funded by Yonggen Zhang, grant number China Agriculture Research System (CARS-36).

Acknowledgments: This study was financially supported by the China Agriculture Research System (CARS-36).

Conflicts of Interest: The authors declare no conflict of interest.

References

1.   Grummer, R.R. Impact of changes in organic nutrient metabolism on feeding the transition dairy cow. *J. Anim. Sci.* **1995**, *73*, 2820–2833. [CrossRef] [PubMed]

2.   Herdt, T.H. Ruminant adaptation to negative energy balance. Influences on the etiology of ketosis and fatty liver. *Vet. Clin. N. Am. Food Anim. Pract.* **2000**, *16*, 215–230. [CrossRef]

3.   Gross, J.; Van Dorland, H.A.; Schwarz, F.J.; Bruckmaier, R.M. Endocrine changes and liver mRNA abundance of somatotropic axis and insulin system constituents during negative energy balance at different stages of lactation in dairy cows. *J. Dairy Sci.* **2011**, *94*, 3484–3494. [CrossRef] [PubMed]

4.   Cook, N.B.; Nordlund, K.V. Behavioral needs of the transition cow and considerations for special needs facility design. *Vet. Clin. N. Am. Food Anim. Pract.* **2004**, *20*, 495–520. [CrossRef]

5.   Seifi, H.A.; Leblanc, S.J.; Leslie, K.E.; Duffield, T.F. Metabolic predictors of post-partum disease and culling risk in dairy cattle. *Vet. J.* **2011**, *188*, 216–220. [CrossRef]

6.   Kimura, K.; Reinhardt, T.A.; Goff, J.P. Parturition and hypocalcemia blunts calcium signals in immune cells of dairy cattle. *J. Dairy Sci.* **2006**, *89*, 2588–2595. [CrossRef]

7.   Ruprechter, G.; Adrien, M.D.L.; Larriestra, A.; Meotti, O.; Batista, C.; Meikle, A.; Noro, M. Metabolic predictors of peri-partum diseases and their association with parity in dairy cows. *Res. Vet. Sci.* **2018**, *118*, 191–198. [CrossRef]

8.   Wathes, D.C.; Cheng, Z.; Bourne, N.; Taylor, V.J.; Coffey, M.P.; Brotherstone, S. Differences between primiparous and multiparous dairy cows in the inter-relationships between metabolic traits, milk yield and body condition score in the periparturient period. *Domest. Anim. Endocrinol.* **2007**, *33*, 203–225. [CrossRef]

9.   Piñeyrúa, J.T.M.; Fariña, S.R.; Mendoza, A. Effects of parity on productive, reproductive, metabolic and hormonal responses of Holstein cows. *Anim. Reprod. Sci.* **2018**, *191*, 9–21. [CrossRef]

10. Roche, J.R.; Friggens, N.C.; Kay, J.K.; Fisher, M.W.; Stafford, K.J.; Berry, D.P. Invited review: Body condition score and its association with dairy cow productivity, health, and welfare. *J. Dairy Sci.* **2009**, *92*, 5769–5801. [CrossRef]

11. Pires, J.A.A.; Delavaud, C.; Faulconnier, Y.; Pomiès, D.; Chilliard, Y. Effects of body condition score at calving on indicators of fat and protein mobilization of periparturient Holstein-Friesian cows. *J. Dairy Sci.* **2013**, *96*, 6423–6439. [CrossRef] [PubMed]

12. Grummer, R.R.; Mashek, D.G.; Hayirli, A. Dry matter intake and energy balance in the transition period. *Vet. Clin. N. Am. Food Anim. Pract.* **2004**, *20*, 447–470. [CrossRef] [PubMed]

13. Rathbun, F.M.; Pralle, R.S.; Bertics, S.J.; Armentano, L.E.; Cho, K.; Do, C.; Weigel, K.A.; White, H.M. Relationships between body condition score change, prior mid-lactation phenotypic residual feed intake, and hyperketonemia onset in transition dairy cows. *J. Dairy Sci.* **2017**, *100*, 3685–3696. [CrossRef] [PubMed]

14. Ingvartsen, K.L.; Andersen, P.H. Free fatty acid in plasma and hepatic lipid concentration around parturition in heifers and cows fed total mixed rations varying in energy concentration prepartum. In Proceedings of the Sixth International Conference on Production Diseases in Farm Animals, Berlin, Germany, 11–14 September 1997.

15. Roche, J.R.; Dominique, B.; Kay, J.K.; Miller, D.R.; Sheahan, A.J.; Miller, D.W. Neuroendocrine and physiological regulation of intake with particular reference to domesticated ruminant animals. *Nutr. Res. Rev.* **2008**, *21*, 207–234. [CrossRef] [PubMed]

16. Zhao, G. *Ruminant Physiology: Digestion, Metabolism, Growth and Reproduction*; Cronjé, P.B., Ed.; CABI Publishing: Wallingford, UK, 2000; p. 474. ISBN 0-85199-463-6.

17. De Koster, J.; Hostens, M.; Van Eetvelde, M.; Hermans, K.; Moerman, S.; Bogaert, H.; Depreester, E.; Van den Broeck, W.; Opsomer, G. Insulin response of the glucose and fatty acid metabolism in dry dairy cows across a range of body condition scores. *J. Dairy Sci.* **2015**, *98*, 4580–4592. [CrossRef]

18. Weber, C.; Hametner, C.; Tuchscherer, A.; Losand, B.; Kanitz, E.; Otten, W.; Singh, S.P.; Bruckmaier, R.M.; Becker, F.; Kanitz, W. Variation in fat mobilization during early lactation differently affects feed intake, body condition, and lipid and glucose metabolism in high-yielding dairy cows. *J. Dairy Sci.* **2013**, *96*, 165–180. [CrossRef]

19. Weber, C.; Hametner, C.; Tuchscherer, A.; Losand, B.; Kanitz, E.; Otten, W.; Sauerwein, H.; Bruckmaier, R.M.; Becker, F.; Kanitz, W. Hepatic gene expression involved in glucose and lipid metabolism in transition cows: Effects of fat mobilization during early lactation in relation to milk performance and metabolic changes. *J. Dairy Sci.* **2013**, *96*, 5670–5681. [CrossRef]

20. Rico, J.; Bandaru, V.; Dorskind, J.; Haughey, N.; Mcfadden, J. Plasma ceramides are elevated in overweight Holstein dairy cows experiencing greater lipolysis and insulin resistance during the transition from late pregnancy to early lactation. *J. Dairy Sci.* **2015**, *98*, 7757–7770. [CrossRef]

21. Kolk, J.H.V.D.; Gross, J.J.; Gerber, V.; Bruckmaier, R.M. Disturbed bovine mitochondrial lipid metabolism: A review. *Vet. Q.* **2017**, *37*, 1–18.

22. Morais, J.P.G.D.; Cruz, A.P.D.S.; Minami, N.S.; Veronese, L.P.; Valle, T.A.D.; Aramini, J. Lactation performance of Holstein cows treated with 2 formulations of recombinant bovine somatotropin in a large commercial dairy herd in Brazil. *J. Dairy Sci.* **2017**, *100*, 5945. [CrossRef]

23. Du, X.; Zhu, Y.; Peng, Z.; Cui, Y.; Zhang, Q.; Shi, Z.; Guan, Y.; Sha, X.; Shen, T.; Yang, Y. High concentrations of fatty acids and β-hydroxybutyrate impair the growth hormone-mediated hepatic JAK2-STAT5 pathway in clinically ketotic cows. *J. Dairy Sci.* **2018**, *101*, 3476–3487. [CrossRef] [PubMed]

24. Koster, J.D.; Urh, C.; Hostens, M.; Broeck, W.V.D.; Sauerwein, H.; Opsomer, G. Relationship between serum adiponectin concentration, body condition score, and peripheral tissue insulin response of dairy cows during the dry period. *Domest. Anim. Endocrinol.* **2017**, *59*, 100. [CrossRef] [PubMed]

25. Ferguson, J.D.; Galligan, D.T.; Thomsen, N. Principal descriptors of body condition score in Holstein cows. *J. Dairy Sci.* **1994**, *77*, 2695–2703. [CrossRef]

26. Roche, J.R.; Berry, D.P.; Kolver, E.S. Holstein-Friesian strain and feed effects on milk production, body weight, and body condition score profiles in grazing dairy cows. *J. Dairy Sci.* **2006**, *89*, 3532–3543. [CrossRef]

27. Berry, D.P.; Veerkamp, R.F.; Dillon, P. Phenotypic profiles for body weight, body condition score, energy intake, and energy balance across different parities and concentrate feeding levels. *Livest. Sci.* **2006**, *104*, 1–12. [CrossRef]

28. Gheise, N.J.E.; Riasi, A.; Shahneh, A.Z.; Celi, P.; Ghoreishi, S.M. Effect of pre-calving body condition score and previous lactation on BCS change, blood metabolites, oxidative stress and milk production in Holstein dairy cows. *Ital. J. Anim. Sci.* **2017**, *16*, 474–483. [CrossRef]

29. Chebel, R.C.; Mendonça, L.G.D.; Baruselli, P.S. Association between body condition score change during the dry period and postpartum health and performance. *J. Dairy Sci.* **2018**, *101*, 4595–4614. [CrossRef]

30. Dechow, C.D.; Rogers, G.W.; Clay, J.S. Heritability and correlations among body condition score loss, body condition score, production and reproductive performance. *J. Dairy Sci.* **2002**, *85*, 3062–3070. [CrossRef]

31. Clempson, A.M.; Pollott, G.E.; Brickell, J.S.; Bourne, N.E.; Munce, N.; Wathes, D.C. Evidence that leptin genotype is associated with fertility, growth, and milk production in Holstein cows. *J. Dairy Sci.* **2011**, *94*, 3618–3628. [CrossRef]

32. Barletta, R.V.; Maturana, F.M.; Carvalho, P.D.; Del Valle, T.A.; Netto, A.S.; Rennó, F.P.; Mingoti, R.D.; Gandra, J.R.; Mourão, G.B.; Fricke, P.M. Association of changes among body condition score during the transition period with NEFA and BHBA concentrations, milk production, fertility, and health of Holstein cows. *Theriogenology* **2017**, *104*, 30. [CrossRef]

33. Proudfoot, K.L.; Veira, D.M.; Weary, D.M.; Keyserlingk, M.A.G.V. Competition at the feed bunk changes the feeding, standing, and social behavior of transition dairy cows. *J. Dairy Sci.* **2009**, *92*, 3116–3123. [CrossRef] [PubMed]

34. Oksbjerg, N.; Gondret, F.; Vestergaard, M. Basic principles of muscle development and growth in meat-producing mammals as affected by the insulin-like growth factor (IGF) system. *Domest. Anim. Endocrinol.* **2004**, *27*, 219–240. [CrossRef] [PubMed]

35. Hayirli, A.; Grummer, R.R.; Nordheim, E.V.; Crump, P.M. Animal and dietary factors affecting feed intake during the prefresh transition period in Holsteins. *J. Dairy Sci.* **2002**, *85*, 3430–3443. [CrossRef]

36. Neave, H.W.; Lomb, J.; Keyserlingk, M.A.G.V.; Behnam-Shabahang, A.; Weary, D.M. Parity differences in the behavior of transition dairy cows. *J. Dairy Sci.* **2017**, *100*, 548–561. [CrossRef] [PubMed]

37. Urh, C.; Denißen, J.; Harder, I.; Koch, C.; Gerster, E.; Ettle, T.; Kraus, N.; Schmitz, R.; Kuhla, B.; Stamer, E. Circulating adiponectin concentrations during the transition from pregnancy to lactation in high-yielding dairy cows: testing the effects of farm, parity, and dietary energy level in large animal numbers. *Domest. Anim. Endocrinol.* **2019**. [CrossRef] [PubMed]

38. Leblanc, S.J.; Lissemore, K.D.; Kelton, D.F.; Duffield, T.F.; Leslie, K.E. Major advances in disease prevention in dairy cattle. *J. Dairy Sci.* **2006**, *89*, 1267–1279. [CrossRef]

39. Leblanc, S.J.; Leslie, K.E.; Duffield, T.F. Metabolic predictors of displaced abomasum in dairy cattle. *J. Dairy Sci.* **2005**, *88*, 159–170. [CrossRef]

40. Ospina, P.A.; Nydam, D.V.; Stokol, T.; Overton, T.R. Association between the proportion of sampled transition cows with increased nonesterified fatty acids and β-hydroxybutyrate and disease incidence, pregnancy rate, and milk production at the herd level. *J. Dairy Sci.* **2010**, *93*, 3595–3601. [CrossRef]

41. Hammon, D.S.; Evjen, I.M.; Dhiman, T.R.; Goff, J.P.; Walters, J.L. Neutrophil function and energy status in Holstein cows with uterine health disorders. *Vet. Immunol. Immunopathol.* **2006**, *113*, 21–29. [CrossRef]

42. Bernabucci, U.; Ronchi, B.; Lacetera, N.; Nardone, A. Influence of Body Condition Score on Relationships Between Metabolic Status and Oxidative Stress in Periparturient Dairy Cows. *J. Dairy Sci.* **2005**, *88*, 2017–2026. [CrossRef]

43. Mulligan, F.J.; Doherty, M.L. Production diseases of the transition cow. *Vet. J.* **2008**, *176*, 3–9. [CrossRef] [PubMed]

44. Zinicola, M.; Bicalho, R.C. Association of peripartum plasma insulin concentration with milk production, colostrum insulin levels, and plasma metabolites of Holstein cows. *J. Dairy Sci.* **2019**, *102*, 1473–1482. [CrossRef] [PubMed]

Article

The Limiting Sequence and Appropriate Amino Acid Ratio of Lysine, Methionine, and Threonine for Seven- to Nine-Month-Old Holstein Heifers Fed Corn–Soybean M-Based Diet

Yuan Li [1], Yanliang Bi [1], Qiyu Diao [1], Minyu Piao [1], Bing Wang [2], Fanlin Kong [1], Fengming Hu [1], Mengqi Tang [3], Yu Sun [3] and Yan Tu [1],*

[1] Feed Research Institute, Chinese Academy of Agricultural Sciences/ Sino-US Joint Lab on Nutrition and Metabolism of Ruminants, Beijing 100081, China; lyqiu0304@163.com (Y.L.); vetbi2008@163.com (Y.B.); diaoqiyu@caas.cn (Q.D.); tmpark729@126.com (M.P.); a895833622@163.com (F.K.); zgdxsjyzd@163.com (F.H.)

[2] College of Animal Science and Technology, China Agricultural University, Beijing 100193, China; wangb@cau.edu.cn

[3] College of Animal Science and Veterinary Medicine, Henan Agriculture University, Zhengzhou 450002, China; tmq0819@163.com (M.T.); sunyu95@163.com (Y.S.)

* Correspondence: tuyan@caas.cn; Tel.: +86-10-8210-6055

Received: 19 April 2019; Accepted: 29 September 2019; Published: 30 September 2019

Simple Summary: Research on the amino acid nutrition of cattle is limited, particularly research on the amino acid patterns of growing heifer. This lack of research has made it difficult to minimize the costs and reduce nitrogen emission of dairy heifers. Lysine might be the first limiting amino acid for seven- to nine-month-old Holstein heifers that are fed a corn–soybean meal-based diet, followed by methionine and threonine. The appropriate ratio of lysine, methionine, and threonine—calculated based on the nitrogen retention of seven- to nine-month-old Holstein heifers—were 100:32:57. We expect to reduce the input of protein feed and nitrogen emissions for dairy farms by using this ratio.

Abstract: An "Amino acid (AA) partial deletion method" was used in this experiment to study the limiting sequences and appropriate ratio of lysine (Lys), methionine (Met), and threonine (Thr) in the diets of 7- to 9-month-old Holstein heifers. The experiment was conducted for three months with 72 Holstein heifers (age = 22 ± 0.5 weeks old; BW = 200 ± 9.0 kg; mean ± standard deviation). Following an initial two weeks adaptation period, heifers were allocated to one of four treatments: a theoretically balanced amino acid diet (positive control [PC]; 1.00% Lys, 0.33% Met, and 0.72% Thr), a 30% Lys deleted diet (partially deleted Lys [PD–Lys]; 0.66% Lys, 0.33% Met, and 0.72% Thr), a 30% Met deleted diet (partially deleted Met, [PD–Met]; 1.00% Lys, 0.22% Met, and 0.72% Thr), and a 30% Thr deleted diet (partially deleted Thr [PD–Thr]; 1.00% Lys, 0.33% Met, and 0.45% Thr). Experimental animals were fed a corn–soybean meal-based concentrate and alfalfa hay. In addition, the animals were provided with supplemental Lys, Met, and Thr (ruminal bypass). The results found no differences in the growth performance and nitrogen retention between PD–Thr treatment and PC treatment ($p > 0.05$). The average daily gain ($p = 0.0013$) and feed conversion efficiency ($p = 0.0057$) of eight- to ninr-month-old heifers were lower in both PD–Lys and PD–Met treatment than those in PC treatment. According to growth performance, Lys was the first limiting AA, followed by Met and Thr. Moreover, nine-month-old Holstein heifers in PD–Lys treatment and PD–Met treatment had higher levels of serum urea nitrogen ($p = 0.0021$), urea nitrogen ($p = 0.0011$) and total excreted N ($p = 0.0324$) than those in PC treatment, which showed that nitrogen retention significantly decreased ($p = 0.0048$) as dietary Lys and Met levels decreased. The limiting sequence based on nitrogen retention was the same as that based on growth performance. The appropriate ratio of Lys, Met, and Thr in the diet based on nitrogen retention was 100:32:57. In summary, the limiting sequence and appropriate amino acid ratio of Lys, Met, and Thr for seven- to nine-month-old Holstein heifers fed a corn–soybean meal-based diet were Lys > Met > Thr and 100:32:57, respectively.

Keywords: amino acid pattern; Holstein heifers; lysine; methionine; threonine

---

1. Introduction

Nitrogen (N) loss is a major source of environmental pollution and causes significant economic losses for dairy farms. Given the high amount of N excretion that occurs in dairy cattle relative to their N intake, it is likely that these heifers were fed unbalanced amino acids and that their amino acid requirements were ignored [1]. A key factor for improving dietary amino acid (AA) utilization is the formulation of diets with appropriate amino acid patterns that meet but do not exceed the requirements. Many attempts have been made to decrease the environmental effects of cattle N excretion by manipulating the metabolizable amino acid levels of rations to increase the capture of dietary N by cattle [2,3]. However, deleterious effects may occur in cattle not only due to over-doses amino acid but also due to amino acid imbalances, where there is a lack of an appropriate amino acid pattern.

The amino acid partial deletion method is the most common method used to develop balanced AA models in animals [4]. This method can be used to determine the sequences needed to limit AA and calculate the optimal ratios. Dorigam et al. [4] estimated the essential AA profile of poultry and determined the ideal pattern for maintenance using this method. Wang et al. [5] also used the deletion method to determine the AA patterns in calf diets. Lysine (Lys), methionine (Met), and threonine (Thr) were found to be the most limiting amino acids, and their concentrations were related to the growth, physiology, and reproductive performance of calves. Ragland et al. [6] also reported that the limiting amino acids for beef cattle were ranked as Lys > Met > Thr, leading us to the conclusion that Lys, Met, and Thr may be the first three limiting amino acids for dairy heifers.

Research on the amino acid patterns of cattle is limited, particularly regarding the amino acid patterns for each growth stage. Several studies [7] have reported on the amino acid patterns in calf diets [8]. However, one "ideal amino acid pattern" cannot, alone, reliably meet the AA requirements at all growth stages. Balanced AA models should account for changes in growth, body protein composition, and physiological requirements throughout life. The costs of growing heifers are the second largest part in the annual operating expenses of a dairy farm. The lack of optimal amino acid patterns made it difficult to minimize the costs and reduce the nitrogen emissions of heifers. The objective of this study was to determine the amino acid limiting sequence and establish an amino acid ratio in corn–soybean meal and alfalfa hay-based diets for Holstein heifers, aged seven to nine months, using the amino acid partial deletion method.

2. Materials and Methods

2.1. Animals, Diets, and Experimental Design

The experimental procedures were approved by the Animal Ethics Committee of the CAAS. Human animal care and handing procedures were followed throughout the experiment (AEC-CAAS-2017-01).

In this experiment, an AA partial deletion method developed by Wang et al. (1989) [9] was used to prepare the different patterns of the Lys, Met, and Thr diets. The AA levels of the total mixed ration (TMR) in the theoretically balanced AA ration were calculated according to the formula proposed by Zinn et al. (1998) [10]: METR = 1.956 + 0.0292 × ADG × [268 − (29.4 × 0.0557 × BW$^{0.75}$ × ADG$^{1.097}$)/ADG] + 0.112 × BW$^{0.75}$ (METR = methionine requirement; ADG = average daily gain; BW = body weight; BW$^{0.75}$ = metabolic weight). Because of the absence of amino acid patterns in cattle at this stage, Lys and Thr were added according to the AA patterns of the growing swine [11] using a Lys: Met: Thr ratio of 100:30:65. Seventy-two Holstein heifers (age = 5.5 ± 0.5 months old; BW = 200 ± 9.0 kg; mean ± standard deviation) were reared at the Third Dairy Farm of Yinxiang Group Company in Shandong Province, China. The basal diet nutrient level is shown in Table 1.

Table 1. Composition and nutrient levels of basal total mixed ration (TMR) (dry matter basis).

Ingredients	Contents, %	Nutrient Levels [2]	Levels
Corn	45.67	Metabolizable energy, (MJ/kg)	10.13
Soybean meal	11.97	Crude protein, %	14.95
Wheat bran	15	Ether extract, %	3.04
Alfalfa hay	25	Ash, %	7.58
Limestone	1.06	Neutral detergent fiber, %	29.22
Salt	0.3	Acid detergent fiber, %	13.99
Premix [1]	1	Calcium, %	1.12
Total	100	Phosphorus, %	0.60
		Lysine, %	0.51
		Methionine, %	0.07
		Threonine, %	0.49

[1] The premix provided the following minerals and vitamins for TMR: Cu, 12.5mg/kg; Fe, 90 mg/kg; Zn, 90 mg/kg; Mn, 30 mg/kg; I, 1.0 mg/kg; Se, 0.3 mg/kg; Co, 0.5 mg/kg; vitamin A, 15,000 IU/kg; vitamin D35,000 IU/kg; vitamin E, 50 mg/kg; [2] nutrient levels were measured values, except for metabolizable energy, which was measured and calculated through digestibility and metabolism trials. The energy of CH_4 was calculated by equation 10.21 (IPCC, 2006), Ym = 5.5%.

A completely randomized design was used for this study. Heifers were randomly allocated to four treatments with 18 heifers each, based on body weight and age, and fed one of the four total mixed rations (TMRs): (1) theoretically balanced AA TMR (Positive control, PC); (2) 30% Lys deleted TMR (partially deleted Lys; PD–Lys); (3) 30% Met deleted TMR (partially deleted Met; PD–Met); and (4) 30% Thr deleted TMR (partially deleted Thr; PD–Thr). Ruminal bypass Lys (Yahe Nutrition Co., Beijing, China, 36% content, 80% bypass rate), Ruminal bypass Met (Adisseo Co., Hebei, China, 44.4% content, 50% bypass rate), and Ruminal bypass Thr (King Technology Co., Hangzhou, China, 40% content, 90% bypass rate) were added to the basal TMR diet. The AA levels in the four treatments are shown in Table 2. The amounts of AA added were adjusted monthly according to BW and dry matter intake (DMI). After an adaptation period of two weeks, each animal was weighed and began the study with an average initial BW of 226 ± 10 kg and age of 6 ± 0.5 months old.

Table 2. Amino acid (AA) levels of TMRs (dry matter basis).

Items	Treatments [1] (%)			
	PC	PD–Lys	PD–Met	PD–Thr
Total AA content				
Lysine	1.00	0.66	1.00	1.00
Methionine	0.33	0.33	0.22	0.33
Threonine	0.72	0.72	0.72	0.45
AA content in basal diet				
Lysine	0.51	0.51	0.51	0.51
Methionine	0.07	0.07	0.07	0.07
Threonine	0.45	0.45	0.45	0.45
Exogenously added AA				
Lysine	0.49	0.15	0.49	0.49
Methionine	0.25	0.25	0.15	0.25
Threonine	0.23	0.23	0.23	0.00

[1] Treatments: PC = theoretical amino acid balance TMR; PD–Lys = 30% Lys deleted TMR; PD–Met = 30% Met deleted TMR; PD–Thr = 30% Thr deleted TMR.

A digestibility and metabolism trial was conducted by selecting four heifers from each treatment during the week before the end of the trial, with a 4-day adaptation period and a 3-day feces and urine collection period. Feces (weight) and urine (volume) outputs were recorded and sampled daily at 07:00, and nitrogen was immediately fixed with 10 mL 10% dilute hydrochloric acid per 100 g feces to determine the N retention (NR). Heifers were housed in individual iron cages (3 × 2.2 m, 6.6 m^2/ head) bedded with rice husks and fermented cow dung. Fresh water was added ad libitum and replaced daily. The animals were fed TMR twice daily at 08:00 and 17:00. Amino acids were supplemented into the TMR during morning feeding. Individual intakes of TMR were recorded daily and collected weekly during the entire experiment to calculate the dry matter intake (DMI). Environmental conditions (including air temperature) were continuously recorded. The mean air temperature was 11.87 ± 7.54 °C.

The experimental feeding periods were 90 days in duration (September to November 2017). All heifers were immunized according to the standard immunization procedure of the farm, with the brucellosis vaccination administered at 7 months of age.

2.2. Sampling and Analyses

BW and body size were measured before the morning feeding period every 30 days. Diet samples were collected weekly before the morning feeding and stored at −20 °C for further analysis. TMR, feces, and urine samples were sent to the Lab of Ruminant Physiology and Nutrition, Feed Research Institute, Chinese Academy of Agricultural Sciences (Beijing, China) for nutrient analysis. The TMR and feces samples were dried in a forced-air oven at 65 °C for 48 h. Then, the DM (105 °C for 5 h), crude protein (CP), ash, and ether extract (EE) contents were analyzed (method 968.08; AOAC, 1990) [12]. Calcium (Ca) content was analyzed using an atomic absorption spectrophotometer (M9W–700; Perkin–Elmer Corp., Norwalk, CT, U.S.A.) (method 968.08; AOAC, 1990) [12]. Total phosphorus (P) content was analyzed by the molybdovanadate colorimetric method (method 965.17; AOAC, 1990) [12] using a spectrophotometer (UV–6100; Mapada Instruments Co., Ltd., Shanghai, China). The neutral detergent fiber (NDF) and acid detergent fiber (ADF) contents were determined with an Ankom A200 apparatus (Ankom Technology, Macedon, NY, USA) with heat-stable amylase (Ankom Technology) and sodium sulfite (Fisher Scientific, Waltham, MA, USA) and an expressed inclusive of residual ash [13].

A blood sample was collected from six heifers in each treatment (at 24 and 36 weeks of age) before morning feeding, by a jugular venipuncture, and transferred into vacuum tubes without anticoagulants. Serum was immediately separated from the blood by centrifugation at 3000× g at 4 °C for 10 min and stored at −20 °C until analysis. Serum urea nitrogen (SUN) was analyzed using blood colorimetric commercial kits (DiaSys Diagnostics Systems GmbH, Frankfurt, Germany).

2.3. Amino Acid Partial Partial Deletion Method

The principle of the amino acid partial partial deletion method is that there is a linear relationship between the first limiting amino acid and the NR. In other words, the NR will decrease greatest after deleting the first limiting animo acid and will result in the largest slope. The model diagram is as follows (Figure 1):

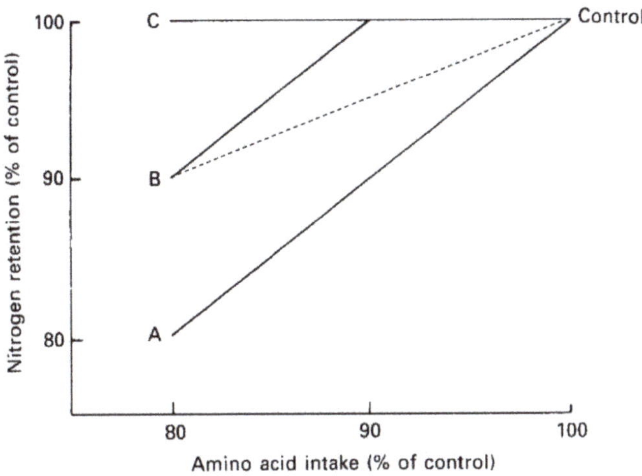

Figure 1. In this model, amino acid intake (AAI) should be presented as the percentage or ratio to control treatment for better distinguishing differences of NR among three AA deleting treatments. To keep the linear relationship between the NR and the first limiting amino acid, the NR should also be converted to the percentage or ratio of the control treatment. The model assumes that deleting the first limiting amino acid (as A) reduces NR to the greatest extent (largest slope); deleting C does not reduce the NR at all (slope = 0), as it remains in excess (over 20%) relative to the first limiting amino acid. Deleting B results in a reduction in the NR intermediate between A and C (0 < Slope B (dashed) < Slope A), and part of B is in excess relative to the first limiting amino acid. In other words, A is the first limiting amino acid while B is second limiting amino acid. According to the principle of the "wooden barrel", all essential AAs can be controlled by the same limitation by adjusting the amount of AAs in the diet. In this model, B is 10% more than A, which means that we should reduce 10% of B from the control treatment to achieve the minimum addition and ensure it is co-limiting with A [9]. Then, we can calculate the ratio of A and B.

2.4. Statistical Analyses

Data on SUN and N retention were analyzed with a one–way ANOVA procedure using the SAS software (SAS version 9.4; SAS Institute Inc., Cary, NC, USA). Least square means were calculated and separated using the PDIFF option, and differences between diets were detected by Duncan's multiple comparison in SAS. A MIXED procedure was used to analyze the growth performance data. Month, treatment, and treatment by month of age interactions were fixed effects, and the heifers within each treatment were random effects. The effect of the month was included as a repeated measure. For the repeated measures analysis, the covariance structure with the lowest Akaike information criterion was used. The results were reported as the least squares. A significance level was declared at $p < 0.05$.

3. Results

3.1. Growth Performance

The results of the growth performance are presented in Table 3. No significant differences (P > 0.05) were observed in the BW and DMI of heifers among the four treatments during the experiment. However, the ADG ($p = 0.0013$) and G/F ($p = 0.0057$) of heifers in the PD–Lys and PD–Met treatment were decreased significantly compared to PC treatment at eight to nine months old.

Table 3. Effects of deleting Lysine (Lys), Methionine (Met), and Threonine (Thr) levels in corn–soybean based TMR on the growth performance of heifers aged seven to nine months old (n = 72).

Items [1]	Treatments [2]				SEM	p Value [3]		
	PD–Lys	PD–Met	PD–Thr	PC		T	M	T × M
BW, kg								
Average	273.7	273.9	276.3	274.8	2.98	0.4798	<0.0001	<0.0001
6 mon	227.5	227.0	229.9	228.5	4.08	0.4647		
7 mon	258.0	257.8	259.3	257.6	4.00	0.7016		
8 mon	282.7	284.3	282.5	280.3	4.52	0.3667		
9 mon	326.4	326.8	333.4	333.0	4.02	0.0997		
ADG, kg								
Average	1.04	1.09	1.09	1.11	0.029	0.1566	<0.0001	<0.0001
6–7 mon	0.98	1.02	0.94	0.95	0.075	0.2848		
7–8 mon	0.95	0.95	0.90	0.89	0.079	0.4946		
8–9 mon	1.21 [c]	1.30 [bc]	1.44 [ab]	1.48 [a]	0.080	0.0013		
DMI, kg								
Average	7.15	7.16	7.16	7.05	0.050	0.1073	<0.0001	<0.0001
6–7 mon	6.25	6.20	6.26	6.28	0.091	0.2942		
7–8 mon	6.99	6.98	6.94	6.84	0.089	0.0916		
8–9 mon	8.21	8.28	8.16	8.25	0.088	0.2297		
Feed conversion rate, G/F								
Average	0.146	0.152	0.152	0.156	0.008	0.1452	0.0001	0.0002
6–7 mon	0.156	0.164	0.151	0.155	0.011	0.2138		
7–8 mon	0.135	0.134	0.127	0.130	0.010	0.5009		
8–9 mon	0.147 [c]	0.158 [bc]	0.176 [ab]	0.181 [a]	0.012	0.0057		

[1] BW=body weight, ADG=average daily gain, DMI=dry matter intake; [2] Treatments: PC = theoretical amino acid balance TMR; PD–Lys = 30% Lys deleted TMR; PD–Met = 30% Met deleted TMR; PD–Thr h = 30% Thr deleted TMR; [3] T = Treatment, M = Month of age, T × M = The interaction between treatment and month of age; [a,b,c] values within the same row with different superscripts are different ($p < 0.05$).

3.2. Serum Urea Nitrogen Levels

The SUN levels of heifers aged eight months old ($p = 0.0013$) and nine months old ($p = 0.0021$) in the PD–Lys and PD–Met treatments were higher than those in the PC treatment (Figure 2). No significant differences of SUN were observed between the PD–Thr treatment and PC treatments ($p > 0.05$).

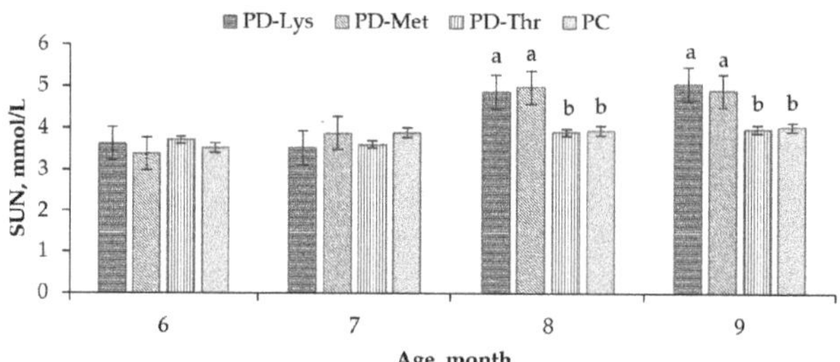

Figure 2. Comparison of serum urea nitrogen levels of seven- to nine-month-old heifers fed corn–soybean based TMRs among the four treatments (n = 24); PD–Lys = 30% Lys deleted TMR diet (diagonal stripes bar), PD–Met = 30% Met deleted TMR diet (vertical stripes bar), PD–Thr = 30% Thr deleted TMR diet (horizonal stripes bar), PC = theoretically balanced amino acid TMR diet (gray bar); The y-axis represents the serum urea nitrogen levels of four treatments; the x-axis was the age of heifers. Error bars indicate SEM. The a,b above the bars indicate the significant differences among treatments ($p < 0.05$).

3.3. Nitrogen Metabolism

There was no difference in N intake among treatments (Table 4). Total excreted N significantly increased (p = 0.0208) when dietary Lys and Met were reduced, as there were significant increases in urine N (p = 0.0011). However, fecal N and Digestible N did not differ among four treatments (p > 0.05). Moreover, the amount of urine N and NR of heifers in the PD–Thr treatment were not significantly different from those in the PC treatment (p > 0.05).

Table 4. Effects of deleting Lysine, Methionine, and Threonine levels in corn–soybean based TMRs on nitrogen metabolism of heifers aged seven to nine months old (n = 16).

Items [1]	Treatments [2]				SEM	p Value
	PD–Lys	PD–Met	PD–Thr	PC		
Intake N, g·(kg^{-1} BW$^{0.75}$) d^{-1}	2.92	2.97	2.90	2.87	0.020	0.2961
Fecal N, g·(kg^{-1} BW$^{0.75}$)·d^{-1}	0.90	0.82	0.80	0.79	0.015	0.1223
Urine N, g·(kg^{-1} BW$^{0.75}$)·d^{-1}	1.06 [b]	1.16 [a]	1.02 [b]	0.88 [c]	0.033	0.0011
Total excrete N, g·(kg^{-1} BW$^{0.75}$)·d^{-1}	1.96 [b]	1.98 [a]	1.82 [ab]	1.67 [ab]	0.032	0.0208
N retention, g·(kg^{-1} BW$^{0.75}$)·d^{-1}	0.96 [b]	0.99 [b]	1.08 [ab]	1.20 [a]	0.034	0.0324
Digestible N, g·(kg^{-1} BW$^{0.75}$)·d^{-1}	2.02	2.15	2.06	2.08	0.020	0.2908
N utilization, %	33.08 [b]	33.26 [b]	34.96 [b]	41.77 [a]	1.210	0.0048
N digestibility, %	69.50	72.38	70.79	72.6	0.512	0.0798

[1] N = nitrogen; Total excrete N=Fecal N + Urine N, Absorbed N =Intake – Total excrete N, NR (N retention) = N intake – fecal N – urinary N, N utilization = (N intake – fecal N excretion)/N intake × 100%, N digestibility = (N intake – fecal N excretion)/N intake × 100%; [2] treatments: PC = theoretical amino acid balance TMR; PD–Lys = 30% Lys deleted TMR; PD–Met = 30% Met deleted TMR; PD–Thr = 30% Thr deleted TMR; [a, b, c] values within the same row with different superscripts differ (p < 0.05).

3.4. Appropriate Amino Acid Model

3.4.1. N Retention and Amino Acid Intake

N retention (NR) and amino acid intake (AAI) based on metabolic weight (Table 5) were converted in proportion to the PC treatment based on the requirements of the "Amino acid partial deletion method model". Then, the proportions of the intakes of Lys, Met, and Thr in the PD–Lys, PD–Met, and PD–Thr treatment to those in the PC treatment were calculated (e.g., the AAI of Lys in the PD–Lys treatment is 0.60, the AAI of Lys in the PC treatment is 0.90, and the ratio of Lys in the PD–Lys treatment to Lys in the PC treatment is 0.60/0.90 = 0.67). After conversion, the ratio of Lys, Met, Thr in amino acid to PC treatment were 0.67, 0.69, and 0.62, respectively. This result differs slightly from 0.7 due to the deferences of the metabolic body weights of the heifers in the four treatments.

Table 5. The proportions of amino acid intake and nitrogen retention in PD–Lys, PD–Met, and PD–Thr to those in the PC treatment.

Items [1]	Based on Metabolic Body Weight, g·(kg^{-1} BW$^{0.75}$)·d^{-1}				The Ratio to PC			
	NR [2]	AAI [2]			NR	AAI		
		Lys	Met	Thr		Lys	Met	Thr
PD–Lys	0.96	0.60	0.29	0.65	0.80	0.67	1.00	1.00
PD–Met	0.99	0.90	0.20	0.65	0.83	1.00	0.69	1.00
PD–Thr	1.02	0.90	0.29	0.40	0.85	1.00	1.00	0.62
PC	1.20	0.90	0.29	0.65	1.00	1.00	1.00	1.00

[1] PD–Lys = 30% Lys deleted treatment; PD–Met = 30% Met deleted treatment; PD–Thr = 30% Thr deleted treatment; PC = theoretical amino acid balanced treatment; [2] NR=N retention; AAI=amino acid intake.

3.4.2. Calculation and Model Diagram of the Effect on Nitrogen Retention

The proportions of the three essential AAs were calculated using the simple linear model based on the amino acid partial deletion method [9]. The model diagram of the effect on NR after deleting 30% of Lys, 30% of Met, and 30% of Thr in corn–soybean based diets is shown in Figure 2.

Figure 3a shows the rate of NR in relation to the daily AA intake. The values of AAI (x-axis) and NR (y-axis) are provided in Table 6. Point "PC" represents the corresponding AAIs and NRs of the three AAs in the PC treatment (all values = 1 only one point). "Lys" is the point of the Lys intake and NR in the PD–Lys treatment (0.67, 0.80). "Met" is the point of the Met intake and NR in the PD–Met treatment (0.69, 0.83). "Thr" is the point of Thr intake and NR in the PD–Thr treatment (0.62, 0.85).

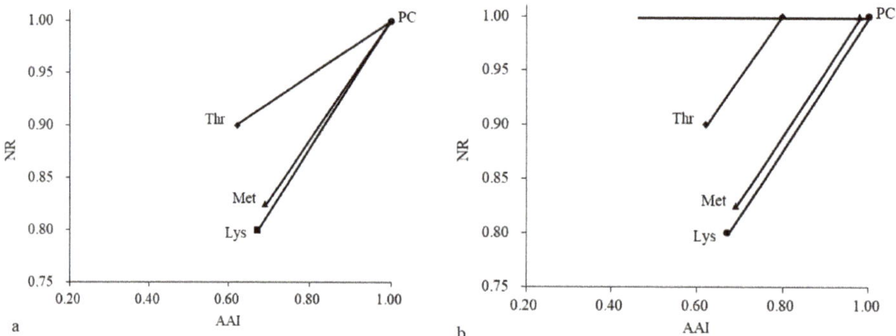

Figure 3. The pattern diagram (**b**) when Met and Thr are converted to an equivalent slope (**a**) with Lys. The y-axis represents the ratio of NR after deleting Lys, Met, Thr to that of the PC treatment; the x-axis is the ratio of the amino acid intake (AAI) in the amino acid deleting treatments to that in the PC treatment. (Lys, ■) = Lys intake and NR level in PD–Lys treatment, (Met, ▲) = Met intake and NR level in PD–Met treatment, (Thr, ♦) = Met intake and NR level in PD–Thr treatment, and (PC, ●) = Lys, Met, Thr intake and NR level in PC treatment; all values =1.

Table 6. The appropiate amino acid ratio of Lysine (Lys), Methionine (Met), and Threonine (Thr) based on the nitrogen retention (NR) of heifers aged seven to nine months old, fed corn–soybean meal-based TMRs.

Items [1]	S	P	C	R
Lys	0.61	1.00	69.99	100.00
Met	0.57	0.98	22.36	31.95
Thr	0.26	0.79	39.78	56.84

[1] S(Slop) = (1-NR)/(1-AAI); P(proportion) = [(1-NR) + S × AAI]/S; C (concentration) = AAI × P; R (ratio) = AA/Lys.

The slope (Table 5) describes the effect of deleting an AA from the PC on the NR (e.g., for Lys, (1–0.80)/(1–0.67) = 0.61). Among the three AAs, a higher slope for the Lys deletion treatment (PD–Lys) means that Lys is the first limiting AA in the PC treatment. The limiting sequence of the three amino acids is ranked as: Lys > Met > Thr.

Figure 3b is the pattern diagram for when Met and Thr are converted to an equivalent slope with Lys. To calculate the proportion of each AA that could be removed from the PC amino acid pattern to make it equally limiting to Lys, it was assumed that when all AAs are equally limiting, they should all have the same slope. Therefore, the required amount of Met can be caculated as: S (Lys) = (1 − 083)/(x − 0.69)—that is, 0.61 = (1 − 0.83)/(x − 0.69), x = 0.98, thence 1.00 − 0.98 = 0.02. In other words, 0.02 of Met should be removed from the PC to make the Met co-limiting with the Lys (the actual requirement of Met is 0.98 × Met in the PC treatment). In the same way, we calculate that the 0.21 of Thr should be removed from the PC treatment when it is equally limited with Lys (the actual requirement of Thr is 0.79 × Thr in the PC treatment).

3.4.3. Appropriate Amino Acid Ratio

An appropriate amino acid model of seven- to nine-month-old Holstein heifers is shown in Table 6 (calculated from Figure 2). S (slope) represents the effect of deleting 30% amino acid on N retention (S = (1 − NR)/(1 − AAI), calculated from Figure 2a). The S value of the Lys deleted treatment was the highest, indicating that Lys was the first limiting amino acid. P (proportion) is the proportion of amino acid (except for Lys) in the PC treatment when it was equally limited to that of Lys (P = [(1−NR) + S × AAI]/S, calculated from Figure 2b). The P value was calculated based on the principle of "equal limitation means equal slope." C (concentration) is the actual concentration of amino acid when it was equally limited with Lys (C = AAI (in PC treatment) × P). R (ratio) is the ratio of the actual amino acid concentration to the Lys concentration (R = AA/Lys). The optimal ratios based on the NR of the three amino acids for seven- to nine-month-old Holstein heifers was 100:32:57.

4. Discussion

4.1. Growth Performance and Body Size

The function of dietary protein is determined by amino acid composition, the nutrient digestive abilities of animals, and how well the composition of absorbed amino acid matches the balance required by the animals. The deficiency and overdose of certain amino acids in the diet will cause an imbalance between amino acids and thus affect the growth and development of the animals. For calves, the addition of Lys and Met in a milk replacer significantly increased the feed conversion efficiency of calves, but the addition of Thr had no significant effect on the growth performance of calves [7]. Ludden et al. [14] observed that supplementation with Lys improved the ADG in the growing cattle. Awawdeh et al. [15] showed that when Met was limiting amino acid, the dietary supplementation of other amino acids increased the utilization efficiency of Met and increase the growth of bulls. In this experiment, deleting 30% of Lys and Met led to a decrease of ADG and G/F. Such a reduction of growth performance might be due to the unbalanced amino acids. Another important observation is that DMI seems not to be affected by treatment. Wang et al. [16] found no significant differences in the DM and N intake of dairy cows after adding Lys and Met to the diet. That is to say, growth performance is affected by limiting amino acid deficiencies rather than feed intake [17]. Lys and Met might be the first two limiting amino acids for growing cattle. Unlike Lys and Met, the growth responses to Thr deletion were not significantly decreased. It remains possible that the theoretical Thr addition in this experiment was relatively higher than the requirement of heifers due to the absence of accurate data on Thr requirements in this trial.

4.2. Serum Urea Nitrogen and Nitrogen Retention

SUN, as an end metabolite of the liver's N metabolism [18], is negatively correlated with the utilization rate of protein [19]. The balance of amino acids is the basic condition needed to improve protein utilization and reduce SUN concentration [19]. Jiang et al. [20] estimated sharp decreases in the content of SUN and the emission of urine nitrogen, as well as an increase of the N retention of cows after adding Met and Lys in their diets. In our study, SUN concentration increased after deleting 30% of dietary Lys and Met, which might indicate an imbalance of amino acids and decrease N utilization. Urine N is the main excretion pathway of SUN, accounting for a large part of the N excretion of heifers. An amino acid balanced diet can improve the N utilization rate and reduce the excretion of fecal and urine nitrogen (about 46%) of dairy cows, especially urine N excretion [21]. Adding Lys and Met to diets can promote a balance of amino acids, reduce urinary nitrogen concentration, and improve the protein utilization rate of dairy cows [22]. Recent research by Lee et al. [2] concluded that the efficiency of feed N absorbed by the small intestine increased when dietary amino acid was balanced. We also observed that urine N was significantly increased as the Lys and Met levels decreased. Therefore, it was further confirmed in this study that deleting Lys and Met led to an imbalance of amino acids, which resulted in an increase of urine nitrogen.

N retention reflects the efficiency of protein deposition and amino acid utilization [23], which is also closely related to the production performance of animals [24]. A balance of amino acids in the diet can enhance the digestion and absorption of N in animals [25]. In particular, the metabolic amount of the first limiting amino acid has a linear relationship with N retention [26]. Balancing a complete amino acid profile increased the efficiency of dietary N utilization in both a low and a high small intestine protein supply [27]. The efficiencies of nitrogen estimated in the current study confirmed that adding Lys to the Lys deficient diet of calves reduced the rate of N excretion and increased the rate of N deposition [28]. Conversely, heifers fed with Lys and Met deficient diets caused an increase of nitrogen retention, indicating inefficiencies in their use of absorbed amino acid for protein accretion [25]. Importantly, the effects of dietary Lys, Met, and Thr levels on the N retention of heifers is not consistent and largely depends on the balance and limiting sequences of these three amino acids. In this study, the decrease of N retention, in combination with the deficiency of Lys and Met, indicated that Lys and Met are the first and second limiting amino acids for heifers , respectively.

In addition, an increase in N retention was commonly reported for cows fed diets with a supplementation of rumen-protected amino acids [29,30], similar to the present study, which indicated that the added ruminal protected amino acids were effectively protected from ruminal degradation and guaranteed amino acids to be released and absorbed in the small intestine for better utilization. Dietary supplementation of rumen-protected Met and Lys could ensure a balance of amino acids, promote the increase of nitrogen deposition, and improve the utilization rate of proteins [31]. In this case, supplementation with rumen-protected amino acids may be a successful strategy for establishing the amino acid pattern of heifers based on dietary amino acids.

4.3. Limiting Sequence and Appropriate Ratio of Amino Acid

The limiting sequence of amino acids in ruminants was affected by the composition of their diets. Maize silage/maize gain based diets can supply adequate protein but do not provide enough Lys to growing cattle, which indicates that Lys is the first limiting amino acid [32]. Klemesrud et al. [33] also reported that Lys is the first limiting amino acid in steers fed with diets containing large amount of maize products. Wang [8] found that Lys was the first limiting amino acid (the second and third were Met and Thr, respectively) due to the large decrease of ADG in calves fed with a milk replacer, starters, and Leymus chinensis after reducing Lys. We found that the limiting sequence of seven- to nine-month-old Holstein heifers was Lys > Met > Thr, based on corn–soybean meal–alfalfa TMRs. Therefore, Lys plays the most important role in the growth of heifers that are fed rations made from corn–soybean meal.

An optimal amino acid pattern is needed as a standard for evaluating the diets of animals. The requirements of amino acids are not well defined for heifers with corn–soybean meal-based diets, and modifying amino acid patterns can increase the bypass protein efficiency [34]. When the Lys ratio is expressed, variation in the estimated requirement of the specific AA is greatly reduced compared to the amino acid ratio of the diets [25]. NRC (2012) [35] pointed out that the ideal amino acid pattern should be expressed as the ratio of amino acid to Lys. In this study, Lys happened to be the first limiting amino acid, so the calculated model is appropriate. The results from this study show that the appropriate pattern of amino acids in the diet based on maximum N retention in seven- to nine-month-old Holstein heifers (fed with corn–soybean meal) was 100:32:57. However, we did not precisely determine the limiting amino acid pattern because of our inability to accurately calculate metabolic proteins, so this pattern cannot be applied to all type of diets. This may offer an explanation for the differences between this pattern and the patterns for calves and cows (Table 7). Of course, the amino acid requirement for growing heifers might change according to age. It is possible that differences in diet type and digestion among heifers, calves, and cows directly affect the profile of delivered amino acids to the intestine.

However, the ruminal bypass amino acid products and microbial metabolism made it difficult to determine whether there was a large difference between intake N and metabolic N. Therefore, further research is needed to determine the precise amino acid patterns based on metabolic protein.

Amino acid is mainly used for the bodily growth and development of heifers. Amino acid requirements are mainly determined by body protein retention and N emission, similar to beef cattle, so the amino acid requirement of heifers may be determined by the composition of the body's amino acid [36]. However, the amino acid pattern in this research was different from that of beef cattle (Table 7). Studies have shown dietary amino acid patterns are not equivalent to carcass amino acid composition. Decomposition and conversion by intestinal bacteria produced a significant difference between amino acid in the diet and amino acid absorbed into the blood. Moreover, different tissues have different uses and metabolic efficiencies for amino acid, which may cause a deviation between a carcass's amino acid composition and dietary amino acid patterns. Considering animal welfare and economic benefits, the calculated amino acid model was not verified by a carcass's amino acid components. Whether a carcass's amino acid components can be used as an appropriate amino acid model for growing heifers needs to be further verified.

Table 7. The amino acid ratio of calves and cows in previous studies.

Stage	Index [1]	Lys: Met: Thr Ratio	Reference
Calves	NR	100:26:66	Gerrits et al., 1997 [37]
Calves	maximum ADG	100:31:77	Hill et al., 2008 [7]
Calves	maximum ADG	100:35:63	Wang et al., 2011 [8]
Beef cattle	body amino acids	100:31:61	NRC (2016) [38]

[1] NR = nitrogen retention; ADG = average daily gain.

5. Conclusions

In this study, there were negative effects on the average daily gain, feed conversion rate, and nitrogen retention of seven- to nine-month-old heifers after deleting 30% dietary Lys and Met. However, deleting Thr content did not affect the growth performance and N metabolism of heifers. The sequence of the three amino acids for seven- to nine-month-old Holstein heifers that were fed a TMR of corn–soybean meal concentrate and alfalfa hay was Lys > Met > Thr. Additionally, the appropriate amino acid ratio calculated from nitrogen retention of this ratio of diet was 100:32:57.

Author Contributions: Conceptualization, Y.T. and Q.D.; methodology, Y.T and Y.L.; software, Y.L.; validation, Y.T; formal analysis, Y.L.; investigation, F.H., M.T., Y.S.; resources, Q.D.; data curation, F.K.; writing—original draft preparation, Y.L.; writing—review and editing, M.P. and B.W.; visualization, Y.T.; supervision, Y.B.; project administration, Y.B.; funding acquisition, Y.T. and Q.D. All authors participated in writing the final draft of the manuscript and agreed on the final format.

Funding: This work was supported by the Beijing Innovation Team for Technology Systems in the Dairy Industry (BAIC06), the Chinese Academy of Agricultural Science and Technology Innovation Project (CAAS-XTCX2016011-01), Fundamental Research Funds for Central Non-profit Scientific Institution of CAAS (Y2019CG08), and Science and Technology Open Cooperation Project of Henan Province (182106000035)—Study on the Amino Acid Pattern of Diet for Different Physiological Stages of Heifers.

References

1. Kim, S.W.; Hurley, W.L.; Wu, G.; Fu, J. Ideal amino acid balance for sows during gestation and lactation. *J. Anim. Sci.* **2009**, *87*, E123–E132. [CrossRef] [PubMed]
2. Lee, C.; Hristov, A.N.; Cassidy, T.W.; Heyler, K.S.; Lapierre, H.; Varga, G.A.; Veth, M.D.; Patton, R.A.; Parys, C. Rumen-protected lysine, methionine, and histidine increase milk protein yield in dairy cows fed a metabolizable protein-deficient diet. *J. Dairy Sci.* **2012**, *95*, 6042–6056. [CrossRef]

3. Lean, I.J.; Ondarza, M.B.; Sniffen, C.J.; Santos, J.P.; Griswold, K.E. Meta-analysis to predict the effects of metabolizable amino acids on dairy cattle performance. *J. Dairy Sci.* **2018**, *101*, 340–364. [CrossRef] [PubMed]

4. Dorigam, J.C.; Sakomura, N.K.; Lima, M.B.; Sarcinelli, M.F.; Suzuki, R.M. Establishing an essential amino acid profile for maintenance in poultry using deletion method. *J. Anim. Physiol.* **2016**, *100*, 884–892. [CrossRef] [PubMed]

5. Wang, J.H.; Diao, Q.Y.; Xu, X.C.; Tu, Y.; Zhang, N.F.; Yun, Q. The limiting sequence and proper ratio of lysine, methionine and threonine for calves fed milk replacers containing soy protein. *Asian-Australas. J. Anim. Sci.* **2012**, *25*, 224–233. [CrossRef] [PubMed]

6. Ragland-Gray, K.K.; Amos, H.E.; Mccann, M.A.; Williams, C.C.; Sartin, J.L.; Barb, C.R.; Kautz, F.M. Nitrogen metabolism and hormonal responses of steers fed wheat silage and infused with amino acids or casein. *J. Anim. Sci.* **1997**, *75*, 3038–3045. [CrossRef]

7. Hill, T.M.; Bateman, H.G.; Aldrich, J.M.; Schlotterbeck, R.L.; Tanan, K.G. Optimal concentrations of lysine, methionine, and threonine in milk replacers for calves less than five weeks of age. *J. Dairy Sci.* **2008**, *91*, 2433–2442. [CrossRef] [PubMed]

8. Wang, J.H.; Diao, Q.Y.; Xu, X.C.; Tu, Y.; Zhang, N.F.; Yun, Q. Effects of Dietary Addition Pattern of Lysine, Methionine and Threonine in the Diet on Growth Performance, Nutrient Digestion and Metabolism, and Serum Biochemical Parameters in Calves at the Ages of 0–2 Months. *Sci. Agric. Sin.* **2011**, *44*, 1898–1907.

9. Wang, T.C.; Fuller, M.F. An Optimal Dietary Amino Acid Pattern for Growing Pigs. *Br. J. Nutr.* **1989**, *62*, 77–89. [CrossRef]

10. Zinn, R.A.; Shen, Y. An evaluation of ruminally degradable intake protein and metabolizable amino acid requirements of feedlot calves. *J. Anim. Sci.* **1998**, *76*, 1280–1289. [CrossRef]

11. Chung, T.K.; Baker, D.H. Ideal amino acid pattern for 10-kilogram pigs [L500, L600], pigs, amino acids, growth, nitrogen metabolism, nitrogen balance, diet IND92069907. *J. Anim. Sci.* **1992**, *62*, 77–89.

12. AOAC (Association of Official Analytical Chemists). *Official Methods of Analysis*, 15th ed.; AOAC: Washington, DC, USA, 1990.

13. Soest, P.J.V.; Robertson, J.B.; Lewis, B.A. Methods for Dietary Fiber, Neutral Detergent Fiber, and Nonstarch Polysaccharides in Relation to Animal Nutrition. *J. Dairy Sci.* **1991**, *74*, 3583–3597. [CrossRef]

14. Ludden, P.A.; Kerley, M.S. Amino acid and energy interrelationships in growing beef steers, II. Effects of energy intake and metabolizable lysine supply on growth. *J. Anim. Sci.* **1998**, *76*, 3157–3168. [CrossRef] [PubMed]

15. Awawdeh, M.S.; Titgemeyer, E.C.; Schroeder, G.F.; Gnad, D.P. Excess amino acid supply improves methionine and leucine utilization by growing steers. *J. Anim. Sci.* **2006**, *84*, 1801–1809. [CrossRef] [PubMed]

16. Wang, C.; Liu, H.Y.; Wang, Y.M.; Yang, Z.Q.; Liu, J.X.; Wu, Y.M.; Yan, T.; Ye, H.W. Effects of dietary supplementation of methionine and lysine on milk production and nitrogen utilization in dairy cows. *J. Dairy Sci.* **2010**, *93*, 3661–3670. [CrossRef]

17. Awawdeh, M.S.; Titgemeyer, E.C.; Mccuistion, K.C.; Gnad, D.P. Effects of ammonia load on methionine utilization by growing steers. *J. Anim. Sci.* **2004**, *82*, 3537–3542. [CrossRef]

18. Kohn, R.A.; Dinneen, M.M.; Russek-Cohen, E. Using blood urea nitrogen to predict nitrogen excretion and efficiency of nitrogen utilization in cattle, sheep, goats, horses, pigs, and rats. *J. Anim. Sci.* **2005**, *83*, 879–886. [CrossRef]

19. Ponnampalam, E.N.; Dixon, R.M. Intake, growth and carcass characteristics of lambs consuming low digestible hay and cereal grain. *Anim. Feed Sci. Technol.* **2004**, *114*, 31–41. [CrossRef]

20. Jiang, S.Z.; Yang, W.R.; Liu, F.X.; Yang, Z.B. Study on Effects of Methionine Medicate into the Rumen on Apparent Digestibility of Amino Acid in the Small Intestine of the Steers. *Acta Vet. Zootech. Sin.* **2006**, *12*, 957–962.

21. Wang, C.; Liu, J.X.; Yuan, Z.P.; Wu, Y.M.; Zhai, S.W.; Ye, H.W. Effect of level of metabolizable protein on milk production and nitrogen utilization in lactating dairy cows. *J. Dairy Sci.* **2007**, *90*, 2960–2965. [CrossRef]

22. Socha, M.T.; Putnam, D.E.; Garthwaite, B.D.; Whitehouse, N.L.; Kierstead, N.A.; Schwab, C.G.; Ducharme, G.A.; Robert, J.C. Improving intestinal amino acid supply of pre- and postpartum dairy cows with rumen-protected methionine and lysine. *J. Dairy Sci.* **2005**, *88*, 1113–1126. [CrossRef]

23. Goodband, B.; Tokach, M.; Dritz, S.; Derouchey, J.; Woodworth, J. Practical starter pig amino acid requirements in relation to immunity, gut health and growth performance. *J. Anim. Sci. Biotechnol.* **2014**, *5*, 251–261. [CrossRef] [PubMed]

24. Lin, X.J.; Jiang, S.Q.; Ding, F.Y.; Zheng, C.T.; Hong, P. Science IOA. Amino Acid Balance Pattern of a Low Protein Diet in Fast-Growing Yellow-Feathered Broilers Aged from 1 to 21 Days. *Chin. J. Anim. Nutr.* **2014**, *26*, 2542–2552.

25. Haque, M.N.; Guinard-Flament, J.; Lamberton, P.; Mustière, C.; Lemosquet, S. Changes in mammary metabolism in response to the provision of an ideal amino acid profile at 2 levels of metabolizable protein supply in dairy cows, Consequences on efficiency. *J. Dairy Sci.* **2015**, *98*, 3951–3968. [CrossRef]

26. Haque, M.N.; Rulquin, H.; Lemosquet, S. Milk protein responses in dairy cows to changes in postruminal supplies of arginine, isoleucine, and valine. *J. Dairy Sci.* **2013**, *96*, 420–430. [CrossRef] [PubMed]

27. Haque, M.N.; Rulquin, H.; Lemosquet, S. Milk protein synthesis in response to the provision of an "ideal" amino acid profile at 2 levels of metabolizable protein supply in dairy cows. *J. Dairy Sci.* **2012**, *95*, 5876–5887. [CrossRef]

28. Schingoethe, D.J. Balancing the amino acid needs of the dairy cow. *Anim. Feed Sci. Technol.* **1996**, *60*, 153–160. [CrossRef]

29. Agle, M.; Hristov, A.N.; Zaman, S.; Schneider, C.; Ndegwa, P.; Vaddella, V.K. The effects of ruminally degraded protein on rumen fermentation and ammonia losses from manure in dairy cows. *J. Dairy Sci.* **2010**, *93*, 1625–1637. [CrossRef]

30. Colmenero, J.J.O.; Broderick, G.A. Effect of dietary crude protein concentration on milk production and nitrogen utilization in lactating dairy cows. *J. Dairy Sci.* **2006**, *89*, 1704–1712. [CrossRef]

31. Abd, P.; Whitehouse, N.L.; Aragona, K.M.; Schwab, C.S.; Reis, S.F.; Brito, A.F. Production and nitrogen utilization in lactating dairy cows fed ground field peas with or without ruminally protected lysine and methionine. *J. Dairy Sci.* **2017**, *100*, 6239–6255.

32. Xue, F.; Zhou, Z.; Ren, L.; Meng, Q. Influence of rumen-protected lysine supplementation on growth performance and plasma amino acid concentrations in growing cattle offered the maize stalk silage/maize grain-based diet. *Anim. Feed Sci. Technol.* **2011**, *169*, 61–67. [CrossRef]

33. Klemesrud, M.J.; Klopfenstein, T.J.; Lewis, A.J. Metabolize methionine and lysine requirements of growing cattle. *J. Anim. Sci.* **2000**, *78*, 199–206. [CrossRef] [PubMed]

34. Doepel, L.; Pacheco, D.; Kennelly, J.J.; Hanigan, M.D.; López, I.F.; Lapierre, H. Milk protein synthesis as a function of amino acid supply. *J. Dairy Sci.* **2004**, *87*, 1279–1297. [PubMed]

35. *NRC (Nutrient Requirements of Swine)*; National Academy Press: Washington, DC, USA, 2012.

36. Christian, W.; Frank, L. Improving the reliability of optimal in-feed amino acid ratios based on individual amino acid efficiency data from N balance studies in growing chicken. *Animals* **2013**, *3*, 558–573.

37. Gerrits, W.J.; France, J.; Dijkstra, J.; Bosch, M.W.; Tolman, G.H.; Tamminga, S. Evaluation of a model integrating protein and energy metabolism in preruminant calves. *J. Nutr.* **1997**, *127*, 1243–1252. [CrossRef] [PubMed]

38. *NRC (Nutrient Requirements of Beef Cattle)*; National Academy Press: Washington, DC, USA, 2016.

Article

An Economic Analysis of the Costs Associated with Pre-Weaning Management Strategies for Dairy Heifers

Anna Hawkins [1], Kenneth Burdine [2], Donna Amaral-Phillips [1] and Joao H.C. Costa [1,*]

[1] Department of Animal & Food Science, University of Kentucky, Lexington, KY 40506, USA
[2] Department of Agricultural Economics, University of Kentucky, Lexington, KY 40506, USA
* Correspondence: costa@uky.edu; Tel.: +1-859-257-7540

Received: 1 June 2019; Accepted: 17 July 2019; Published: 23 July 2019

Simple Summary: Rearing of replacement female calves on a dairy farm is of critical importance to maintain herd sizes, improve the genetic quality of the herd, and remain economically sustainable. A 2-year investment period is needed for replacement female heifers to grow before entering the milking herd. The management of replacements over this 2-year period can vary greatly among operations, making it difficult to compare producers' cost to benchmark. The objective of this project was to develop a model to calculate the cost of rearing a replacement heifer from birth to weaning under different housing, milk source, allotments, and labor and health management decisions to be used as a dairy farm decision support tool. We calculated the cost for management options with general cost values. We found that the average feed cost represented 46% of the total cost while labor, and fixed and variable costs represented 33%, 9%, and 12%, respectively. The total cost increased as milk allotment increased, but cost per Kg of gain decreased. The ranges in total cost within each management scenario often exceed the difference in cost from one scenario to the next. In conclusion, variable costs have the potential to vary among operations, playing a major role in the total cost of rearing replacements from birth to weaning.

Abstract: Dairy calves are raised in various housing and feeding environments on dairy farms around North America. The objective of this study was to develop a simulation model to calculate the cost of raising replacement dairy heifers using different inputs that reflect different management decisions and evaluate their influence on the total cost. In this simulation, 84 calves were modeled between 0–2 months of age to reflect a 1000 heifer herd. The decisions associated with housing, liquid diet source and allowance, labor utilization, and health were calculated. Costs and biological responses were reflective of published surveys, literature, and market conditions. A 10,000-iteration economic simulation was used for each management scenario using @Risk and PrecisionTree add-ons (Palisade Corporation, Ithaca, NY, USA) to account for variation in pre-weaning mortality rate, weaning age, and disease prevalence. As milk allotment increased, total feed cost increased. Feeding calves a higher allowance of milk resulted in a lower cost per kg of gain. Average feed cost percentage of the total cost was 46% (min, max: 33%, 59%) while labor, and fixed and variable cost represented 33% (20%, 45%), 9% (2%, 12%), and 12% (10%, 14%), respectively. Total pre-weaning costs ranged from $258.56 to $582.98 per calf across all management scenarios and milk allotments.

Keywords: calf economics; replacement; ADG; cost per kg

---

1. Introduction

Heifer availability is critical for the dairy operation to maintain a consistent herd size and remain economically sustainable in most cases [1]. Improved fertility and increased use of sexed semen have made replacement heifers more available for dairy operations [2]. Some producers keep all newborn replacement heifers in case more replacements are needed than anticipated, which can create a heavy financial burden for producers when raising excess heifers. Heifer raising expenses are often lumped into broad farm-wide expenses such as feed, labor, and health costs, making it difficult to accurately calculate heifer raising costs [3]. In addition, failing to identify the on-farm cost to raise a replacement heifer can allow for inefficiencies in feed, labor, housing, or health costs to go unnoticed, which accumulate unanticipated replacement female costs.

Previously reported replacement heifer rearing costs are variable and can be explained in part by differences in rearing management systems. For example, the average total cost to raise replacement heifers to wean was found to only increase by $82.88 per heifer from 2000 to 2015 but ranges within each study can exceed $350 per heifer [4–6]. Heinrichs [6] found a range in feed cost on 44 farms of $29.06 to $259.17 per calf; total cost per calf ranged $89.00–$442.78 during the pre-weaning period. In a 2014 survey of 2545 heifer calves in the United States, individual housing was the dominant form of housing pre-weaned heifers at 86.6% and 13.4% were managed in group housing, yet 8 different housing types were reported [7]. Little research has examined the cost between housing types, although the University of Wisconsin has conducted surveys of producers in an automatic and conventional housing scenario. The average cost (min, max) of producers utilizing individual housing was $363.69 ($195.06, $530.76) and those with group housing was $401.58 ($138.39, $585.52), a difference in average cost of $37.89 per calf, but with a difference range of over $300 for individual and $400 for group housing [8].

Housing is the first of many decisions a producer makes on how pre-weaned calves will be managed. Utilization of labor and milk source requires additional decisions based on resources and availability. While gaining in popularity, only 1.9% of the calves were fed through an automatic feeder while almost half of the surveyed calves were fed using a bottle or a bucket [9]. On average, one calf requires 7–12 labor hours during the pre-weaning period, or 7–10 mins per day [8]. Unpasteurized whole milk was the most common milk source utilized by producers but close to 50% of those also utilized milk replacer [7]. More recent surveys show a similar trend, with 40.1% of calves fed whole or waste milk, 34.8% fed milk replacer, and 25.1% fed a combination of the two. Calf starter was provided, starting on average at 5 days old, to all calves surveyed [9].

Thus, it is important to understand the costs associated with the myriad of rearing systems for dairy calves in the United States. The objective of this paper was to evaluate the economic impact of different calf raising management decisions, especially housing, liquid diet and allowance, and health expenses on the total pre-weaning cost of rearing heifer replacements.

2. Materials and Methods

A cost simulation model was developed at the University of Kentucky Dairy Science program during 2018. This economic model was developed in Excel 2013 (Microsoft, Redmond, WA, USA) utilizing @RISK and PrecisionTree add-ons (Palisade Corporation, Ithaca, NY, USA). The base herd used included 1500 milking cows, 1000 replacement heifers in total and 84 heifer calves in the pre-weaning period, assuming a 30% replacement rate and an average age at first calving of 25 months. Costs were calculated on a per head basis for housing, feed, labor, mortality, and health. All remaining variables are static. Interest was accounted for on infrastructure and mortality as well as the depreciation of assets related to replacement females. Costs associated with herd-wide parameters, such as disease prevalence and mortality rates, were distributed across all remaining calves in the pre-weaning phase. The model required a management decision at 3 points: housing type, milk source, and labor shown in Figure 1. Three main housing types were modeled: individual housing outside (IHO), individual housing inside (IHI) [10], and group housing (GH). Three milk sources were built into the model: whole milk (WM), pasteurized whole milk (PWM), or milk replacer (MR). Four possible liquid feeding plans were modeled: 6, 8, 10, and 12 L of milk per calf per day. Labor was modeled for two categories: conventional, where a person was assigned to feeding and caring for the calves; or automatic, where an automatic calf feeder was utilized in addition to human labor. Totals costs were reported per calf for each management decision, the entire pre-weaning period per calf, and per day per calf. Per day cost was calculated by dividing days of age at weaning by the total cost per calf during the pre-weaning period.

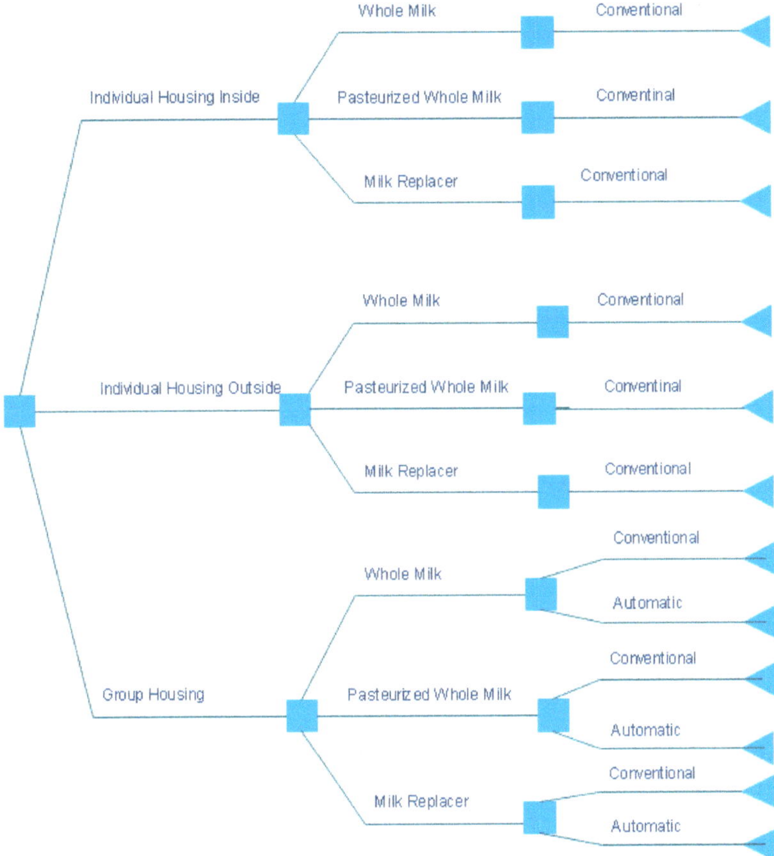

Figure 1. Decision tree of possible management decisions for housing, milk source, and labor for pre-weaned calves.

2.1. Housing

Housing systems that required a barn (IHI and GH) used values found from Table 1 to determine barn value and monthly payment. Barn cost was derived from the Dairy Calf and Heifer Association Gold Standard recommendation of 3.3 M^2 per calf, with an additional 15% of space to account for walkways and feed areas. Thus, replacement heifers were assumed to require 3.7 M^2 per calf. Construction cost [11] varied based on the infrastructure required for each situation, ranging from $10.00 to $15.50 per M^2. Estimated barn value (BV) was then calculated with Equation (1).

$$BV = CC \times 3.7 \text{ M}^2 \times \text{number of pre-weaned calves} \qquad (1)$$

Table 1. Model inputs were adapted from the published literature, the latest USDA reports, and heifer raising surveys.

Variable	Value	Source
Number of pre-weaned calves in 2 months	84	Based on rearing 500 heifers annually
Employee Labor (/h)	$14.00	Based on National Dairy Labor Survey, 2014
Management Labor (/h)	$22.00	Based on National Dairy Labor Survey, 2014
Interest Rate	7%	
Barn construction per M^2 Frame	$10.00	Adkins, 2017
Barn construction per M^2 Frame and Group Pens	$15.50	Adkins, 2017
Individual hutch	$300.00	Based on average market price
Value of newborn calf	$100.00	Based on USDA market reports
Whole milk value (cwt)	$15.00	Based on USDA, 2016
Milk replacer value (22.7 kg)	$65.00	Based on average market price
Calf Starter (mt)	$550.00	Based on average market price
Automatic calf feeder value	$15,000	Based on (Adkins, 2017)
Pasteurizer value	$10,000	Based on average market price
Diarrhea prevalence	21.4%	Urie, 2018
Respiratory illness prevalence	12.7%	Urie, 2018
Pre-weaning mortality rate	5%	NAHMS, 2011
Water cost per calf pre-weaning	$0.50	Based on water price Jan. 2019
Electrical cost per calf pre-weaning	$0.50	Based on electrical price Jan. 2019
Bedding per calf pre-weaning	$11.00	Heinrichs, 2013
Weaning Age	65 days	Adkins, 2017

Barn cost per heifer (BC) was calculated using the payment function in excel with 7% interest, 20 years useful life and BV. BC was included in IHI and GH situations. Calves housed in individual housing outside followed the same payment function. Housing calves year-round in individual housing with an average occupancy time of 2 months ± rest period would allow 5 calves per hutch per year. Days of age at weaning was used as the length of time a heifer was incurring cost during the pre-weaning period. Housing costs also included utility costs, such as water and electricity. Electricity was only factored for housing systems that included a barn (IHI and GH). The bedding was included at a flat evaluation of $11.00 per calf. For pasture scenarios, a cash value price per acre was used as the value of land to try to account for the opportunity cost of a specific acre being used for other purposes. An additional annual maintenance cost of $31.50 per acre was assumed.

2.2. Feed

Milk replacer was mixed at a concentration of 0.11 kg per liter of water. A pasteurizer was depreciated over all calves over the 15-year useful life. The model accounted for four possible feeding milk allotments: 6, 8, 10, and 12 L per calf per day. A 2016 survey of producers in the United States showed over half of the farms were feeding calves between 4–6 L per day [7]. Recent studies have shown that increasing milk allotment can increase average daily gain (ADG) pre-weaning, result in larger skeletal measurements at weaning, and decrease vocalizations caused by milk deprivation [12,13]. Milk allotments and starter intakes per calf for this model were reflective of experimental data [14]. In this study, calves were randomly assigned to 6, 8, 10, or 12 L feeding treatments of pasteurized whole milk with *ad libitum* access to calf starter. A step-down weaning program was performed: milk was fed at maximum allotment until weaning at 42 days. Milk allotment was decreased by 50% until day 50, where allotment was decreased daily by 20% until weaned. Calves were assumed to be consuming at least 2.25 kg of calf starter at weaning. An additional 20% was assumed to be fed to account for waste and loss. This additional expense was added to daily calf starter cost. Milk and calf starter costs were calculated on a daily basis for the entire pre-weaning period, from day 0 to 65. ADG was determined using the dry matter intake requirements and resulting gain from NRC, 2001. Calf weight was modeled daily to determine appropriate weaning weights based on dry matter intake from milk replacer or whole milk.

The assumed birth weight was 40 kg for each calf. Assumed average daily gains on each feeding plan (6, 8, 10, and 12 L) are described in Table 2, following the equation presented in NRC, 2001. The weaning weight was calculated by multiplying ADG by 65 days and adding the weight gain to BW. Feed cost was reported for three variables: total cost during the pre-weaning period, daily feed cost, and feed cost per kg of gain. Total feed cost included milk replacer or whole milk expenses and feeding equipment. Daily feed cost was derived from dividing total feed cost by weaning age (65 days). Daily feed cost was then used to calculate feed cost per kg of gain. Daily cost under each feeding plan was divided by ADG to determine the cost per kg of gain.

Table 2. Birth weight and weaning weight were a result of milk allotted and calf starter intake per calf following experimental data from Rosenberger et al. 2017. Average daily gain (ADG) followed the equation presented in NRC 2001.

Milk Allotment (L/day)	Starter Intake (kg)	ADG (kg)	BW (kg)	WW (kg)
6	64	0.3	40	77.4
8	63.7	0.3–0.6	40	87.6
10	63.4	0.6–0.9	40	98.4
12	60.3	0.9–1.2	40	108.3

Birth weight (BW) was assumed at 40 kg, weaning weight (WW) was calculated based on ADG for 65 day weaning age.

2.3. Labor

Labor to care for calves and the number of employees working were adapted from published surveys of producers employing individual and group housing (Table 1). Because of the lack of data on group housing without automatic feeder labor time requirements, we assumed the median of an automatic feeder and individually housed heifers (5.5 mins/calf/day). Management labor was calculated separately to represent additional labor required from owners, managers, and/or family. Management followed the trend of 10% of the paid labor, creating the assumption of 0.55 mins/calf/day for group housing without an automatic feeder. Minutes per calf could be input directly or total time per all pre-weaned calves could be used to calculate total labor cost using Equation (2).

$$\text{((Daily Paid Labor Hours/Number of Calves)} \times \text{Hourly Paid Labor)} + \text{((Daily Management Labor Hours/Number of Calves)} \times \text{Hourly Management Pay)} \tag{2}$$

The expenses related to buying and using an automatic calf feeder were included in the labor section. Justified by the change in labor demands, the use of an automatic calf feeder can be viewed as an additional autonomous employee. The cost of the feeder was assumed at $15,000 value, 10 years useful life and $200.00 annual maintenance. These values were assumed based on market prices and a routine maintenance program. Equation (3) represents the calculation of daily feeder cost per calf using the payment (PMT) function in excel.

$$(-\text{PMT (interest rate, 120, initial value))/number of pre-weaned calves} \tag{3}$$

2.4. Mortality and Health

The cost of each calf was calculated daily and, therefore, monthly cost to raise one calf in each management style was determined. All calf mortality events were assumed to occur at the end of the first month of life, accruing the additional monthly cost plus interest. This additional cost is divided over the remaining number of calves. Equation (4) explains how calf mortality was added as an additional cost to each remaining calf.

$$\text{(Value of Newborn Calf} + \text{(Cost up to death} \times \text{((Interest Rate/365)} \times 60)) \times \text{Mortality Rate)/Remaining Calves} \tag{4}$$

Health costs are reflective of a standard vaccination protocol including fly control, respiratory vaccine, vitamin A, D, and E, selenium, and a vaccine for rotavirus and coronavirus scours, and *E. Coli*. Labor costs related to health tasks were compiled into a "working heifer" labor expense. The total health cost was figured at $9.22 per calf. Fair market prices were assumed on all vaccine and health-related equipment through averaging online prices obtained in January 2019.

The prevalence of respiratory illness and diarrhea was determined by the 2014 Heifer Raiser Survey conducted by the USDA, 18% for respiratory illness and 25% for diarrhea on average. Because of the variation in this measure from farm to farm it was made stochastic to account for variation between farms. The minimum incidence was 16% for respiratory illness with a maximum of 19%; the minimum of diarrhea was 22% with a maximum of 28%. Based on the selected prevalence, there was a direct relationship to the additional treatment cost for each calf. We modeled a protocol that would include electrolytes and 3 days of antibiotics. We assumed an 85.6% improvement rate and culled the remaining heifers at the end of that week.

2.5. Statistical Simulation

A simulation model was developed in Excel 2013 (Microsoft, Redmond, WA, USA) utilizing @RISK and PrecisionTree add-ons (Palisade Corporation, Ithaca, NY, USA) to evaluate the cost of raising an individual heifer from birth to weaning under different management styles and systems. 10,000 simulations of the model were performed for each of the situations. Stochastic simulations allowed for variation of inputs values which are reflected in ranges of potential outcomes, unlike a static model which will always produce the same outcome. Modeling variables stochastically, such as weaning age, mortality rates, and disease prevalence, we can simulate different outcomes. All variables were modeled following a Pert distribution set with minimum, most likely, and maximum value. Assumptions were made based on published literature, surveys, and market assumptions were also used to calculate the total cost (Table 1). A month in the cost spreadsheet was considered 30 days. This model is available online at https://afs.ca.uky.edu/dairy/decision-tools, wherein all variables and assumptions can be modified to reflect different situations and individual farms.

3. Results and Discussion

3.1. Housing

The total cost to house calves in individual housing outside, individual housing inside and group housing were $21.12, $70.52, $94.30, respectively. All of these costs were within 1 SD of the average found in published literature. For housing that included a barn, the barn payment per heifer was the largest contributor to cost, while bedding was the largest contributing cost per calf for individual housing outside (Table 3).

Table 3. Percentage breakdown of hutch/barn infrastructure, bedding and, water and electric on total housing cost per housing management decision.

Housing System	Individual Housing Outside	Individual Housing Inside	Group Housing
Hutch or Barn *	32%	83%	87%
Bedding	52%	16%	12%
Water & Electric	2%	1%	1%

* includes interest and depreciation of infrastructure.

3.2. Feed

Feed cost was heavily dependent upon the amount of milk allotted per day. Table 4 shows the total cost of each milk source with 6, 8, 10, and 12 L allotments. As milk allotment per calf increased, the cost of milk increased.

Table 4. Cost of milk replacer, whole milk, and pasteurized whole milk as a milk source for calves with 6, 8, 10, and 12 L milk allowances.

Milk Source	Milk Allotment (L)			
	6	8	10	12
Milk Replacer	$81.52	$107.02	$132.53	$158.04
Whole Milk	$81.41	$108.38	$135.36	$162.33
Pasteurized Whole Milk	$99.79	$126.76	$153.73	$180.71

The cost of pasteurizing whole milk ranged from 10 to 18% of the total cost of feeding calves in applicable scenarios. This model assumed the same nutritional value and gain from milk replacer and whole milk, creating a limitation in the model. However, calves fed pasteurized or unpasteurized whole milk have been shown to increase model-produced ADG by at least 0.03 kg/day with the potential to be over 0.25 kg/day of gain in comparison to milk replacer [15]. The additional cost to feed whole milk has the potential to be offset by an increase in weight gain.

The estimated cost per kg of gain decreased as milk allowance increased, and with increasing ADG, shown in Table 5. For example, group-housed calves on milk replacer with an automatic feeder fed 6 L will cost $3.50 per kg of gain. When these same calves are increased to 12 L the cost decreases to $2.67 per kg of gain. The minimum decrease in cost was from feeding 10 L of milk replacer to 12 L of milk replacer at $0.01 difference per kg of gain, and the maximum savings per kg of gain was $0.41 increasing from 10 to 12 L of pasteurized whole milk. If birth weights were 44 kg with a goal of weaning calves at 100 kg, we could assume a minimum of $0.56 to $22.96 in feed efficiency savings per calf alone. Modeling cost per kg of gain following experimental data presented in the (NRC, 2001) equations indicates that feeding calves a higher allowance of milk decreases the cost per kg of gain. The cost of milk and calf starter, with our current assumptions in inputs and ADG, decrease cost per kg of gain.

Table 5. Feed cost per kg of gain of pre-weaned calves fed milk replacer, pasteurized whole milk and whole milk.

Milk Source	Milk Allotment (L)			
	6	8	10	12
ADG (kg)	0.3	0.3–0.6	0.6–0.9	0.9–1.2
Milk Replacer	$3.50	$2.75	$2.68	$2.67
Pasteurized Whole Milk	$3.60	$3.45	$3.31	$2.90
Whole Milk	$2.98	$2.96	$2.92	$2.60

3.3. Labor

The labor decisions depended on the housing system selected. Hourly wages for management are higher than those for paid employees as shown in Table 1. Employees contributed more to the total cost than management in conventional and automatic systems even though their hourly rate is lower. Labor costs associated with the automatic calf feeder were responsible for 23% of the total labor cost. Labor cost of individual housing and group housing contributed 33% and 26%, respectively. The minutes and total cost per hourly laborer were decreased from inside individual housing to group housing by 36% per calf for a value of 2.4 minutes or $0.50 per calf per day. This shows a reduction in overall labor cost but an increased demand for fixed and variable expenses. These include the paying for the feeder, annual maintenance and a barn to house calves.

This breakdown of cost follows the same trends of Wisconsin surveys of conventional and automatic calf raisers. The paid labor cost alone was reduced by 39% for farms utilizing an automatic calf feeder, and paid management decreased by 14%. The total pre-weaning cost decreased 6% from conventional to automatic labor; the cost difference was recovered in an additional fixed variable cost of the automatic calf feeders.

3.4. Health

Mortality rate and prevalence of diarrhea or respiratory illness, which were included in variable costs, impacted the total cost. The average cost, including the risk of each calf being healthy or experiencing diarrhea, totaled (mean ± SD) $5.39 ± 14.42 per calf. The average cost per BRD case was $0.70 ± 7.33 per calf. Preventative health costs added an additional $9.22 to each calf.

The change in total cost per calf, accounting for additional expenses with fewer calves as mortality rate increases (2%, 8%, 10%, and 15%), are reported in Table 6. As mortality rate increased, the cost of infrastructure and higher cost management systems showed a larger increase in the dollar amount added for each calf. Across management styles, decreasing the mortality rate from 15% to 2% reduced overall cost from $39.47 to $36.84 per calf. For a farm raising 500 pre-weaned calves annually, potential savings by decreasing mortality 10% alone could be over $18,000.

Table 6. Total cost under each management pathway per calf when mortality rate is set at 2, 8, 10, and 15%.

Management Pathway	Mortality Rate			
	2%	8%	10%	15%
Individual Housing Outside				
Milk Replacer-Conventional	$283.03	$298.74	$304.44	$319.87
Pasteurized Whole Milk-Conventional	$291.27	$307.26	$313.06	$328.75
Whole Milk-Conventional	$287.98	$303.85	$309.61	$325.20
Individual Housing Inside				
Milk Replacer-Conventional	$303.03	$319.40	$325.34	$341.42
Pasteurized Whole Milk-Conventional	$311.28	$327.92	$333.96	$350.30
Whole Milk-Conventional	$307.98	$324.51	$330.51	$346.75
Group Housing				
Milk Replacer-Conventional	$312.06	$328.73	$334.78	$351.15
Pasteurized Whole Milk- Conventional	$320.31	$337.24	$343.39	$360.03
Whole Milk-Conventional	$317.01	$333.84	$339.95	$356.48
Milk Replacer-Automatic	$293.10	$309.15	$314.97	$330.72
Pasteurized Whole Milk-Automatic	$301.35	$317.66	$323.58	$339.60
Whole Milk-Automatic	$298.05	$314.25	$320.14	$336.05

It has been found that management practices specific to a housing type can change illness prevalence. For example, calves housed in groups of 12–18 had a higher incidence of respiratory illness and lower daily gains than calves housed in groups of 6–9 [9]. We assume a constant square footage per calf, therefore the barn square footage increases as the number of calves increase and this may not always be reflective of true management practices. A limitation to the model is the same probabilities in averages and ranges in mortality for all management pathways for calculated cost.

3.5. Total Cost of Management Scenarios

All possible combinations of management decisions (each combination of housing type, milk source, and labor type) and for each of the 4 milk allotments were analyzed for total cost, daily cost, and percentage of feed, labor, and fixed and variable costs (Table 7). Fixed costs included barn and housing infrastructure, depreciation of assets, and interest. Variable costs included health-related expenses, mortality, and utilities for electricity and water. Feed represented the largest factor in all management scenarios, followed by labor, then variable and fixed costs. This follows the same results found in previously published models where 57% of total cost were due to feed costs [6].

Table 7. Total cost mean, SD, min and max of each management pathway under 6, 8, 10, and 12 L milk allowances.

Management Scenario	Milk Allotment (L)															
	6				8				10				12			
	Mean	SD	Min	Max	Mean	SD	Min	Max	Mean	SD	Min	Max	Mean	SD	Min	Max
Individual Hutches Outside																
Milk Replacer-Conventional	$276.03	$16.77	$259.92	$407.56	$301.71	$16.82	$285.07	$433.73	$327.39	$16.87	$310.22	$459.90	$353.07	$16.93	$335.37	$486.08
Pasteurized Whole Milk-Conventional	$295.55	$16.81	$279.04	$427.45	$323.38	$16.86	$306.30	$455.82	$351.22	$16.92	$333.56	$484.19	$379.06	$16.99	$360.82	$512.57
Whole Milk-Conventional	$274.63	$16.77	$258.56	$406.13	$302.47	$16.82	$285.82	$434.51	$330.30	$16.88	$313.08	$462.88	$358.14	$16.94	$340.34	$491.25
Individual Housing Inside																
Milk Replacer-Conventional	$301.11	$16.82	$284.49	$433.12	$326.79	$16.87	$309.63	$459.29	$352.47	$16.93	$334.78	$485.47	$378.15	$16.98	$359.93	$511.64
Pasteurized Whole Milk-Conventional	$320.63	$16.86	$303.60	$453.01	$348.46	$16.92	$330.86	$481.39	$376.30	$16.98	$358.12	$509.76	$404.13	$17.04	$385.39	$538.13
Whole Milk-Conventional	$299.71	$16.81	$283.12	$431.70	$327.55	$16.87	$310.38	$460.07	$355.38	$16.93	$337.64	$488.44	$383.22	$16.99	$364.90	$516.81
Group Housing																
Milk Replacer-Conventional	$345.11	$16.91	$327.58	$477.97	$370.79	$16.97	$352.73	$504.14	$396.47	$17.03	$377.88	$530.32	$422.15	$17.09	$403.03	$556.49
Pasteurized Whole Milk-Conventional	$364.63	$19.95	$346.70	$497.87	$392.47	$17.02	$373.96	$526.24	$420.30	$17.08	$401.22	$554.61	$448.14	$17.15	$428.48	$582.98
Whole Milk-Conventional	$343.72	$19.91	$326.21	$476.55	$371.55	$16.97	$353.48	$504.92	$399.39	$17.03	$380.74	$533.29	$427.22	$17.10	$408.00	$561.66
Milk Replacer-Automatic	$339.97	$16.90	$322.54	$472.73	$365.65	$16.95	$347.69	$498.90	$391.33	$17.01	$372.84	$525.07	$417.01	$17.08	$397.99	$551.25
Pasteurized Whole Milk-Automatic	$359.49	$16.94	$341.66	$492.62	$387.32	$17.00	$368.92	$520.99	$415.16	$17.07	$396.18	$549.36	$442.99	$17.14	$423.44	$577.74
Whole Milk-Automatic	$338.57	$16.89	$321.18	$471.30	$366.41	$16.96	$348.44	$499.68	$394.24	$17.02	$375.70	$528.05	$422.08	$17.09	$402.96	$556.42

Using the assumptions in Table 1, on average, the most expensive management style was the one utilizing group housing, feeding pasteurized whole milk with conventional labor. The least expensive management pathway was the one utilizing individual housing outside, feeding whole milk with conventional labor. The main difference in cost can be attributed to the larger infrastructure needs for group housing and the additional cost of a pasteurizer. Total and daily cost for all management scenarios with 6, 8, 10, and 12 L allotments are shown in Table 3.

The mean for total cost ranged between $258.56 to $582.98 per calf across all management pathways. As seen in previous literature, the mean cost in each milk allotment has less variation than when looking at the range of projected costs per management scenario. This can be attributed to variation in health and mortality rates. Increasing the mortality rate and disease prevalence increased the cost for the remaining calves by spreading infrastructure costs, the loss of the calf and incurred expenses, and additional illness treatments over fewer calves. Variation in costs is not always related to efficiency on-farm but instead related to trade-offs in management styles.

The least expensive pathways were the 3 combinations for individual housing outside. In these scenarios, housing cost contributed 7–8% of the total cost compared to other management pathways utilizing more infrastructure, where housing accounted for 21–30% of the total cost. The addition of barns with individual housing inside and in group housing was the contribution of the additional 14–23% of housing cost.

When costs were broken down by day, assuming a 65-day weaning age, the cost ranged from $3.83 to $6.19 per calf per day. The average daily charge for a contract raiser from birth to weaning was $1.88/day [16]. Based on our calculated total cost for rearing pre-weaning calves, this would create a significant loss for the contract raiser. But in the Wisconsin heifer raising survey the cost per day of fixed and variable costs, which most closely matches our model, $2.05–$8.73 for minimum and maximum daily cost [8]. This simulation model can be compared to surveys to validate the results are reflective of on-farm total values. Finally, this model can be used to estimate other housing situations other than the ones presented in the current survey, and herd numbers could be used to estimate heifer costs for individual herds. In this survey, we chose the most common management practices to raise dairy heifers in North America, but there are many other options of housing and feeding dairy heifer calves that require further investigation.

4. Conclusions

Raising calves from birth to weaning contributes to a major portion of the total heifer raising cost. Milk and calf starter contributed over half the cost to raise a calf from birth to weaning. Costs calculated by this model are based on currently available data; it is likely some of our assumptions will under or overestimate total and specific costs of calf raising practices across the US. More data are needed to improve accurate assumptions for farms. However, no model will be able to accurately describe all situations of calf rearing in various locations. Calculating pre-weaning cost for each individual farm is critical in making management decisions and remaining sustainable.

Author Contributions: Conceptualization, A.H. and J.H.C.C.; methodology, A.H. and J.H.C.C.; formal analysis, A.H., K.B., and J.H.C.C.; investigation, A.H.; resources, J.H.C.C.; writing—original draft preparation, A.H. and J.H.C.C.; writing—review and editing, A.H., K.B., D.A.-P., and J.H.C.C.; visualization, A.H. and K.B.; supervision, K.B. and J.H.C.C.; project administration, J.H.C.C.

Funding: The authors would like to thank Dannon for their continued support of this project.

Acknowledgments: The authors would like to acknowledge the dairy producers in Kentucky, Indiana and Ohio, USA that provided feedback and insight into our model design. We are thankful for the time and effort they spent with us consulting on this project. The authors would like to thank Dr. Jeffrey Bewley and Dr. Michael Overton for the input and assistance during the development of this economic model. This open model can be found on https://afs.ca.uky.edu/dairy/decision-tools.

Conflicts of Interest: The authors declare no conflict of interest.

References

1. Zanton, G.I.; Heinrichs, A.J. Meta-Analysis to Assess Effect of Prepubertal Average Daily Gain of Holstein Heifers on First-Lactation Production. *J. Dairy Sci.* **2005**, *88*, 3860–3867. [CrossRef]
2. De Vries, A. Economic trade-offs between genetic improvement and longevity in dairy cattle. *J. Dairy Sci.* **2017**, *100*, 4184–4192. [CrossRef] [PubMed]
3. Mohd Nor, N.; Steeneveld, W.; Mourits, M.C.M.; Hogeveen, H. The optimal number of heifer calves to be reared as dairy replacements. *J. Dairy Sci.* **2015**, *98*, 861–871. [CrossRef] [PubMed]
4. Boulton, A.; Rushton, J.; Wathes, D.C. A Study of Dairy Heifer Rearing Practices from Birth to Weaning and Their Associated Costs on UK Dairy Farms. *Open J. Anim. Sci.* **2015**, *5*, 185–197. [CrossRef]
5. Gabler, M.T.; Tozer, P.R.; Heinrichs, A.J. Development of a Cost Analysis Spreadsheet for Calculating the Costs to Raise a Replacement Dairy Heifer. *J. Dairy Sci.* **2000**, *83*, 1104–1109. [CrossRef]
6. Heinrichs, A.J.; Jones, C.M.; Gray, S.M.; Heinrichs, P.A.; Cornelisse, S.A.; Goodling, R.C. Identifying efficient dairy heifer producers using production costs and data envelopment analysis. *J. Dairy Sci.* **2013**, *96*, 7355–7362. [CrossRef] [PubMed]
7. USDA. *Dairy 2014, Dairy Cattle Management Practices in the United States*; USDA: Washington, DC, USA, 2016.
8. Akins, M.S.; Cavitt, M.; Hagedorn, M.A.; Mills-Lloyd, S.; Kohlman, T.L.; Sterry, R. *Economic Costs and Labor Efficiencies Associated with Raising Dairy Calves for Operations Using Individual or Automated Feeding*; University of Wisconsin Extension: Madison, WI, USA, 2017; Volume 8.
9. Urie, N.J.; Lombard, J.E.; Shivley, C.B.; Kopral, C.A.; Adams, A.E.; Earleywine, T.J.; Olson, J.D.; Garry, F.B. Preweaned heifer management on US dairy operations: Part, I. Descriptive characteristics of preweaned heifer raising practices. *J. Dairy Sci.* **2018**, *101*, 9168–9184. [CrossRef] [PubMed]
10. Peña, G.; Risco, C.; Kunihiro, E.; Thatcher, M.J.; Pinedo, P.J. Effect of housing type on health and performance of preweaned dairy calves during summer in Florida. *J. Dairy Sci.* **2016**, *99*, 1655–1662. [CrossRef] [PubMed]
11. McCullock, K.; Hoag, D.L.K.; Parsons, J.; Lacy, M.; Seidel, G.E.; Wailes, W. Factors affecting economics of using sexed semen in dairy cattle. *J. Dairy Sci.* **2013**, *96*, 6366–6377. [CrossRef] [PubMed]
12. Kiezebrink, D.J.; Edwards, A.M.; Wright, T.C.; Cant, J.P.; Osborne, V.R. Effect of enhanced whole-milk feeding in calves on subsequent first-lactation performance. *J. Dairy Sci.* **2015**, *98*, 349–356. [CrossRef] [PubMed]
13. Thomas, T.J.; Weary, D.M.; Appleby, M.C. Newborn and 5-week-old calves vocalize in response to milk deprivation. *Appl. Anim. Behav. Sci.* **2001**, *74*, 165–173. [CrossRef]
14. Rosenberger, K.; Costa, J.H.C.; Neave, H.W.; von Keyserlingk, M.A.G.; Weary, D.M. The effect of milk allowance on behavior and weight gains in dairy calves. *J. Dairy Sci.* **2017**, *100*, 504–512. [CrossRef] [PubMed]
15. Shivley, C.B.; Lombard, J.E.; Urie, N.J.; Kopral, C.A.; Santin, M.; Earleywine, T.J.; Olson, J.D.; Garry, F.B. Preweaned heifer management on US dairy operations: Part VI. Factors associated with average daily gain in preweaned dairy heifer calves. *J. Dairy Sci.* **2018**, *101*, 9245–9258. [CrossRef] [PubMed]
16. Wolf, C.A. Custom Dairy Heifer Grower Industry Characteristics and Contract Terms. *J. Dairy Sci.* **2003**, *86*, 3016–3022. [CrossRef]

Article

Mortality-Culling Rates of Dairy Calves and Replacement Heifers and Its Risk Factors in Holstein Cattle

Hailiang Zhang [1], Yachun Wang [1,*], Yao Chang [1], Hanpeng Luo [1], Luiz F. Brito [2], Yixin Dong [1], Rui Shi [1], Yajing Wang [3], Ganghui Dong [4] and Lin Liu [5]

[1] Key Laboratory of Animal Genetics, Breeding and Reproduction, MARA, National Engineering Laboratory of Animal Breeding, College of Animal Science and Technology, China Agricultural University, Beijing 100193, China; zhl108@cau.edu.cn (H.Z.); changyao@cau.edu.cn (Y.C.); luohanpeng@cau.edu.cn (H.L.); dyx18634593316@163.com (Y.D.); srandeffy@163.com (R.S.)

[2] Department of Animal Sciences, College of Agriculture, Purdue University, West Lafayette, IN 47907, USA; britol@purdue.edu

[3] State Key Laboratory of Animal Nutrition, College of Animal Science and Technology, China Agricultural University, Beijing 100193, China; yajingwang@cau.edu.cn

[4] Beijing Sunlon Livestock Development Co. Ltd, Beijing 100176, China; ganghui89@126.com

[5] Beijing Dairy Cattle Center, Beijing 100192, China; liulin@bdcc.com.cn

* Correspondence: wangyachun@cau.edu.cn

Received: 23 July 2019; Accepted: 25 September 2019; Published: 26 September 2019

Simple Summary: High mortality and involuntary culling rates cause great economic losses to the dairy industry around the world, and the survival of dairy calves and replacement heifers is paramount in modern dairy breeding. However, little has been done to genetically improve mortality rates of dairy calves and replacement heifers in Chinese Holstein cattle. In this study, we investigated population parameters (descriptive statistics) of mortality rates of dairy calves and replacement heifers and risk factors affecting mortality and involuntary culling rates in Chinese Holstein cattle. The mortality rate of dairy calves and replacement heifers from day 3 to 60, 61 to 365, and 366 to first calving was 5.5%, 7.4%, and 8.7%, and an unfavorable increasing trend has been observed in the Chinese Holstein population. Health events associated with digestive and respiratory or circulatory systems were the main reasons for deaths. Herd-birth year, birth season, and dam parity had significant effects on survival. Our findings will help farmers to better manage dairy calves and replacement heifers and highlight the need to include these survival traits as part of the national genetic evaluation schemes.

Abstract: The rates of mortality and involuntary culling of dairy calves and replacement heifers have great economic implications on the dairy cattle industry around the world. The main objectives of this study were: (1) to obtain population parameters of mortality and involuntary culling rates of dairy calves and replacement heifers; and, (2) to investigate the factors affecting mortality and involuntary culling rates in Chinese Holstein cattle. Two datasets containing records of birth, calving, and culling events from 142,833 Holstein cattle born between 1991 and 2018 were used in this study. The population parameters were obtained using dataset 1, which consisted of dairy calves and replacement heifers that died or were involuntarily culled. Three survival traits were defined in dataset 2, which consisted of females born from 1999 to 2018. A binomial logistic regression was used to analyze the risk factors on the survival traits. The mortality rate of dairy calves and replacement heifers from day 3 to 60, 61 to 365, and 366 to first calving was 5.5%, 7.4%, and 8.7%, and an unfavorable increasing trend was observed. Health events associated with digestive and respiratory or circulatory systems were the main death reasons. Herd-birth year, birth season, and dam parity had significant effects on survival traits. The results from this study will help farmers to better manage calves and replacement heifers and highlight the need to include survival traits in dairy calves and replacement heifers as part of national genetic evaluation schemes.

Keywords: dairy calf; involuntary culling; mortality; replacement heifer; survival rate

---

1. Introduction

The large majority of dairy cattle herds are divided into two groups: milking cows and replacement heifers, in which the latter does not generate any direct income to the producers until the first calving [1,2]. Estimates of expenses associated with rearing replacement heifers range from 15% to 20% of the total milk production costs [3]. However, many potential replacement heifers do not reach their first lactation due to premature death or involuntary culling [2]. Therefore, in addition to welfare issues, high mortality and culling rates cause great economic losses to the dairy industry around the world [4,5]. Various breeding programs include indicator traits of health and longevity measured in dairy cows [6]. However, less importance has been given to survivability of dairy calves and replacement heifers.

The mortality rates of calves and replacement heifers vary across countries, production systems, and populations. For instance, in the United States, annual calf and heifer mortality rates were estimated to be around 9.6%, with pre-weaning calves accounting for 7.8% [7]. In Danish Holstein, the frequency of pre-pubertal mortality was estimated to be 5% to 6% [8], and the most frequent diseases affecting calves were diarrhea and respiratory diseases. Other studies showed that the main causes of replacement heifer mortality or involuntary culling were different depending on the life stage of the animal. Scours, diarrhea, and other digestive problems were the most important death causes of pre-weaning calves, while respiratory diseases were the largest death cause of weaned calves [9]. Furthermore, many herd- or animal-level risk factors affecting dairy calf and heifer mortality have been identified in various populations. These factors include dystocia, sex, twinning rate, dam parity, herd size, and birth season [10–13].

The impacts of mortality rates of calf and replacement heifer on dairy cattle herds should not be neglected. However, there is a lack of literature reporting on this issue in calves and replacement heifers in Chinese Holstein population, especially studies based on individual records. In this context, the main objectives of this study were: (1) to estimate population parameters (descriptive statistics) of mortality and involuntary culling rates of dairy calves and replacement heifers; and, (2) to investigate the risk factors affecting mortality and involuntary culling of dairy calves and heifers in Chinese Holstein cattle, using individual records. The findings of this study will help farmers to design better management strategies for the dairy calves and replacement heifers, and provide a reference for further investigation on the genetic background of survival traits in dairy calves and replacement heifers.

2. Materials and Methods

The records of birth, calving and culling/death events from 1999 to 2018 in female Holstein from 31 herds located in Beijing, Tianjin, Yunnan, Hebei, Henan, Heilongjiang, Jilin, and Inner Mongolia were extracted from the farm management software (AfiFarm, http://www.afimilk.com.cn). The free-stall barn system was used in all herds, and the herds' sizes ranged from 1000 to 10,000 animals. The test-day milk yield in these herds ranged from 30 kg to 40 kg. The herd records before using management software (before 2005) were added into software from herdbook records, and thus, early records might be incomplete or less accurate. Two datasets were defined using event records: Dataset 1 and dataset 2. Dataset 1 included records of dairy calves and replacement heifers that left herds (before the first calving) between 2006 and 2018, which was used to obtain population parameters of involuntary culling/death age on dairy calves and replacement heifers in Chinese Holstein cattle, including average involuntary culling/death age and culling/death reason. The dataset 2 included records from all animals (that left herds either before or after first calving) born from 1999 to 2018, which was only used to investigate risk factors affecting survivability of dairy calves and replacement heifers using logistic regression. In Chinese Holstein herds, most calves left the herd due to premature death, while heifers can also be culled for reproduction disorders, severe disease and other reasons. In this study, we are interested in both mortality and involuntary culling. The records of calves that died within 2 days after birth, replacement heifers that died after 1800 days of age (60 months) and censored records (alive dairy calves and replacement heifers, and sold and transferred individuals) were removed from both dataset 1 and dataset 2. The death records before first 48 h were considered as stillbirth, which is usually a separate dam trait and thus is not part of the current study. In dataset 1, the involuntary culling/death reasons were grouped in a total of 10 categories: digestive system diseases, diseases of respiratory or circulatory systems, reproduction disorders, death without clear reasons, infectious diseases, developmental disorders (e.g., abnormalities and dysplasia), feet and leg diseases, accidental injury, other diseases (e.g., septicemia and meningitis) and unknown reason. Only females were kept in the datasets. After editing, records for 18,077 culled dairy calves and replacement heifers remained in dataset 1 and 113,218 records of all animals in dataset 2. Death/culling age (days) was calculated for each animal in dataset 1 and referred to the interval from birth to death/culling on both dairy calves and replacement heifers. A total of 3 survival traits were defined for females in dataset 2, including survival from 3 to 60 days (Sur1), 61 to 365 days (Sur2), and from day 366 to first calving (Sur3). Survival traits were analyzed as binary traits, in which a value of "0" was assigned to animals that left the herds and "1" to those that survived up to next life stage.

A binomial logistic regression was used to evaluate the risk factors affecting survivability of dairy calves and replacement heifers using the LOGISTIC procedure of SAS software (version 9.1; SAS Institute, 2004 [14]). A total of 4 risk factors associated with survival traits were analyzed using the dataset 2. These factors were herd-birth year (402 levels), birth season (divided into Spring: March to May, Summer: June to August, Fall: September to November, and Winter: December to February), dam parity (defined as 0 = unknown, 1 = first parity, 2 = second parity, 3 = third and greater parities), calving ease score (defined as 0 = unknown, 1 = unassisted, 2 = easy pull, and 3 = hard pull or surgery). The factor of herd-birth year represented the combined effect of herd and birth year of calf, and 402 levels were combined into 5 levels using logit (p) of each level. The statistical model for Sur1, Sur2, and Sur3 can be described as follow:

$$\text{logit}(p) = \ln\left(\frac{p}{1-p}\right) = \beta_0 + \beta_1 x_1 + \beta_2 x_2 + \beta_3 x_3 + \beta_4 x_4,$$

where p is the culling probability of dairy calves and replacement heifers in each life stage; β_0 is the overall mean (intercept); β_1 to β_4 are the regression coefficients of ranked factors; x_1, x_2, x_3, and x_4 correspond to the herd-birth year, birth season, dam parity and calving ease score, respectively, associated with each observation.

3. Results

3.1. Descriptive Statistics

3.1.1. Mortality-Culling Frequency of Dairy Calves and Replacement Heifers

The combined mortality-culling rate of female dairy calves and replacement heifers was 21.2%, in which the mortality-culling rate from day 3 to 60, 61 to 365, and 366 to first calving was 5.5%, 7.4%, and 8.7%, respectively. The variability in mortality-culling within each life stage over time is shown in Figure 1. From 2006 to 2008, the mortality-culling rate of dairy calves and replacement heifers increased from 15.2% to 25.9% (an increase rate of 70.4%).

3.1.2. Death/Culling Age of Dairy Calves and Replacement Heifers

Descriptive statistics of death/culling age for 18,077 dairy calves and replacement heifers born from 2006 to 2018 were calculated. The average death/culling age was 399 days; the median 296 days and the lower and upper quartile were equal to 84 and 658 days, respectively. The death/culling age did not follow a normal distribution as the large majority of calves died early in life (Figure 2). The highest mortality-culling risk on dairy calves was within the first 100 days after birth. Mortality-culling tended to decrease with the increase of animal age. The mean and median of death/culling age over time are presented in Figure 3. Over the past 13 years, there was a large difference on death/culling age of dairy calves and replacement heifers and the average death/culling age fluctuated around 400 days. Furthermore, the variation range of median death/culling age was larger than the means among different years, and the maximum difference of median death/culling age was 337 days (between 2006 and 2012).

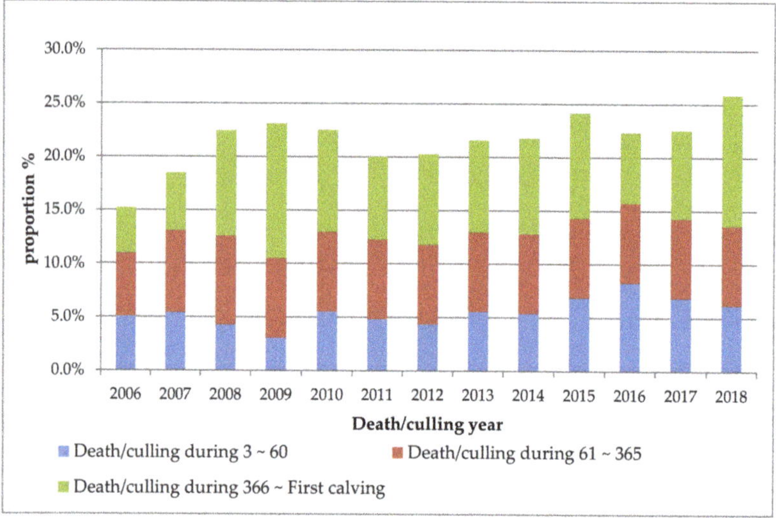

Figure 1. Mortality-culling rates of dairy calves and replacement heifers in different life stages over time.

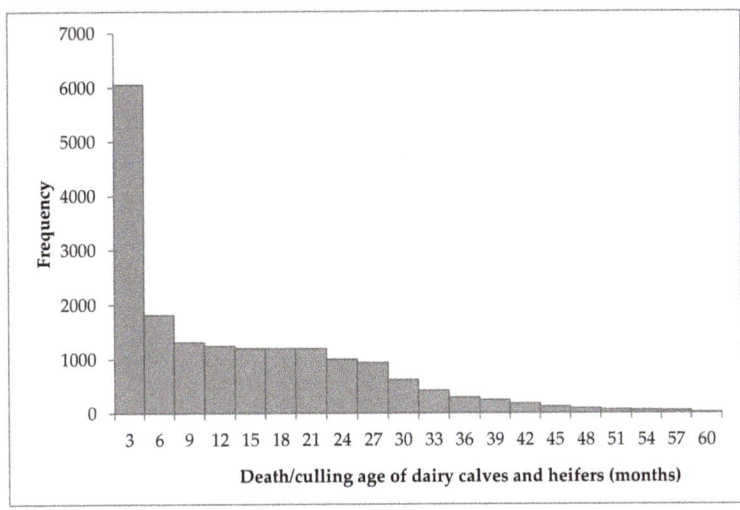

Figure 2. Histogram of death/culling age (months) on dairy calves and replacement heifers.

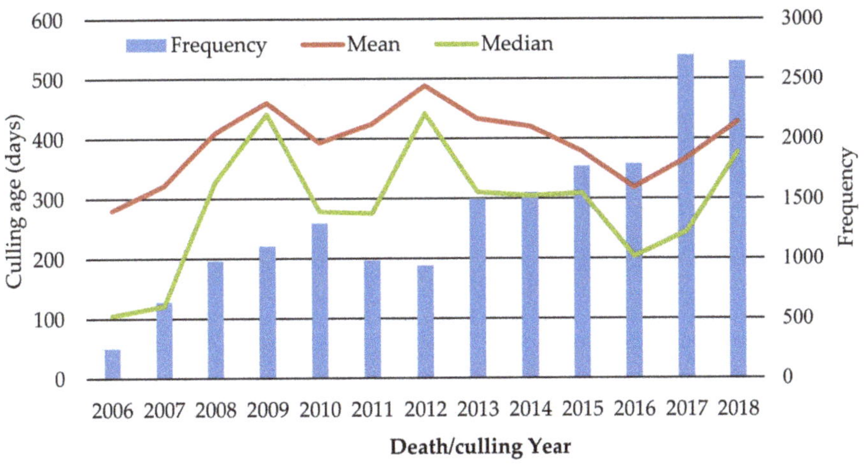

Figure 3. Average death/culling age (days) and number of dairy calves and replacement heifers in different death years.

3.2. Death/Culling Reasons of Dairy Calves and Replacement Heifers

The overall death proportion of each reason category in period from day 3 to first calving in different years is presented in Figure 4. "Unknown reason" was not included in Figure 4, which was reached 8.99%–43.78% over these years. Diseases related to the digestive, respiratory and circulatory systems, and reproductive disorders were the main causes of death. These categories accounted for 58.6% of the known causes of death. Diarrhea, pneumonia, and infertility (based on non-return rate) were the main specific reasons of mortality. Over the past 13 years, the death proportion due to diseases related to the respiratory and circulatory systems and reproductive disorders gradually increased, in contrast with both infectious diseases and other diseases that showed a decrease trend. The average involuntary culling or death ages of dairy calves and replacement heifers based on different death reason categories are presented in Table 1. The individuals with digestive system diseases, diseases of respiratory or circulatory systems, and death without clear reason were culled in early life (mean: up to 226.6 days; median: up to 101.0 days). As expected, the average death age of individuals with reproductive disorders (mean: 937.4; median: 891.0) was greater compared to the other categories (mean range: 165.1–476.7 days; median range: 84.0–450.5 days).

Table 1. The death age (days) of dairy calves and replacement heifers caused by different reason categories.

Death Reasons	Proportion (%)	Death Age (Days)				
		Mean	Median	Lower Quartile	Upper Quartile	SD
Unknown reason	26.0	449.8	430.0	259.0	638.0	262.3
Digestive system diseases	18.9	226.6	84.0	20.0	227.0	350.3
Diseases of respiratory or circulatory systems	13.6	165.1	101.0	51.3	173.0	200.9
Reproductive disorders	10.3	937.4	891.0	776.0	1085.0	259.1
Death without clear reason	6.9	200.6	90.0	3.0	234.0	264.4
Infectious diseases	4.8	444.1	437.0	237.0	635.0	250.4
Developmental disorders	4.2	396.9	413.0	67.3	601.5	334.3
Feet and leg diseases	2.4	470.8	413.5	98.0	743.5	400.7
Accidental injury	2.1	476.7	450.5	266.0	668.5	289.9
Other diseases	10.8	417.1	310.0	82.0	678.0	390.7

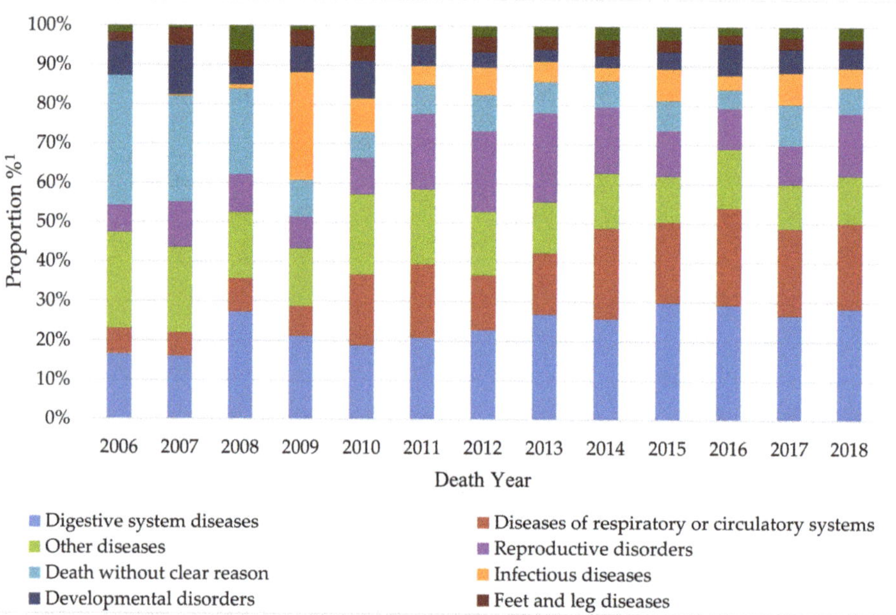

Figure 4. The death proportions of dairy calves and replacement heifers caused by different reason categories over time.[1] The category of "unknown reason" was not included in Figure 4.

3.3. Analyses of the Factors Influencing Survivability of Dairy Calves and Replacement Heifers

According to the Wald test (Chi-square), both herd-birth year and birth season significantly ($p < 0.01$) influenced the mortality of dairy calves and replacement heifers during the stages of 3–60 days (Sur1), 61–365 days (Sur2), and from 366 days to first calving (Sur3). The dam parity significantly influenced Sur1 and Sur2 ($p < 0.01$), and did not significantly influence Sur3 ($p = 0.19$). However, dam calving ease score did not significantly impact any of the 3 survival traits. The results of the binomial logistic regression on Sur1, Sur2, and Sur3 in dairy calves and replacement heifers are presented in Table 2.

The dairy calves and replacement heifers born in Spring had the lowest mortality risk in any of the 3 life stages. Across the 3 life stages (Sur1, Sur2, and Sur3), the mortality risk of dairy calves and replacement heifers born in Fall was between 1.13 and 1.53 times greater than those animals born in the Spring season. The calf birth season had larger impact on survivability of animals during 3–365 days (Sur1 and Sur2) compared to 366 days to first calving. In terms of dam parity, calves born from second parity cows had the lowest culling risk in any of the 3 life stages. From 3 to 60 days, calves born from first parity cows had the highest culling risk, i.e., 1.16 times greater than those animals born from second parity cows. However, the dairy calves and replacement heifers born from cows with 3 or more parities had the highest mortality risk during 61–365 and 366–first calving. Animals born by hard pull or surgery had the highest mortality risk, which was not significant compared with unassisted calves.

Table 2. Associations between different levels of birth season, dam parity and calving ease score on the odds ratio of mortality [1].

Variable	Level	Odds Ratio	95% Confidence Interval		*p*-Value
		Survival from 3 to 60 days			
	Summer vs Spring	1.48	1.34	1.64	<0.01
Birth season	Autumn vs Spring	1.56	1.41	1.73	<0.01
	Winter vs Spring	1.60	1.44	1.77	<0.01
	Unknown vs 2	0.45	0.28	0.71	<0.01
Dam parity	1 vs 2	1.16	1.02	1.33	0.03
	"3 and above" vs 2	1.14	0.98	1.32	0.09
	Unknown vs 1	1.46	0.93	2.30	0.10
Calving ease score	2 vs 1	1.39	0.97	1.99	0.07
	3 vs 1	1.83	0.77	4.34	0.17
		Survival from 61 to 365 days			
	Summer vs Spring	1.53	1.41	1.65	<0.01
Birth season	Autumn vs Spring	1.64	1.52	1.77	<0.01
	Winter vs Spring	1.18	1.08	1.28	<0.01
	Unknown vs 2	0.54	0.33	0.87	0.01
Dam parity	1 vs 2	1.02	0.93	1.12	0.68
	"3 and above" vs 2	1.13	1.01	1.25	0.03
	Unknown vs 1	1.00	0.62	1.60	0.98
Calving ease score	2 vs 1	0.87	0.61	1.23	0.42
	3 vs 1	1.36	0.67	2.76	0.39
		Survival from 366 to first calving			
	Summer vs Spring	1.13	1.06	1.21	<0.01
Birth season	Autumn vs Spring	1.08	1.01	1.16	0.02
	Winter vs Spring	1.08	1.01	1.15	0.03
	Unknown vs 2	0.90	0.52	1.57	0.72
Dam parity	1 vs 2	1.04	0.95	1.14	0.39
	"3 and above" vs 2	1.11	1.01	1.23	0.04
	Unknown vs 1	0.75	0.43	1.30	0.31
Calving ease score	2 vs 1	0.99	0.72	1.36	0.95
	3 vs 1	1.20	0.60	2.40	0.61

[1] The Spring season, second parity and calving ease score 1 were the base classes of birth season, dam parity and dam calving ease score, respectively. The results of herd-birth year are not shown.

4. Discussion

The involuntary culling and mortality rates of dairy calves and replacement heifers have been reported to vary across countries and dairy cattle populations. In the population used for the current study, dairy calves were usually weaned at 2 months of age, and the pre-weaning mortality rate was 5.5%, which is within the range reported in the literature. For instance, the calf mortality within the first month of life were 3.1%–3.4% in Danish [15] and UK [2] Holstein populations, while in the US, the mortality of pre-weaning Holstein calves was 7.8% [7]. The mortality rate of 12.9% for dairy calves up to yearling age is also within the ranges reported in the literature for worldwide dairy populations (3.7%–22.5%) [16,17]. Approximately 21.2% of dairy calves and heifers failed to reach first calving, which is substantially higher compared to other reports (e.g., 14.5%) [18]. Furthermore, there was an unfavorable increase trend on mortality of dairy calves and replacement heifers over time in Chinese Holstein population. The calf and replacement heifer survival between day 3 and the start of productive life should be given more attention, especially for genetically select animals with better genetic merit for survival traits.

Many factors have caused mortality of dairy calves and replacement heifers, including calf-related diseases, heifer fertility disorders and farm management factors. In this study, censored and voluntary culling records were removed from the datasets. Therefore, the mortality rates reported here represent involuntary culling in Chinese Holstein calves and replacement heifers. Due to poor data management and insufficient attention paid on data recording, culling/death reason were not always available for each animal, especially in early records. In general, the 2 most frequent causes of mortality are digestive [19,20] and respiratory diseases [21], in which diarrhea and pneumonia accounts for the majority of death cases [9,20,22]. In this study, diseases associated with the digestive, respiratory and circulatory systems were the main culling reasons, which is consistent with the findings reported in other studies. In the US Holstein cattle population, scours, diarrhea, and other digestive problems were the key causes of pre-weaning calf mortality, followed by respiratory diseases. For weaned calves, respiratory disease was the largest mortality reason in the US population [9]. Pritchard et al. [1] reported that a large number of heifers were culled due to been considered unsuitable as breeding replacement, failure to conceive and other reproduction disorders, which is in agreement with our findings. The median death age of calves or replacement heifers due to digestive system diseases, diseases of respiratory or circulatory systems, and reproductive disorders were 84, 101, and 891 days, respectively. Furthermore, the main causes of mortality were different over these years in calves and replacement heifers. The fertility recession and more attention on epidemic prevention may respectively result in increase/decrease trends of culling/death proportion of reproductive disorders/infectious diseases over these years.

Considering the impacts of management differences across herds and a likely interaction with birth year, the effect of herd was included in the statistical model as a combined effect (herd-birth year) with birth year of the calf in current study. Herd-birth year significantly impacted all survival traits consistent with Norberg et al. [8]. Birth season and dam parity significantly impacted all survival traits, which is in agreement with results reported by Norberg et al. [8] and Gulliksen et al. [16]. During Summer and Fall, animals can be under heat stress in the main dairy farming areas in China (including the herds in the current study). The calves that experience maternal heat stress during late gestation have been reported to have reduced survival rate before puberty [23]. Dairy producers can plan the calving accordingly in order to reduce calving mortality rates and/or implement other mitigation approaches. Furthermore, Henderson et al. [24] and Ring et al. [25] reported that dam calving ease score was an important risk factor of mortality, especially within the first 182 days of life. According to them, the calves and heifers born from increased calving ease score were more likely to die in early life stages. This is likely due to the stress suffered by calves during birth. The dam calving ease score had no statistically significant impact on the survival traits analyzed here, which may be related a small data size in current study. These information from risk factors will help farmers to reduce mortality rate of calves by implementing better management practices in their herds. In addition, the influence of the risk factors identified here will be important effects to be included in the statistical models for genetic evaluation for survival traits in dairy calves and replacement heifers.

Survival traits defined at different life periods during replacement heifer development may enable selection against certain diseases commonly prevalent during those life stages [1]. Three survival traits were defined in this study aiming to describe the likelihood of death (or survival) during pre-weaning period, day 61 to yearling and yearling to the first calving. Survival trait at early life of the calf (Sur1) may enable indirect selection against diseases related to the digestive, respiratory, and circulatory systems, which were the two most frequent mortality reason categories. The survival traits at different life stages, defined in this study, can be used to genetically improve the mortality rates of dairy calf and replacement heifers, and the results from the current study laid the foundation for establishing the statistical models for genetic evaluations. The next step will be the estimation of genetic parameters (heritability and genetic correlations) for the survival traits defined in this study.

5. Conclusions

The combined mortality rate of dairy calves and replacement heifers in Chinese Holstein cattle was 21.2% and an unfavorable trend on dairy calves and replacement heifer' mortality was observed. Diseases related with digestive (e.g., diarrhea), respiratory (e.g., pneumonia) and circulatory systems, and reproductive disorders (infertility based on non-return rate) were the main death reason categories. Herd-birth year, birth season, and parity of dam had significant effects on the survival traits of dairy calves and replacement heifers. Survival traits in dairy cattle from birth to first calving are important breeding goals to be incorporated into dairy genetic selection schemes.

Author Contributions: Data curation, H.L. and Y.D.; Formal analysis, H.Z.; Methodology, H.Z. and Y.C.; Project administration, Y.W.; Resources, G.D. and L.L.; Supervision, Y.W.; Visualization, R.S.; Writing—original draft, H.Z.; Writing—review & editing, L.B. and Y.W.

Funding: This research was funded by Modern Agro-industry Technology Research System (CARS-36); Beijing Dairy Industry Innovation Team (BAIC06-2019, Beijing, China); Beijing Sciences and Technology Program (D171100002417001); the Program for Changjiang Scholar and Innovation Research Team in University (IRT_15R62); and Postgraduate Internationalization Training Promotion Project of CAU (31051521, Beijing, China).

Conflicts of Interest: The authors declare no conflict of interest. The funding providers had no role in the design, execution, interpretation, or writing of the study.

References

1. Pritchard, T.; Coffey, M.; Mrode, R.; Wall, E. Understanding the genetics of survival in dairy cows. *J. Dairy Sci.* **2013**, *96*, 3296–3309. [CrossRef] [PubMed]
2. Wathes, D.C.; Brickell, J.S.; Bourne, N.E.; Swali, A.; Cheng, Z. Factors influencing heifer survival and fertility on commercial dairy farms. *Animal* **2008**, *2*, 1135–1143. [CrossRef] [PubMed]
3. Heinrichs, A.J. Raising dairy replacements to meet the needs of the 21st century. *J. Dairy Sci.* **1993**, *76*, 3179–3187. [CrossRef]
4. Meyer, C.L.; Berger, P.J.; Koehler, K.J.; Thompson, J.R.; Sattler, C.G. Phenotypic trends in incidence of stillbirth for Holsteins in the United States. *J. Dairy Sci.* **2001**, *84*, 515–523. [CrossRef]
5. Ortiz-Pelaez, A.; Pritchard, D.G.; Pfeiffer, D.U.; Jones, E.; Honeyman, P.; Mawdsley, J.J. Calf mortality as a welfare indicator on British cattle farms. *Vet. J.* **2008**, *176*, 177–181. [CrossRef] [PubMed]
6. Miglior, F.; Fleming, A.; Malchiodi, F.; Brito, L.F.; Martin, P.; Baes, C.F. A 100-Year Review: Identification and genetic selection of economically important traits in dairy cattle. *J. Dairy Sci.* **2017**, *100*, 10251–10271. [CrossRef] [PubMed]
7. NAHMS. *Heifer and Cow Mortality*; USDA APHIS VS: Fort Collins, CO, USA, 2007; p. 3.
8. Norberg, E.; Pryce, J.E.; Pedersen, J. Short communication: A genetic study of mortality in Danish Jersey heifer calves. *J. Dairy Sci.* **2013**, *96*, 4026–4030. [CrossRef] [PubMed]
9. US Department of Agriculture. *Dairy 2007—Heifer Calf Health and Management Practices on US Dairy Operations, 2007*; USDA APHIS VS CEAH: Fort Collins, CO, USA, 2010.
10. Wells, S.J.; Dargatz, D.A.; Ott, S.L. Factors associated with mortality to 21 days of life in dairy heifers in the United States. *Prev. Vet. Med.* **1996**, *29*, 9–19. [CrossRef]
11. Meyer, C.L.; Berger, P.J.; Koehler, K.J. Interactions among factors affecting stillbirths in Holstein cattle in the United States. *J. Dairy Sci.* **2000**, *83*, 2657–2663. [CrossRef]
12. Lombard, J.E.; Garry, F.B.; Tomlinson, S.M.; Garber, L.P. Impacts of dystocia on health and survival of dairy calves. *J. Dairy Sci.* **2007**, *90*, 1751–1760. [CrossRef]

13. Stull, C.L.; McV, M.L.; Collar, C.A.; Peterson, N.G.; Castillo, A.R.; Reed, B.A.; Andersen, K.L.; VerBoort, W.R. Precipitation and temperature effects on mortality and lactation parameters of dairy cattle in California. *J. Dairy Sci.* **2008**, *91*, 4579–4591. [CrossRef] [PubMed]

14. SAS Institute. *SAS/Graph 9.1 Reference*; SAS Institute Inc.: Cary, NC, USA, 2004.

15. Fuerst-Waltl, B.; Sørensen, M.K. Genetic analysis of calf and heifer losses in Danish Holstein. *J. Dairy Sci.* **2010**, *93*, 5436–5442. [CrossRef] [PubMed]

16. Gulliksen, S.M.; Lie, K.I.; Løken, T.; Østerås, O. Calf mortality in Norwegian dairy herds. *J. Dairy Sci.* **2009**, *92*, 2782–2795. [CrossRef] [PubMed]

17. Yalew, B.; Lobago, F.; Goshu, G. Calf survival and reproductive performance of Holstein–Friesian cows in central Ethiopia. *Trop. Anim. Health Prod.* **2011**, *43*, 359–365. [CrossRef] [PubMed]

18. Brickell, J.S.; McGowan, M.M.; Pfeiffer, D.U.; Wathes, D.C. Mortality in Holstein-Friesian calves and replacement heifers, in relation to body weight and IGF-I concentration, on 19 farms in England. *Animal* **2009**, *3*, 1175–1182. [CrossRef] [PubMed]

19. Menzies, F.D.; Bryson, D.G.; McCallion, T.; Matthews, D.I. Mortality in cattle up to two years old in Northern Ireland during 1992. *Vet. Rec.* **1996**, *138*, 618–622. [CrossRef] [PubMed]

20. Svensson, C.; Linder, A.; Olsson, S.O.; Sveriges, L. Mortality in Swedish Dairy Calves and Replacement Heifers. *J. Dairy Sci.* **2006**, *89*, 4769–4777. [CrossRef]

21. McCorquodale, C.E.; Sewalem, A.; Miglior, F.; Kelton, D.; Robinson, A.; Koeck, A.; Leslie, K.E. Short communication: Analysis of health and survival in a population of Ontario Holstein heifer calves. *J. Dairy Sci.* **2013**, *96*, 1880–1885. [CrossRef]

22. Virtala, A.; Mechor, G.D.; Grohn, Y.T.; Erb, H.N. Morbidity from nonrespiratory diseases and mortality in dairy heifers during the first three months of life. *J. Am. Vet. Med. Assoc.* **1996**, *208*, 2043–2046.

23. Tao, S.; Dahl, G.; Laporta, J.; Bernard, J. Effects of heat stress during late gestation on the dam and its calf. *J. Anim. Sci.* **2018**, *963*, 351. [CrossRef]

24. Henderson, L.; Miglior, F.; Sewalem, A.; Kelton, D.; Robinson, A.; Leslie, K.E. Estimation of genetic parameters for measures of calf survival in a population of Holstein heifer calves from a heifer-raising facility in New York State. *J. Dairy Sci.* **2011**, *94*, 461–470. [CrossRef] [PubMed]

25. Ring, S.C.; McCarthy, J.; Kelleher, M.M.; Doherty, M.L.; Berry, D.P. Risk factors associated with animal mortality in pasture-based, seasonal-calving dairy and beef herds. *J. Anim. Sci.* **2018**, *96*, 35–55. [CrossRef] [PubMed]

Article

The Effect of Heat Stress on Autophagy and Apoptosis of Rumen, Abomasum, Duodenum, Liver and Kidney Cells in Calves

Ruina Zhai [1],[†], Xusheng Dong [2],[†], Lei Feng [2], Shengli Li [3],[*] and Zhiyong Hu [2],[*]

[1] College of Animal Science, Xinjiang Agricultural University, Urumqi 830052, China; 18562301385@163.com
[2] Ruminant Nutrition and Physiology Laboratory, College of Animal Science and Technology, Shandong Agricultural University, Taian 271018, China; 17863801700@163.com (X.D.); Flei0921@163.com (L.F.)
[3] State Key Laboratory of Animal Nutrition, College of Animal Science and Technology, China Agricultural University, Beijing 100193, China
*   Correspondence: lisheng0677@163.com (S.L.); huzhiyong@sdau.edu.cn (Z.H.);
     Tel.: +86-1333-116-8629 (S.L.); +86-1856-236-0982 (Z.H.)
†   These authors contributed equally to this work.

Received: 17 September 2019; Accepted: 21 October 2019; Published: 22 October 2019

Simple Summary: Heat stress causes significant negative responses in the dairy industry. The objective of this study was to assess the effect of heat stress on the autophagy and apoptosis of the rumen, abomasum, duodenum, liver and kidney in calves. The results showed that heat stress could increase the autophagy and apoptosis of the kidney, duodenum and abomasum. However, heat stress had no effect on the autophagy and apoptosis of the liver. In cows, most studies of autophagy and apoptosis have only focused on mammary remodeling. Our results provide new knowledge regarding autophagy and autophagy in calf heat stress management.

Abstract: The objective of this study was to assess the effect of heat stress on the autophagy and apoptosis of the rumen, abomasum, duodenum, liver and kidney in calves. Two groups of Holstein male calves were selected with similar birth weights and health conditions. Heat stress (HT): Six calves (birth weight 42.2 ± 2.3) were raised from July 15 to August 19. Cooling (CL): Six calves (birth weight 41.5 ± 3.1 kg) were raised from April 10 to May 15. All the calves were euthanized following captive bolt gun stunning at 35 d of age. The expression of protein 1 light chain 3-II (LC3-II) and caspase3 in the rumen, abomasum, duodenum, liver and kidney were determined by western blotting. In addition, other possible relevant serum biochemical parameters were evaluated. Significant differences were observed in alkaline phosphatase (ALP), albumin (ALB) and glucose (Glu). The results showed that heat stress could increase the autophagy and apoptosis of the kidney, duodenum and abomasum. However, heat stress had no effect on the autophagy and apoptosis of the liver. Additionally, the expression of caspase-3 in the rumen in HT was significantly lower than that in CL. In conclusion, the effects of heat stress on autophagy and apoptosis are organ-specific. The results provide knowledge regarding autophagy and autophagy in calf heat stress management.

Keywords: autophagy; apoptosis; heat stress; calf

---

1. Introduction

Heat stress causes significant negative responses in the dairy industry. Heat stress is caused by excessive temperature conditions that cannot be compensated for by the temperature regulation mechanism of cows. The temperature regulation ability is weaker for young calves than for adult cows, with an upper end of about 29 °C, and heat stress is considered to occur at temperatures greater than 32 °C and 60% humidity [1]. Extensive research has shown the diminution of liver enzyme activities and kidney functions in cows during heat stress [2,3]. In rats and pigs, heat stress apparently promotes intestinal mucosal damage due to reduced intestinal blood flow and tissue hyperthermia [4,5]. The reduced blood flow of the digestive tract is probably harmful to the barrier function integrity of rumen [6].

Autophagy and apoptosis are universal mechanisms that regulate gut homeostasis and reduce digestive tract damage [7,8]. When cells are under stress, autophagy and apoptosis are activated [9]. Autophagy is a specific protein degradation process that has been recognized as an important mechanism for cell survival under stress conditions [10,11]. The protein 1 light chain 3-II (LC3-II) is a useful marker of autophagic membranes and is essential for the expansion of the early autophagosome during cellular house-keeping and autophagic cell death [12,13]. In elegans, autophagy-related genes are transcriptionally upregulated in response to heat shock [14]. Heat stress also triggers autophagy in different types of cells such as human alveolar basal epithelial cells and rat cardiomyocytes [15]. In vivo apoptosis and autophagy are two forms of physiological and conserved programmed cell death [16]. Apoptosis is characterized by a series of morphological changes, including plasma membrane blebbing, nuclear condensation, and fragmentation, all of which lead to the formation of apoptotic bodies [9]. In general, autophagy is activated first and maintains cell homeostasis [17]. When stress is prolonged or exceeds a threshold, apoptosis is activated [9,18]. In cows, heat stress can induce the apoptosis of granulosa cells, as evidenced by the activation of caspase-3 [19]. Caspase-3 is an executioner caspase which plays an important role in apoptosis [9].

In cows, most studies of autophagy and apoptosis have only focused on mammary remodeling [20–22]. Thus far, very little attention has been paid to the role of autophagy and apoptosis in calves. The objective of this study was to assess the effect of heat stress on the autophagy and apoptosis of the rumen, abomasum, duodenum, liver and kidney in calves. We hypothesized that autophagy and autophagy would be stimulated to relieve heat stress. We hope to provide knowledge regarding autophagy and autophagy in calf heat stress management.

2. Materials and Methods

The study was conducted at the Shandong high-speed modern dairy farm in Ji Ning, Shandong, China in 2018. Animal care and use were approved and conducted under established standards of the Ethics Committee on animals of Shandong Agricultural University (SDAUA-2018-012).

Two groups of Holstein male calves were selected with similar birth weights and health conditions. Heat stress (HT): Six calves (birth weight 42.2 ± 2.3) were raised from July 15 to August 19. Cooling (CL): Six calves (birth weight 41.5 ± 3.1 kg) were raised from April 10 to May 15. Calves were individually housed in 1.5 × 3.4 m pens inside a naturally ventilated barn with free-choice water and solid feed. The HT calves were housed in the same pens and were only provided with shade. The relative humidity and air temperature of each pen were recorded at 7 days before euthanasia. The temperature humidity index (THI) of the HT and CL groups were calculated as described previously [23]. In the HT group, the average THI was 85.08, whereas the average the THI was 63.49 in the CL group. All the calves were provided with 4 L of colostrum within 2 hours from birth. From the next day, calves were fed 6 L of whole milk once daily until being euthanized. The ingredient and nutrient composition of the calf starter is given in Table 1.

The rectal temperature and respiratory rate of each calf were recorded 3 times per day. At 35 ± 2 d of age, fifteen milliliters of blood were collected from the caudal vein using 20 ml syringes. The samples were collected in the procoagulant tube, centrifuged at 1000× g for 15 min, and then the serum was collected in a 1.5 ml Eppendorf tube and stored at −20 °C. All the calves were euthanized following captive bolt gun stunning at 35 ± 2 d of age. After the opening of the body cavity, the samples of the rumen, abomasum, duodenum (entire wall from 6 cm distal to the pylorus), liver and left kidney were washed with normal saline. Then, these samples were frozen in liquid nitrogen and stored at −80 °C until western blotting was performed.

Table 1. Ingredient and nutrient composition of the experimental calves' starter.

Items	Content (% of DM)
Ingredients, % of DM	
Corn grain	48
Wheat bran	12.6
Soybean meal	18.8
Extruded soybean	7
Corn gluten meal	9
Salt	0.55
Calcium carbonate	2
Dicalcium phosphate	1.15
Vitamin and trace mineral premix [1]	0.9
Nutrients % of DM	
DM, %	89.3
CP, %	22.13
Crude fat, %	4.32
NDF, %	17.14
ADF, %	6.62
Ca, %	1.07
P, %	0.56
ME, Mcal/kg	2.83

DM: dry matter; CP: crude protein; NDF: neutral detergent fiber; ADF: acid detergent fiber; ME: metabolizable energy.[1] Premix contained (mg/kg): vitamin A, 4,035; vitamin D, 1,740; vitamin E, 39; Fe, 18; Zn, 37; Cu, 10.6; Mn, 15.3; Co, 0.12; I, 0.47; and Se, 0.35.

Briefly, the tissue samples blocks were washed with PBS (Solarbio, P1020-500ml, Beijing, China), cut into small pieces, homogenized in PBS at 4 °C using a Servicebio KZ-II homogenizer, kept on ice for 0.5 h, oscillated to ensure complete tissue cracking every 5 min, and then centrifuged (3000 × g, 10 min, 4 °C). Protein concentration was detected by a bicinchoninic acid (BCA) Protein Assay Kit (G2026, Servicebio, Wuhan, China). A Laemmli sample buffer (Bio-Rad, 1610737, Shanghai, China) was used to dilute the sample and then boiled for 5 min. Immunoblotting was performed as previously described [24,25]. The LC3 (Sigma-Aldrich, L8918, Shanghai, China), caspase-3 (Sigma-Aldrich, C8487, Shanghai, China) and β-actin (Sigma-Aldrich, A2066, Shanghai, China) antibodies and appropriate secondary antibodies (Servicebio, GB23303, Wuhan, China) were applied according to the manufacturer's guidelines. The chemiluminescence of the bands of interest was detected with a digital G: Box imager (Syngene, Frederick, MD, USA). Image J software (National Institutes of Health, Bethesda, MD, USA) was used to quantify the band density. Alanine aminotransferase (ALT), aspartate aminotransferase (AST), alkaline phosphatase (ALP), total protein (TP), albumin (ALB), glucose (Glu) and total cholesterol (TCHO) of the serum were determined with an automatic biochemical analyzer (Type 7020, Hitachi, Tokyo, Japan).

The data were analyzed with a completely randomized design using a one-way ANOVA of SAS 8.2 (SAS Institute Inc., Cary, NC, USA). The individual calf was considered as the experimental unit. The means were compared using Duncan's multiple range test. Significance was declared at $p < 0.05$.

3. Results

In the HT group, the average THI was 85.08, whereas the average THI was 63.49 in the CL group. It could also be seen that the rectal temperature (39.36 ± 0.26 vs. 38.31 ± 0.19 °C, respectively; $p < 0.01$) and respiratory rate (55.17 ± 3.49 vs. 34.17 ± 2.48 breaths per minute; respectively; $p < 0.01$) of the HT calves were significantly higher than the CL calves. Those data indicated that the HT calves were exposed to heat stress, while the CL calves were not subjected to heat stress.

No differences were observed in the concentrations of ALT, AST, TP and TCHO in plasma of two groups (Table 2). ALP and ALB in the CL group were significantly higher than that in the HT group ($p < 0.05$), while the CL calves had a lower amount of serum Glu ($p < 0.05$). Compared with the CL calves, the HT calves had a lower expression of LC3-II in the kidney ($p < 0.05$) and tended to have a lower expression in the duodenum, abomasum and rumen (Figure 1). The expressions of caspase-3 in the kidney, duodenum and abomasum were elevated in the HT calves relative to the CL calves ($p < 0.01$). No significant differences were found in caspase-3 expression of the liver between the two groups (Figure 2). Interestingly, the expression of caspase-3 in the rumen in the HT group was significantly lower than that of the CT group.

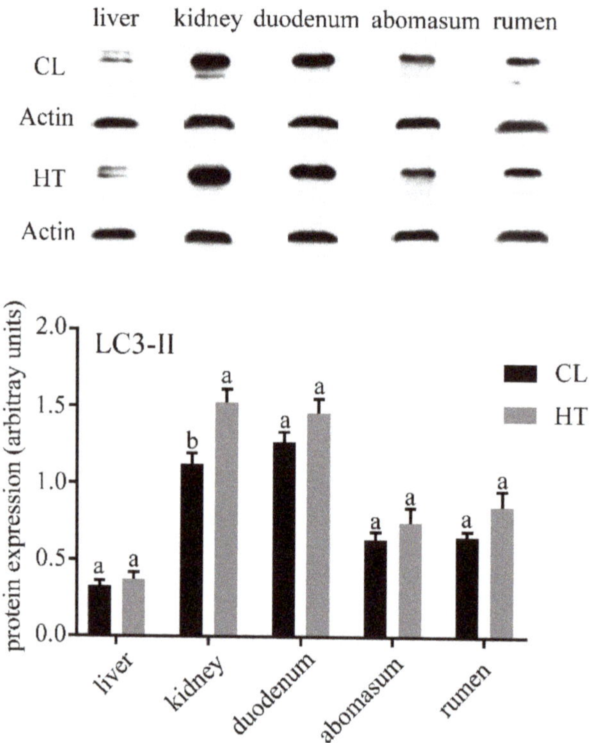

Figure 1. Effects of the temperature humidity index (THI) on the microtubule-associated protein 1 light chain 3-II (LC3-II) expression of the liver, kidney, duodenum, abomasum and rumen in the HT and CL calves. Treatment was as follows: (1) HT: Calves were fed from July 15 to August 19; (2) CL: Calves were fed from April 10 to May 15. Insets depict representative blots. Values represent means ± standard deviation Response from statistical result, $p < 0.05$. β-Actin was used to normalize the expression of target proteins. The letters below the bar graph indicate different organs. Different letters above the bar indicate differences between different groups ($p < 0.05$).

Table 2. Serum concentrations of alanine aminotransferase (ALT), aspartate aminotransferase (AST), alkaline phosphatase (ALP), total protein (TP), albumin (ALB), glucose (Glu) and total cholesterol (TCHO) of the calves in the CL (cooling) and HT (heat stress) groups.

Item	CL	HT	*p*-Value	SEM
ALT, U/L	17.00	13.00	0.063	1.094
AST, U/L	44.78	44.98	0.959	1.908
ALP, U/L	274.99	191.16	0.082	24.229
TPB, g/L	59.48	62.79	0.490	2.261
ALB, g/L	24.36	22.75	0.007	0.334
Glu, mmol/L	4.97	6.29	0.014	0.291
TCHO, mmol/L	2.44	2.28	0.211	0.0631

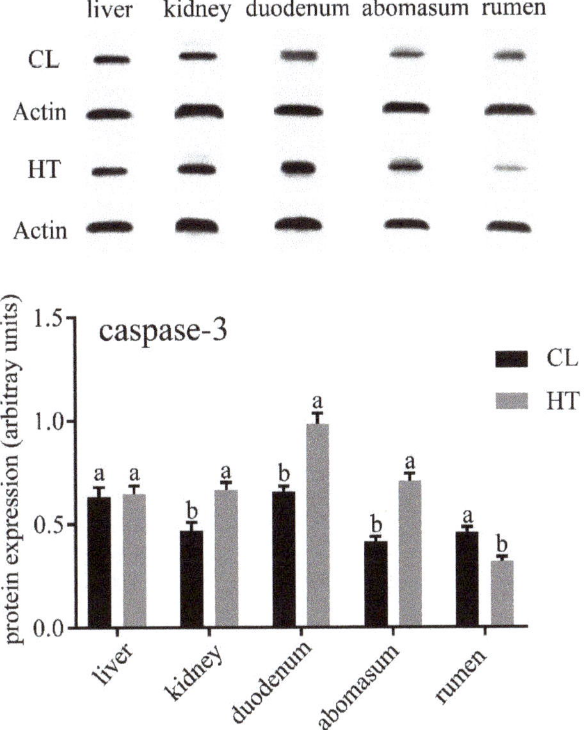

Figure 2. Effects of the THI on the caspase3 expression of the liver, kidney, duodenum, abomasum and rumen in the HT and CL calves. Treatment was as follows: (1) HT: Calves were fed from July 15 to August 19; (2) CL: Calves were fed from April 10 to May 15. Insets depict representative blots. Values represent means ± standard deviation. Response from statistical result, $p < 0.05$. β-Actin was used to normalize the expression of target proteins. The letters below the bar graph indicate different organs. Different letters above the bar indicate differences between different groups ($p < 0.05$).

4. Discussion

In our study, the THI, rectal temperature and respiratory rate indicated that the HT calves were exposed to significant environmental heat stress, while the CL calves did not suffer from heat stress. The current study found that the HT group had a higher concentration of Glu than the CL group. This result agrees with an earlier study, which showed that Glu tended to have a higher concentration in an HT group [26]. However, a previous study found that pancreatic insulin response to Glu stimulation and the concentration of insulin in calves were not affected by the heat stress [27,28]. It seems possible that the high concentration of serum Glu was connected to the impaired cell metabolism and osmolarity caused by heat stress without the change of insulin. It has been suggested that ALP plays a vital role in bone mineralization and hepatobiliary diseases [29,30]. The low concentration of ALP in the HT calves may impact the development of bone and hepatobiliary. ALB may be able to regulate oncotic pressure and modulate inflammatory or immunological responses [31]. In our study, the concentration of serum ALB was lower in the HT calves than the CL calves. Our result agrees with a previous study, indicating that heat stress could decrease the concentration of serum ALB, which perhaps, in turn, could impact oncotic pressure [31].

We analyzed the expression of LC3-II and caspase-3 of the HT and CL calves to assess the effect of the THI on autophagy and apoptosis in calves. In our study, we found that different organs have different levels of autophagy and apoptosis. All the organs in our study except liver were affected by heat stress. In the duodenum and abomasum, especially the kidney, the level of autophagy and apoptosis were increased by heat stress. A previous study found that heat stress could induce autophagy in hepatocellular carcinoma [32]. In the skeletal muscle of *Sus scrofa*, the markers of autophagosome formation and autophagic activation were increased by heat stress [33]. These results were similar to those of our study. The enhanced autophagy may be a self-protection mechanism of organs under heat stress.

Hyperosmolarity, which is one of the consequences of heat stress, could activate several mediator systems that may cause renal injury [34]. Calves extensively sweating results in a serious loss of water and salt, which could lead to an increase of urine-specific gravity and osmolarity [34]. In our study, the high concentration of Glu in the HT calves may have been due to the increase of osmotic pressure. Additionally, the activity of aldose reductase, which can convert Glu into sorbitol and is increased by hyper osmolarity [35]. Sorbitol can protect kidney cells from the hyperosmotic environments under the conditions of plasma hyperosmolarity and dehydration [36,37]. Previous studies have found that hyperosmotic stress could induce apoptosis and suppressed mammalian target of rapamycin complex 1 (mTORC1) which could inhibit autophagy [38,39]. However, sorbitol dehydrogenase could convert sorbitol into fructose. In the small intestine, the metabolism of fructose is associated with local inflammation and increased intestinal permeability [40,41]. These physiological duodenal changes probably caused the high level of apoptosis in the duodenum seen in our study.

One interesting finding was that the level of autophagy and apoptosis in the liver were not affected by heat stress. Autophagy, which is a response to stressful conditions of the liver, could eliminate damaged mitochondria and accumulated lipid droplets in liver [42]. Both autophagy and apoptosis play important roles in liver injury [42]. It seems possible that the liver has a strong regulatory ability to reduce the damage caused by heat stress. Additionally, the apoptosis level was decreased by heat stress in the rumen. It seems that the effects of heat stress on autophagy and apoptosis are organ-specific. Further work is required to evaluate the effect of osmotic pressure on autophagy and apoptosis. In addition, further study should focus on the effects of heat stress on the liver.

5. Conclusions

In conclusion, heat stress could increase the level of autophagy and apoptosis of the kidney, duodenum and abomasum. However, heat stress has no effect on the autophagy and apoptosis of the liver. Heat stress decreased serum ALP and ALB and increased Glu concentration. In conclusion, the effects of heat stress on autophagy and apoptosis are organ-specific. These results provide knowledge regarding autophagy and autophagy in calf heat stress management.

Author Contributions: Z.H. and S.L. conceived and designed the experiments; X.D. and R.Z. performed the experiments; X.D. and L.F. analyzed the data; X.D. and R.Z. wrote the paper.

Funding: This research was funded by the National Natural Science Foundation of China, grant number 31772624, National Key Research and Development Program of China, grant number 2018YFD0501600.

Conflicts of Interest: The authors declare no conflict of interest.

References

1. Moore, D.A.; Duprau, J.L.; Wenz, J.R. Short communication: Effects of dairy calf hutch elevation on heat reduction, carbon dioxide concentration, air circulation, and respiratory rates. *J. Dairy Sci.* **2012**, *95*, 4050–4054. [CrossRef]
2. Nardone, A.; Ronchi, B.; Lacetera, N.; Ranieri, M.S.; Bernabucci, U. Effects of climate changes on animal production and sustainability of livestock systems. *Livest. Sci.* **2010**, *130*, 57–69. [CrossRef]
3. Schneider, P.L.; Beede, D.K.; Wilcox, C.J. Nycterohemeral patterns of acid-base status, mineral concentrations and digestive function of lactating cows in natural or chamber heat stress environments. *J. Anim. Sci.* **1988**, *66*, 112–125. [CrossRef]

4. Lambert, G.P.; Gisolfi, C.V.; Berg, D.J.; Moseley, P.L.; Oberley, L.W.; Kregel, K.C. Selected contribution: Hyperthermia-induced intestinal permeability and the role of oxidative and nitrosative stress. *J. Appl. Physiol. (1985)* **2002**, *92*, 1750–1761; discussion 1749. [CrossRef]

5. Pearce, S.C.; Sanz-Fernandez, M.V.; Hollis, J.H.; Baumgard, L.H.; Gabler, N.K. Short-term exposure to heat stress attenuates appetite and intestinal integrity in growing pigs. *J. Anim. Sci.* **2014**, *92*, 5444–5454. [CrossRef]

6. Kadzere, C.T.; Murphy, M.R.; Silanikove, N.; Maltz, E. Heat stress in lactating dairy cows: A review. *Livest. Prod. Sci.* **2002**, *77*, 59–91. [CrossRef]

7. Negroni, A.; Cucchiara, S.; Stronati, L. Apoptosis, Necrosis, and Necroptosis in the Gut and Intestinal Homeostasis. *Mediat. Inflamm.* **2015**, *2015*, 250762. [CrossRef]

8. Randall-Demllo, S.; Chieppa, M.; Eri, R. Intestinal epithelium and autophagy: Partners in gut homeostasis. *Front. Immunol.* **2013**, *4*, 301. [CrossRef]

9. Marino, G.; Niso-Santano, M.; Baehrecke, E.H.; Kroemer, G. Self-consumption: The interplay of autophagy and apoptosis. *Nat. Rev. Mol. Cell Biol.* **2014**, *15*, 81–94. [CrossRef]

10. Sakiyama, T.; Musch, M.W.; Ropeleski, M.J.; Tsubouchi, H.; Chang, E.B. Glutamine increases autophagy under Basal and stressed conditions in intestinal epithelial cells. *Gastroenterology* **2009**, *136*, 924–932. [CrossRef]

11. Huang, H.; Li, X.; Zhuang, Y.; Li, N.; Zhu, X.; Hu, J.; Ben, J.; Yang, Q.; Bai, H.; Chen, Q. Class A scavenger receptor activation inhibits endoplasmic reticulum stress-induced autophagy in macrophage. *J. Biomed. Res.* **2014**, *28*, 213–221. [CrossRef]

12. Klionsky, D.J.; Abdalla, F.C.; Abeliovich, H.; Abraham, R.T.; Acevedo-Arozena, A.; Adeli, K.; Agholme, L.; Agnello, M.; Agostinis, P.; Aguirre-Ghiso, J.A.; et al. Guidelines for the use and interpretation of assays for monitoring autophagy. *Autophagy* **2012**, *8*, 445–544. [CrossRef]

13. Tanida, I.; Ueno, T.; Kominami, E. LC3 conjugation system in mammalian autophagy. *Int. J. Biochem. Cell Biol.* **2004**, *36*, 2503–2518. [CrossRef]

14. Kumsta, C.; Chang, J.T.; Schmalz, J.; Hansen, M. Hormetic heat stress and HSF-1 induce autophagy to improve survival and proteostasis in C. elegans. *Nat. Commun.* **2017**, *8*, 14337. [CrossRef]

15. Dokladny, K.; Zuhl, M.N.; Mandell, M.; Bhattacharya, D.; Schneider, S.; Deretic, V.; Moseley, P.L. Regulatory coordination between two major intracellular homeostatic systems: Heat shock response and autophagy. *J. Biol. Chem.* **2013**, *288*, 14959–14972. [CrossRef]

16. Yi, H.Y.; Yang, W.Y.; Wu, W.M.; Li, X.X.; Deng, X.J.; Li, Q.R.; Cao, Y.; Zhong, Y.J.; Huang, Y.D. BmCalpains are involved in autophagy and apoptosis during metamorphosis and after starvation in Bombyx mori. *Insect Sci.* **2018**, *25*, 379–388. [CrossRef]

17. Liu, G.; Pei, F.; Yang, F.; Li, L.; Amin, A.D.; Liu, S.; Buchan, J.R.; Cho, W.C. Role of Autophagy and Apoptosis in Non-Small-Cell Lung Cancer. *Int. J. Mol. Sci.* **2017**, *18*, 367. [CrossRef]

18. Kroemer, G.; Marino, G.; Levine, B. Autophagy and the integrated stress response. *Mol. Cell* **2010**, *40*, 280–293. [CrossRef]

19. Li, L.; Wu, J.; Luo, M.; Sun, Y.; Wang, G. The effect of heat stress on gene expression, synthesis of steroids, and apoptosis in bovine granulosa cells. *Cell Stress Chaperones* **2016**, *21*, 467–475. [CrossRef]

20. Wohlgemuth, S.E.; Ramirez-Lee, Y.; Tao, S.; Monteiro, A.P.A.; Ahmed, B.M.; Dahl, G.E. Short communication: Effect of heat stress on markers of autophagy in the mammary gland during the dry period. *J. Dairy Sci.* **2016**, *99*, 4875–4880. [CrossRef]

21. Zarzynska, J.; Motyl, T. Apoptosis and autophagy in involuting bovine mammary gland. *J. Physiol. Pharmacol. Off. J. Polish Physiol. Soc.* **2008**, *59* (Suppl. 9), 275–288.

22. Teplova, I.; Lozy, F.; Price, S.; Singh, S.; Barnard, N.; Cardiff, R.D.; Birge, R.B.; Karantza, V. ATG proteins mediate efferocytosis and suppress inflammation in mammary involution. *Autophagy* **2013**, *9*, 459–475. [CrossRef]

23. Tao, S.; Bubolz, J.W.; do Amaral, B.C.; Thompson, I.M.; Hayen, M.J.; Johnson, S.E.; Dahl, G.E. Effect of heat stress during the dry period on mammary gland development. *J. Dairy Sci.* **2011**, *94*, 5976–5986. [CrossRef]

24. Nestal de Moraes, G.; Carvalho, E.; Maia, R.C.; Sternberg, C. Immunodetection of caspase-3 by Western blot using glutaraldehyde. *Anal. Biochem.* **2011**, *415*, 203–205. [CrossRef]

25. Wohlgemuth, S.E.; Seo, A.Y.; Marzetti, E.; Lees, H.A.; Leeuwenburgh, C. Skeletal muscle autophagy and apoptosis during aging: Effects of calorie restriction and life-long exercise. *Exp. Gerontol.* **2010**, *45*, 138–148. [CrossRef]

26. Guo, J.R.; Monteiro, A.P.A.; Weng, X.S.; Ahmed, B.M.; Laporta, J.; Hayen, M.J.; Dahl, G.E.; Bernard, J.K.; Tao, S. Short communication: Effect of maternal heat stress in late gestation on blood hormones and metabolites of newborn calves. *J. Dairy Sci.* **2016**, *99*, 6804–6807. [CrossRef]

27. Tao, S.; Monteiro, A.P.; Hayen, M.J.; Dahl, G.E. Short communication: Maternal heat stress during the dry period alters postnatal whole-body insulin response of calves. *J. Dairy Sci.* **2014**, *97*, 897–901. [CrossRef]

28. Min, L.; Cheng, J.B.; Shi, B.L.; Yang, H.J.; Zheng, N.; Wang, J.Q. Effects of heat stress on serum insulin, adipokines, AMP-activated protein kinase, and heat shock signal molecules in dairy cows. *J. Zhejiang Univ. Sci. B* **2015**, *16*, 541–548. [CrossRef]

29. Sharma, U.; Pal, D.; Prasad, R. Alkaline phosphatase: An overview. *Indian J. Clin. Biochem.* **2014**, *29*, 269–278. [CrossRef]

30. Zierk, J.; Arzideh, F.; Haeckel, R.; Cario, H.; Fruhwald, M.C.; Gross, H.J.; Gscheidmeier, T.; Hoffmann, R.; Krebs, A.; Lichtinghagen, R.; et al. Pediatric reference intervals for alkaline phosphatase. *Clin. Chem. Lab. Med.* **2017**, *55*, 102–110. [CrossRef]

31. Baldassarre, M.; Naldi, M.; Domenicali, M.; Volo, S.; Pietra, M.; Dondi, F.; Caraceni, P.; Peli, A. Simple and rapid LC-MS method for the determination of circulating albumin microheterogeneity in veal calves exposed to heat stress. *J. Pharm. Biomed. Anal.* **2017**, *144*, 263–268. [CrossRef]

32. Jiang, J.; Chen, S.; Li, K.; Zhang, C.; Tan, Y.; Deng, Q.; Chai, Y.; Wang, X.; Chen, G.; Feng, K.; et al. Targeting autophagy enhances heat stress-induced apoptosis via the ATP-AMPK-mTOR axis for hepatocellular carcinoma. *Int. J. Hyperthermia* **2019**, *36*, 499–510. [CrossRef]

33. Ganesan, S.; Pearce, S.C.; Gabler, N.K.; Baumgard, L.H.; Rhoads, R.P.; Selsby, J.T. Short-term heat stress results in increased apoptotic signaling and autophagy in oxidative skeletal muscle in Sus scrofa. *J. Therm. Biol.* **2018**, *72*, 73–80. [CrossRef]

34. Johnson, R.J.; Rodriguez-Iturbe, B.; Roncal-Jimenez, C.; Lanaspa, M.A.; Ishimoto, T.; Nakagawa, T.; Correa-Rotter, R.; Wesseling, C.; Bankir, L.; Sanchez-Lozada, L.G. Hyperosmolarity drives hypertension and CKD–water and salt revisited. *Nat. Rev. Nephrol.* **2014**, *10*, 415–420. [CrossRef]

35. Bardoux, P.; Bichet, D.G.; Martin, H.; Gallois, Y.; Marre, M.; Arthus, M.F.; Lonergan, M.; Ruel, N.; Bouby, N.; Bankir, L. Vasopressin increases urinary albumin excretion in rats and humans: Involvement of V2 receptors and the renin-angiotensin system. *Nephrol. Dial. Transplant.* **2003**, *18*, 497–506. [CrossRef]

36. Burg, M.B. Molecular basis of osmotic regulation. *Am. J. Physiol.* **1995**, *268*, F983–F996. [CrossRef]

37. Schmolke, M.; Schilling, A.; Keiditsch, E.; Guder, W.G. Intrarenal distribution of organic osmolytes in human kidney. *Eur. J. Clin. Chem. Clin. Biochem. J. Forum Eur. Clin. Chem. Soc.* **1996**, *34*, 499–501.

38. Burg, M.B.; Ferraris, J.D.; Dmitrieva, N.I. Cellular response to hyperosmotic stresses. *Physiol. Rev.* **2007**, *87*, 1441–1474. [CrossRef]

39. Su, K.H.; Dai, C. mTORC1 senses stresses: Coupling stress to proteostasis. *Bioessays* **2017**, *39*. [CrossRef]

40. Ishimoto, T.; Lanaspa, M.A.; Le, M.T.; Garcia, G.E.; Diggle, C.P.; Maclean, P.S.; Jackman, M.R.; Asipu, A.; Roncal-Jimenez, C.A.; Kosugi, T.; et al. Opposing effects of fructokinase C and A isoforms on fructose-induced metabolic syndrome in mice. *Proc. Natl. Acad. Sci. USA* **2012**, *109*, 4320–4325. [CrossRef]

41. Johnson, R.J.; Rivard, C.; Lanaspa, M.A.; Otabachian-Smith, S.; Ishimoto, T.; Cicerchi, C.; Cheeke, P.R.; Macintosh, B.; Hess, T. Fructokinase, Fructans, Intestinal Permeability, and Metabolic Syndrome: An Equine Connection? *J. Equine Vet. Sci.* **2013**, *33*, 120–126. [CrossRef]

42. Wang, K. Autophagy and apoptosis in liver injury. *Cell Cycle* **2015**, *14*, 1631–1642. [CrossRef]

Article

Influence of Farm Management for Calves on Growth Performance and Meat Quality Traits Duration Fattening of Simmental Bulls and Heifers

Denis Kučević [1], Tamara Papović [1,*], Vladimir Tomović [2], Miroslav Plavšić [1], Igor Jajić [1], Saša Krstović [1] and Dragan Stanojević [3]

[1] Faculty of Agriculture, University of Novi Sad, Trg D. Obradovića 8, 21000 Novi Sad, Serbia;
    denis.kucevic@stocarstvo.edu.rs (D.K.); plavsic@gmail.com (M.P.); igor.jajic@stocarstvo.edu.rs (I.J.);
    sasa.krstovic@stocarstvo.edu.rs (S.K.)
[2] Faculty of Technology, University of Novi Sad, Bulevar cara Lazara 1, 21000 Novi Sad, Serbia;
    tomovic@uns.ac.rs
[3] Faculty of Agriculture, University of Belgrade, Nemanjina 6, 11080 Beograd, Zemun, Serbia;
    stanojevic@agrif.bg.ac.rs
* Correspondence: tamarapapovic@uns.ac.rs

Received: 28 August 2019; Accepted: 29 October 2019; Published: 9 November 2019

Simple Summary: Cattle have been selected for their adaptation to a specific environment and productive system, in which they show, in theory, their best economical results. With appropriate nutrition, the calf's performance enhances during early life and improve the production limit providing distinctive opportunities to optimize feeding strategies and increase the profitability of beef production. There is considerable variation in fattening protocols as well as in farm conditions. Meat quality parameters and carcass traits are the main objectives of most research carried out in the beef production area. Optimizing meat quality parameters and carcass traits are important for farmer profits and consumer satisfaction. According to that, at the phenotypic level, growth performance and traits could be observed. Rearing practices are known to have an impact on cattle carcasses and meat characteristics. The rearing practices applied after calving have an influence on the animal's performance at the growth period and can involve different animal properties at the beginning of the fattening period.

Abstract: This study assessed the effects of farm management during rearing practices in the first months of a calf's life on growth performance and meat quality traits during the fattening period. A total of 48 Simmental calves were divided into two groups at a commercial cattle feedlot. In the first group were calves from the same farm and herd (n = 12 male and n = 12 female). The second group included calves from several different herds and farms (n = 12 male and n= 12 female). Calves were transferred to a feedlot and fed with a commercial feedlot ration at three to four months of age. The aim was to determine if identical fattening conditions at feedlot can reduce initial calf rearing differences between cattle during the fattening period. Bulls grew faster than heifers reaching higher total gain and showed significantly higher slaughter weight than heifers. Meat samples of heifers from the same herd had the highest intramuscular fat content and reddest color with significant differences among cattle groups. The most abundant fatty acid was oleic acid (C18:1), followed by palmitic (C16:0), stearic (C18:0), linoleic (C18:2), and myristic acid (C14:0). Meat samples of heifers from different herds were darkest with highest content of iron (Fe) with significant differences among cattle groups.

Keywords: heifer; bull; Simmental; fattening; management; carcass and meat quality

---

1. Introduction

There is wide variation in meat production and productivity levels. Variations in these production traits can be attributed to differences in genetic composition, nutrition, slaughter endpoints, and gender [1,2]. The bulls grow faster and more efficiently, had a higher slaughtering proportion, and produce leaner carcasses with a higher proportion of total meat than heifers. Therefore, the meat from heifers compared to bulls have more dry matter and intramuscular fat, and is more tender and acceptable [3,4]. Many studies showed that different rearing factors applied during the fattening period have an impact on carcass or meat properties [4–6]. Further, it has been shown that rearing management before the fattening period could impact both carcass [7–9] and meat quality traits. Hence, the consideration of a wider period rather than the fattening period alone could be of great interest to improve the prediction power of carcass and meat quality traits. There is considerable variation in fattening protocols as well as in farm conditions [10]. The rearing practices applied after calving have an influence on the animal's performance at the growth period. These differences in performance involve different animal properties at the beginning of the fattening period [11].

Constant dynamic changes in industry demand experts with multidisciplinary knowledge and skills with the need to find faults in the production processes in a short time but also to react preventively in order to enable continual process workflow [12]. Currently developed cattle identification systems are based on electronic technologies that allow automation, instead of traditional systems based on visual identification [13]. An automated system can work autonomously, and, if required, can be easily integrated into the new or existing complex farm management system [13] and also improve consumer confidence and provide assurance to buyers regarding the animal's life history [14]. According to topics of interest, developers of new products and services need to do thorough analysis of information available in patent databases and to use collected information for defining future research and development plans and market strategies [15]. Producing a product that delivers a consistently high-quality eating experience is paramount to the beef industry to ensure consumer satisfaction [16].

Simmental cattle, a dual purpose worldwide breed common in central Europe, is usually slaughtered between 16–18 months and 600–700 kg live weight [17]. Considering that Simmental is the most widespread breed in Serbia (more than 70%) and because of the agro-climatic conditions, intensive systems of fattening based on concentrates ad libitum and cereal straw, with young animals, are the most common type of beef production systems. Calves from intensive systems are housed indoors, weaned at an early age (two to four months) and reared with concentrate and cereal straw ad libitum, when their diet is switched to concentrate [18].

Male and female calves from different farms for this research were considered together during the fattening period under identical conditions and also their expression of the observed parameters at the phenotypic level. We hypothesized that the different rearing practices from the first three to four months of calf's life can influence the characteristics of the beef carcasses and quality of produced meat. Moreover, differences between Simmental bulls and heifers in relation to growth performance, carcasses, and meat quality traits were significant and in agreement with our expectations that calves from the identical rearing conditions have more similar final results and with those reported in the literature.

2. Materials and Methods

2.1. Animals and Growth Performance, Slaughter Procedures, and Carcass Quality

The investigation was conducted on 48 calves of Simmental breed produced under an intensive rearing system at commercial beef feedlot. A total of 24 calves came from the same herd (from one farm 12 male and 12 female) from intensive system. They were weaned early and started with four weeks of age to be fed with concentrate (corn middlings 43%; limestone flour 25%; sunflower meal 19%; soybean meal 10%; premix 1.5%; limestone 1%; monocalcium phosphate 1%; animal feed salt 0.5%), and oats straw. The other 24 calves were from several different herds (from different farms including the same number of male and female) reared in semi-intensive system with different rearing practices.

For the fattening period, the two groups were housed at the commercial beef feedlot. Calves from the same herd previously carried out as the first group and the second group included calves from several other herds. They all were up from three to four months of age when transferred to feedlot and fed a commercial feedlot ration. The adaption period was three weeks. During that, animals started to consume ad libitum the same diet and reared under the same environmental and production regime. The fattening period ends when bulls reached up 568 to 613 kg and heifers reached up 517 to 547 kg of body weight.

During the fattening period, the rearing system was free, and food consisted of concentrated feeds, hay and corn grain silage locally produced and were formulated to meet the nutrient requirements [18] for the different growth phases. Animals had ad libitum access to water during the whole fattening period. Changing the concentrate composition at body weight of 250–300 kg (from all the way through to and finish phase of fattening) was a correction associated with declining ratio of protein to energy connected with age. The cattle were fed ad libitum a total mixed ration (TMR) composed of corn grain and maize silage (70%) and concentrate (30% in total, including: Corn middlings 4.3%; sunflower meal 70%; limestone flour 15%; premix 3%; limestone 3%; monocalcium phosphate 3%; animal feed salt 1.5%).

Data for each animal included initial weight (kg), total gain (kg), slaughter weight (kg), fattening period (days), and slaughter age (days) which were recorded systematically. Individual calves weights were measured using a heavy duty scale with accuracy ± 0.5 kg (initial weight) at the beginning and the end of the fattening period prior to slaughter. An estimated total gain during the fattening period was calculated between the initial weight and at slaughter weight. When the target slaughter age was achieved, the cattle were slaughtered in the slaughterhouse.

2.2. Slaughter Procedures

From feedlot to slaughterhouse, cattle were transported unmixed in early morning hours and after transport, which took about 3 h (farms are 60 km far from the slaughterhouse), animals were rested for about 2 h in the abattoir. The animals were rested by isolating them from noise and human activity during the lairage period. All the cattle were slaughtered according to routine procedures of the slaughterhouse. Carcasses were conventionally chilled for 24 h in a chiller at 0–4 °C. After chilling, *M. longissimus lumborum* (LL) was removed from the right side of each carcass, in the area between the sixth and seventh rib to determine meat quality. The meat samples were trimmed of visible adipose and connective tissue. Physical and sensory characteristics were measured on fresh or cooked beef. Samples for chemical analysis (approximately 250 g) were taken after the homogenization of the LL; they were vacuum packaged in polyethylene bags and stored at −40 °C until analysis.

2.3. Carcass Quality Traits Evaluation

The carcass quality was characterized by: Hot carcass weight (HCW), dressing percentage (ratio between hot carcass weight and live weight before slaughter, in %), and conformation score. Carcass conformation was graded under the EU beef carcass classification (SEUROP) scheme. After slaughtering of the animals, their carcass weighing and muscle development evaluation was done [19]. Beef carcass conformations are defined with the EUROP scale, represented by the letters E, U, R, O, and P (class S is used only in countries where there is a basis for its use—double muscled cattle). The scoring consists of a visual assessment of carcass muscling where carcasses graded as E have the most muscularity, and this decreases through to P which have the least muscularity (muscle development). At the same time, the degree of fat cover of the carcasses was based on visual evaluation numerically scored from 1 = very low to 5 = very high, according to the same European classification [19].

2.4. Meat Quality Measurements

2.4.1. Physical and Sensory Quality Measurements

The pH value was measured in the center of LL muscles at 24 h (pH$_{24}$ h) *post-mortem* [20,21]. Samples for color measurements were taken from the central part of all muscles, perpendicularly to the long axis of LL, after 60 min of blooming [22]; the minimum thickness of samples was 2.5 cm. The instrumental color was determined using Minolta Chroma Meter CR-400 (Minolta Co., Ltd., Osaka, Japan) using D-65 lighting, a 2° standard observer angle and an 8 mm aperture in the measuring head. The CIE *L*a*b** color coordinates [23] were lightness (*L**), redness (*a**), yellowness (*b**), *C** (chroma—saturation index; *C** = (*a**2 + *b**2)1/2), *h* (hue angle; *h* = arctangent (*b*/a**)), and λ (dominant wavelength (nm)) [23–25]. Water-holding capacity (WHC) was determined as free water (exudative juice) using the filter paper press method [21,26,27]. The cooking loss was determined by the method as described by Tomović et al. [28]. Samples of cooked meat, after cooking loss determination, were used for objective determination of tenderness [28,29]. Tenderness was measured as the shear force (N) using Warner–Bratzler shear machine (Model SD—50 of 50 lb or 222 N capacity, John Chatillon & Sons, New York, NY, USA) as described by Senk et al. [12]. The sensory analyses were performed by an eight-member panel. Samples for sensory evaluation were taken perpendicularly to the long axis of LL; the minimum thickness was 2.54 cm. Panelists evaluated color using sets of [25] official color (1 = extremely bright cherry-red to 8 = extremely dark red) and marbling [30] (1 = slight to 7 = moderately abundant) standards.

2.4.2. Proximate and Mineral Composition

Moisture [31], protein (nitrogen × 6.25; [32]), total fat [33], and total ash [34] contents of muscles were determined according to methods recommended by the International Organization for Standardization. The minerals contents of the meat (calcium (Ca), sodium (Na), magnesium (Mg), iron (Fe), zinc (Zn), and copper (Cu)) were determined by the flame atomic absorption spectrometry as described in detail described by Tomović et al. [35] after mineralization by dry ashing [34]. Phosphorus (P) was determined by the standard spectrophotometric method [36]. All analyses were performed in duplicate.

2.4.3. Fatty Acids Composition

Meat samples of 5 g were dried at the temperature of 105 °C. Then, samples were quantitatively transferred into an extraction cartridge, and petroleum ether extraction was run for 5 h in the Soxhlet extractor [37,38]. The methyl esters of the fat extracted were formed according to the method described by Yurchenko et al. [39]. Fatty acids methyl esters were identified by comparing the retention times of fatty acid methyl ester peaks from samples with those of standards obtained from Supelco (Supelco C4-C24 Even Carbon for saturated: C14, C16, C18, and Supelco Fame Mix GLC-10 for unsaturated fatty acids: C18:1, C18:2). Chromatographic analysis of the methyl esters was carried out with a gas chromatograph GC-2010 Plus, Shimadzu, equipped with a flame ionization detector and autosampler AOC-20i, Capillary Column InterCap WAX (length 30 m, inner diameter 0.25 mm, film thickness 0.25 μm). Analysis of the standard mixture of methyl esters was carried out using reference probe sample of 0.6 μL at split ratio 40:1. The injector and detector temperatures were 260 °C, and the analysis was performed in isothermal conditions at 200 °C. Helium was applied as carrier gas with flow rate of 3 mL/min.

2.5. Statistical Analysis

All data are presented as mean and standard deviation (SD). Data were studied by two-way factorial ANOVA (gender and group) and Post-Hoc test (Duncan's multiple range test) was used to characterize statistically significant differences at the level $p < 0.05$ between analyzed groups within the Statistica software package (ver. 13 StatSoft, Inc. 2016, Kraków, Poland). The two-way factorial model equation used for the evaluation was as follows:

$$Y_{ijkl} = \mu + F_i + G_j + I_k + e_{ijk} \tag{1}$$

where: Y_{ijkl}, the value of the tested traits (dependent variable); μ, average mean value of the dependent variable; F_i, fixed effect of the group (i = 1,2); G_j, fixed effect of the gender (j = 1,2); I_k, interaction group x gender; e_{ijk}, other random effects.

3. Results and Discussion

3.1. Growth Performance

In this study, weights were recorded at the beginning and at the end of the fattening period. Despite the fact that calves came from different herds, there were no significant differences between two groups in initial weight at the start of the fattening period which is presented in Table 1. The average of days spent in a feedlot for the second group of cattle from different herds was significantly longer compared to the first group of cattle which were from the same herd. It is well known that the optimal slaughter ages and weights vary widely among cattle breed types depending on how rapidly they mature, which is characterized by fat deposition during the "finishing" period [1]. In this research, the group had significant influence at the slaughter age ($p < 0.001$). Considering slaughter age, the bulls and heifers from the second group were older (512.2 and 530.3 days, respectively) than those from the first group (491.6 days). Moreover, cattle from the second group spent a longer period in the feedlot, which can be explained by the fact that the calves from different herds brought to the same feedlot took a longer period to adapt at the beginning, especially the females.

There was interaction between gender and group for total gain during the fattening period ($p < 0.001$), with bulls achieving higher total gain than heifers. In our study, heifers from the second group achieved lowest total gains and slaughter weight (383 and 518 kg) during the fattening period in comparison with the rest of the animals. Likewise, heifers from the second group spent the longest period at the feedlot (456.8 days) which corresponded to the above-mentioned claim that calves from different herds with different rearing practices should take a longer period to adapt.

Table 1. Growth performance per groups and gender of Simmental cattle fattened in a feedlot.

Parameter	1st Group (Same Herd)		2nd Group (Different Herds)		p-Values		
	Male	Female	Male	Female	Group	Gender	Group × Gender
IW (kg)	142.4 ± 23.2	145.0 ± 9.4	148.3 ± 17.1	135.1 ± 11.6	0.672	0.262	0.099
TG (kg)	426.3 [b] ± 33.8	402.5 [c] ± 25.9	465.4 [a] ± 23.2	382.8 [c] ± 31.0	0.250	<0.001	<0.001
SW (kg)	568.8 [b] ± 35.3	547.5 [b] ± 24.8	613.8 [a] ± 31.78	517.9 [c] ± 24.2	0.369	<0.001	<0.001
DIF (days)	416.3 [b] ± 19.9	410.7 [b] ± 16.7	421.8 [b] ± 17.5	456.8 [a] ± 12.9	<0.001	<0.001	<0.001
SA (days)	491.6 [b] ± 27.6	491.7 [b] ± 14.5	512.2 [ab] ± 43.3	530.3 [a] ± 9.9	<0.001	0.253	0.257

IW = initial weight (at start of the fattening period); TG = total gain during the fattening period; SW = slaughter weight; DIF = days in feedlot; SA = slaughter age. [a,b,c] Row means with different superscript differ in significance at $p < 0.05$.

A higher total gain of bulls compared to heifers here resulted in higher slaughter weight of bulls. Similarly, Bureš et al. [3] found a higher slaughter weight for bulls compared to heifers 18 months old, fattened in quite identical husbandry conditions. These results are in accordance with data obtained by Kaminiecki et al. [40] for Charolais x Simmental crossbreeds bulls.

3.2. Carcass Quality Traits Evaluation

The carcass quality traits of cattle are shown in Table 2. Hot carcass weights from bulls (354 and 379 kg) were significantly higher than from heifers (327 and 309 kg) in the first and second group, respectively ($p < 0.001$). Our results for the carcass weight were lower than those [41] published for Simmental bulls and higher for the dressing percentage [1,17]. The effect of nutrition efficiency increased with slaughter weight due to the interaction between the total gain during the fattening period and the slaughter weight which resulted in higher values of the carcass weight and dressing percentage. Moreover, Herva et al. [42] concluded that carcass fat content was increased when carcasses were heavier, and when a daily gain was higher.

Table 2. Carcass quality traits evaluation for investigated groups of Simmental cattle.

Parameter	1st Group (calves from the Same Herd)		2nd Group (Calves from the Different Herds)		p-Values		
	Male	Female	Male	Female	Group	Gender	Group × Gender
HCW (kg)	354.0 [b] ± 18.5	327.9 [c] ± 16.7	379.4 [a] ± 23.4	309.7 [c] ± 39.1	0.634	<0.001	<0.001
Dressing (%)	62.3 ± 1.7	59.9 ± 1.5	61.8 ± 1.2	59.7 ± 6.2	0.726	0.271	0.877
Conformation	2.8 ± 0.4	2.6 ± 0.4	2.8 ± 0.4	2.5 ± 0.4	0.869	0.141	0.620
Fat cover	4.0 [a] ± 0.4	3.8 [ab] ± 0.2	4.0 [a] ± 0.2	3.6 [b] ± 0.4	0.394	<0.001	0.204

HCW = hot Carcass weight; Dressing = dressing percentage; Conformation = conformation scores, EUROP classification scales from E = 5 excellent; U = 4 very good; R = 3 good; O = 2 fair; P = 1 poor; Fat cover = fat cover scores, EUROP classification scales from 1 = low; 2 = slight; 3 = average; 4 = high and 5 = very high. [a,b,c] Row means with different superscript differ in significance at $p < 0.05$.

Group, gender, and their interaction did not significantly affect the dressing and conformation traits evaluation ($p > 0.27$). Kaminiecki et al. [40] found that Simmental × Charolais crossbreeds produced a dressing percentage of 58.5% while [43] reported that carcass dressing percentage was higher in heavier animals, which could result from higher carcass fatness. Both studies were in accordance with our results. Higher final weights of bulls in our trial resulted higher hot carcass weight compared to heifers, however the dressing percentage was not affected. Fat cover scores were significantly influenced by gender. Usually females start to deposit fat earlier than males. In addition, the males were intact (with their testicles), so they should be leaner than heifers. The results regarding fat cover evaluated indicate that most animals belonged between score three and four. Regarding conformation, the majority of cattle carcasses were classified as class R. Bulls showed significantly higher scores of fat cover (4) than heifers (3.8 and 3.6). Our results for bulls were in accordance with results obtained by Chambaz et al. [44] for Simmental steers (conformation score 3.7 which present U class and fatness score 4.1). According to Monteils et al. [8] irrespective of a cattle category, the higher carcass conformation and higher carcass fat cover were found related to increased hot dressing percentage. Interestingly, in each analyzed cattle group in our research, among the carcasses classified to a higher slaughter weight, a higher grade (conformation, fat cover) was recorded.

3.3. Physical and Sensory Quality Measurements

The data of the Simmental cattle showed variations in the properties of interest referring to physical and sensory traits depending on the examined effects (Table 3). In the present study, pH_{24} value was significantly influenced by the gender, but all mean pH_{24} values fell in a very narrow range with 5.44 (heifers) to 5.50 (bulls) which was in accordance with the results obtained by Pilarcyk [45] for Simmental bulls (pH_{24} 5.52). Meat of high quality has pH at the range of 5.4–5.6, but meat of a higher pH value can be characterized by gummy structure, increased water-holding capacity, and decreased specific taste [4]. We found that an interaction effect between the group and gender was found for all instrumental color parameters ($p < 0.001$). All instrumental color parameters showed significant differences between cattle. Significantly paler (lightest color, higher *L**) numerical CIE*L** mean values were found in meat samples from bulls on the second group (39.76) and the lowest (darkest color) was found in meat samples from heifers at the same group (37.73).

Table 3. Physical and sensory quality measurements of fresh and cooked *M. longissimus lumborum* from investigated groups of Simmental cattle.

Parameter	1st Group (Same Herd)		2nd Group (Different Herds)		*p*-Values		
	Male	Female	Male	Female	Group	Gender	Group × Gender
pH$_{24}$	5.50 [a] ± 0.04	5.45 [ab] ± 0.02	5.50 [a] ± 0.04	5.44 [b] ± 0.10	0.817	<0.001	0.817
*L**	38.22 [bc] ± 1.32	39.02 [ab] ± 1.57	39.76 [a] ± 1.89	37.73 [c] ± 1.04	0.780	0.157	<0.001
*a**	19.60 [c] ± 1.02	22.13 [a] ± 1.11	20.79 [b] ± 1.00	19.81 [bc] ± 2.01	0.158	0.054	<0.001
*b**	8.84 [b] ± 0.83	10.16 [a] ± 0.89	9.90 [a] ± 0.70	8.42 [b] ± 1.05	0.192	0.756	<0.001
*C**	21.51 [c] ± 1.23	24.36 [a] ± 1.37	23.03 [b] ± 1.16	21.53 [c] ± 2.23	0.158	0.141	<0.001
h	24.20 [b] ± 1.29	24.54 [ab] ± 0.86	25.42 [a] ± 0.94	22.93 [c] ± 1.13	0.523	<0.001	<0.001
λ (nm)	609.16 [b] ± 1.90	609.34 [b] ± 1.08	607.60 [c] ± 1.35	611.47 [a] ± 1.97	0.546	<0.001	<0.001
WHC-M (cm^2)	4.10 [b] ± 0.45	3.98 [b] ± 0.19	4.64 [a] ± 0.46	4.25 [b] ± 0.39	<0.001	<0.001	0.244
WHC–T (cm^2)	11.37 [a] ± 0.40	11.15 [ab] ± 0.33	11.42 [a] ± 0.42	11.03 [b] ± 0.33	0.744	<0.001	0.432
WHC–RZ (cm^2)	7.26 [a] ± 0.59	7.18 [ab] ± 0.44	6.78 [b] ± 0.52	6.79 [b] ± 0.53	<0.001	0.805	0.763
WHC-M/RZ	0.57 [b] ± 0.10	0.56 [b] ± 0.06	0.70 [a] ± 0.12	0.63 [ab] ± 0.10	<0.001	0.190	0.396
WHC-M/T	0.36 [b] ± 0.04	0.36 [b] ± 0.02	0.41 [a] ± 0.04	0.39 [ab] ± 0.04	<0.001	0.232	0.413
Cooking loss (%)	38.34 [a] ± 1.75	33.93 [b] ± 1.46	37.17 [a] ± 1.83	33.30 [b] ± 2.24	0.099	<0.001	0.610
WBSF (N)	56.03 [b] ± 6.65	52.98 [bc] ± 3.96	61.02 [a] ± 6.76	50.13 [c] ± 5.34	0.526	<0.001	<0.001
Color sensoric (1–8)	4.50 ± 1.15	4.50 ± 0.60	4.50 ± 0.83	4.54 ± 0.58	0.930	0.930	0.930
Marbling scores (1–7)	4.08 [a] ± 1.00	3.25 [b] ± 0.45	3.00 [b] ± 0.74	4.17 [a] ± 0.83	0.713	0.464	<0.001

*L** = a measure of darkness/lightness (higher value indicates a lighter color); *a** = a measure of redness (higher value indicates a redder color); *b** = a measure of yellowness (higher value indicates a more yellow color); *C** = saturation index (higher values indicates greater saturation of red); *h* = hue angle (lower values indicates a redder color); λ = dominant wavelength; WHC-M = surface of the pressed meat film; WHC-T = surface of the wet area on the filter paper; WHC-RZ = WHC-T–WHC-M, a bigger WHC-M/T = ratio indicates a better WHC; CL = cooking loss; WBSF = Warner–Bratzler shear force; [a,b,c] Row means with different superscript differ in significance at *p* < 0.05.

Furthermore, meat of heifers from the first group had the reddest color (CIE*a** value was 22.13). As well, heifers from the first group also had the significantly highest CIE*b** value (10.16). Brighter color of meat from heifers as compared with meat from bulls could be due to the increased fat disposition content of heifers as fat increases brightness of meat color and fiber type as well [4]. Concomitant, heifers from the first group had significantly highest values of CIE*C** (24.36). Bulls from the second group had significantly higher value for the *h* (hue angle) (25.42) and the lower value of λ (dominant wavelength) (607.60 nm) than the rest of the animals. A lower *L** value and yellowness *b** were found in the meat of older cattle (heifers from the second group) whereas hue angle (*h*) was similar for all animals, which was in accordance with [46].

WHC (M/T cooking losses) was influenced by the group. Cattle from the second group had better WHC (M/T = 0.41 for bulls and 0.39 for heifers, a bigger M/T ratio indicating a better WHC) than cattle from the first group (M/T = 0.36, for both). If more water is retained in the muscle/myofibrillar structure, generally a product with a higher sensory tenderness and juiciness is obtained [14].

Gender significantly affected the cooking loss (*p* < 0.001). Bulls showed higher content of cooking loss (38.34% and 37.17%) than heifers (33.93% and 33.30%). Moreover, cattle from the first group had higher content than cattle from the second group comparing in total. Values for cooking loss in our study were similar to the results of Scollan et al. [41] for crossbred Charolais × White Holstein-Friesian bulls (34.53%). Significant effect of gender and interaction between group and gender was found for WBSF. Bulls showed significantly higher WBSF value (56.03 and 61.02 N) than heifers (52.98 and 50.13 N) for the first and second group, respectively. Weglarz [4] found that comparing meat from bulls and heifers, heifer meat appeared slightly more tender, which must have been related to the higher content of intramuscular fat. A slightly lower Warner–Blatzer shear force values than those in our study, were reported by Bureš and Barton [47] for Fleckvieh bulls (49.8 N) and for Simmental bulls (48.19 N) [48]. Beef from cattle with a high intramuscular fat level often has a lower shear force [49], which is in accordance with results from our study. The color sensory attribute of the meat samples did not differ significantly between the cattle groups (*p* > 0.05). Scollan et al. [41] demonstrated that the meat from lighter and younger animals was significantly more tender, however with larger variation within WBSF values. Marbling score is being used as an indirect mean for meat sensory quality assessment [50]. There was an interaction effect between group and gender for marbling score. Marbling score was significantly highest ($p < 0.05$) for the heifers at the second group (4.17) than the other animals.

3.4. Proximate and Mineral Composition

The proximate composition of meat samples from Simmental cattle are shown in Table 4. We found an interaction effect between the group and gender for moisture content ($p < 0.001$) where the bulls had higher moisture contents (73.21% to 74.54%) than heifers (up 72.11% to 72.24%) for the first and second group, respectively. Proximate composition, except protein was influenced mainly by the gender. No differences were found in the content of protein among meat samples from two groups. As expected, the protein content was in agreement with some earlier investigations [39,43].

Table 4. Proximate composition (%) of fresh *M. longissimus lumborum* from investigated groups of Simmental cattle.

Parameter	1st Group (same herd)		2nd Group (different herds)		*p*-Values		
	Male	Female	Male	Female	Group	Gender	Group × Gender
Moisture	73.21 [b] ± 0.94	72.24 [bc] ± 0.99	74.54 [a] ± 1.32	72.11 [c] ± 1.43	0.086	<0.001	<0.001
Protein	21.32 ± 0.45	21.07 ± 0.72	21.18 ± 0.27	21.38 ± 0.68	0.625	0.868	0.165
Total fat (IMF)	4.38 [ab] ± 1.30	5.40 [a] ± 1.59	3.00 [b] ± 1.37	5.19 [a] ± 1.68	0.071	<0.001	0.185
Total ash	1.04 [c] ± 0.04	1.14 [a] ± 0.03	1.08 [b] ± 0.06	1.13 [ab] ± 0.07	0.254	<0.001	0.088

IMF = intramuscular fat. [a,b,c] Row means with different superscript differ in significance at $p < 0.05$.

Gender significantly affected ($p < 0.001$) content of total fat and total ash. However, the content of total fat was the most variable inside the investigated groups. Total fat content was significantly higher for heifers (ranged between 5.19% to 5.40%) and lower for bulls (3.00% to 4.38%) at the first and second group, respectively. According to the results of Weglarz [10] that are comparable to ours, meat from bulls had higher moisture and significantly lower fat and total ash content in comparison with meat from heifers. Content of total ash was significantly higher for heifers (1.13%) than for bulls (from 1.04% to 1.08%) which was in accordance with total ash content reported by Pilarczyk et al. [45] and Monteils et al. [43].

An overview of obtained results for the mineral composition of meat samples are presented in Table 5. Gender affected the content of phosphorus, calcium, iron, and zinc in the meat samples ($p < 0.001$). Phosphorous was the most abundant mineral in fresh meat samples. As shown in Table 6, the content of phosphorous was significantly higher for the bulls (152.28 and 157.97 mg/100g) than for heifers (106.91 and 110.26 mg/100g) from the first and second group, respectively. Accordingly, bulls showed significantly higher content of calcium than heifers. Interaction effect between group and gender was found to be significant for magnesium content. The highest magnesium content was found in the meat samples for bulls from the first group (24.61 mg/100g). All investigated effects (group, gender, and their interaction) significantly affected ($p < 0.001$) the content of iron and zinc. Heifers showed significantly higher content of iron compared to bulls with significant differences among cattle groups. Heifers from the second group had a significantly higher content of iron in meat samples (2.46 mg/100g) in regard to rest animals. According to Domaradzki et al. [50] a similar variation those to ours in the content of minerals of young Simmental bulls is reported.

There were noticeable significant differences between investigated groups for the content of zinc, where the highest content of zinc was found for bulls from the first group (6.26 mg/100g) and the lowest for bulls from the second group (5.21 mg/100g). Investigated effects did not significantly affect sodium and copper content in meat samples, and there were no differences between groups. Nogalski et al. [51] said that breed is a significant factor determining the content of minerals in the muscles of cattle raised under the same conditions.

Table 5. Mineral composition (mg/100g) of fresh *M. longissimus lumborum* from investigated groups of Simmental cattle.

Parameter	1st Group (Same Herd)		2nd Group (Different Herds)		p-Values		
	Male	Female	Male	Female	Group	Gender	Group × Gender
P	152.28 [a] ± 14.92	106.91 [b] ± 5.37	157.97 [a] ± 8.15	110.26 [b] ± 11.96	0.152	<0.001	0.708
Ca	4.99 [a] ± 0.64	4.02 [c] ± 0.86	4.76 [ab] ± 0.73	4.23 [bc] ± 1.05	0.978	<0.001	0.366
Na	51.90 [a] ± 4.01	48.06 [b] ± 5.62	47.67 [b] ± 2.56	47.50 [b] ± 3.73	0.050	0.099	0.130
Mg	24.61 [a] ± 1.98	22.07 [bc] ± 1.33	21.03 [c] ± 3.20	23.78 [ab] ± 1.94	0.153	0.869	<0.001
Fe	1.89 [b] ± 0.18	2.09 [b] ± 0.36	1.91 [b] ± 0.24	2.46 [a] ± 0.34	<0.001	<0.001	<0.001
Zn	6.26 [a] ± 0.89	5.26 [b] ± 0.50	5.21 [b] ± 0.62	5.35 [b] ± 0.57	<0.001	<0.001	<0.001
Cu	0.03 ± 0.01	0.03 ± 0.01	0.03 ± 0.01	0.03 ± 0.01	0.921	0.370	0.728

P = phosphorus; Ca = calcium; Na = sodium; Mg = magnesium; Fe = iron; Zn = zinc; Cu = copper. [a,b,c] Row means with different superscript differ in significance at $p < 0.05$.

Table 6. Fatty acid composition (g/100g fat) of fresh *M. longissimus lumborum* from investigated groups of Simmental cattle.

Parameter	1st Group (same herd)		2nd Group (different herds)		p-Values		
	Male	Female	Male	Female	Group	Gender	Group × Gender
C14:0	2.56 [a] ± 0.39	2.18 [b] ± 0.26	2.29 [ab] ± 0.29	2.54 [ab] ± 0.65	0.723	0.594	<0.001
C16:0	25.03 [a] ± 1.33	24.23 [ab] ± 1.18	23.32 [b] ± 1.68	25.36 [a] ± 1.43	0.483	0.137	<0.001
C18:0	19.28 [a] ± 1.92	16.37 [c] ± 1.69	18.50 [ab] ± 3.52	17.06 [bc] ± 1.52	0.944	<0.001	0.274
C18:1	42.21 [bc] ± 1.45	44.98 [a] ± 2.86	40.26 [c] ± 2.14	42.75 [b] ± 2.97	<0.001	<0.001	0.844
C18:2	4.01 [a] ± 0.37	3.09 [b] ± 0.70	4.08 [a] ± 0.53	3.13 [b] ± 0.41	0.722	<0.001	0.885
∑SFAs	46.88 [a] ± 1.59	42.68 [b] ± 1.91	44.11 [b] ± 4.39	44.00 [b] ± 2.10	0.362	<0.001	<0.001
∑UFAs	46.39 [ab] ± 1.33	47.83 [a] ± 2.36	44.84 [b] ± 1.55	45.38 [b] ± 3.04	<0.001	0.123	0.476
∑OFAs	6.74 [b] ± 1.75	9.49 [a] ± 3.41	11.05 [a] ± 3.98	10.62 [a] ± 3.57	<0.001	0.226	0.101

SFAs = saturated fatty acids (myristic acid—C14:0, palmitic acid—C16:0, stearic acid—C18:0); UFAs = unsaturated fatty acids (oleic acid—C18:1, linoleic acid—C18:2); OFAs = other fatty acid. [a,b,c] Row means with different superscript differ in significance at $p < 0.05$.

3.5. Fatty Acids Composition

The results for the fatty acid profile of meat samples for investigated groups are presented in Table 6. In general, the most abundant fatty acid was oleic acid (C18:1) with g/100g fat up 40.26 to 42.21 for bulls and 42.75 to 44.98 g/100g fat for heifers, followed by palmitic (C16:0), stearic (C18:0), linoleic (C18:2), and myristic acid (C14:0). Gender significantly affected oleic acid (C18:1) content ($p < 0.001$) where the heifers had significantly higher oleic acid (C18:1) content (44.98 and 42.75 g/100g fat) than bulls (42.21 and 40.26 g/100g fat). Gender also significantly affected stearic acid (C18:0) composition, where bulls had significantly higher content of stearic acid (C18:0) (19.28 and 18.50 g/100g fat) than heifers (16.37 and 17.06 g/100g fat). Results reported by Monteils et al. [43] for oleic acid (C18:1) and linoleic acid (C18:2) were lower compared to our results. In the research of [43], a higher IMF content of meat was associated with a considerable increase in MUFAs concentrations and a decrease in PUFAs levels which could result from feeding grass silage ad libitum. The interaction between the group and gender had significant influence on saturated fatty acids ($p < 0.001$), such as myristic acid (C14:0) and palmitic acid (C16:0). Content of myristic acid (C14:0) was significantly higher for bulls at the first group (2.56 g/100g fat) and significantly lower for the bulls at the same group (2.18 g/100g fat). A difference between investigated animals for palmitic acid (C16:0) content was significant. Similarly, the heifers from the second group had the highest palmitic acid (25.36 g/100g fat) content, while the lowest palmitic acid content was obtained for bulls from the same group (23.32 g/100g fat). Results obtained in studies by [6,43,52] were similar to ours.

De Smet et al. [53] found that an increased fat content of bovine meat was paralleled by increased proportions of SFAs and MUFAs, and a decreased proportion of PUFAs. Therefore, it is known that FA composition is mainly affected by rearing and feeding conditions. Feed composition is known to be one of the most important factors influencing fatty acids composition in beef. Some researchers [34,35] demonstrated that when animals were grown at the same rate, muscles from cattle which had a high grass intake had a higher PUFA/SFA ratio and a lower n-6/n-3 PUFA ratio in comparison with muscles from cattle fed concentrates. Cattle with a high potential for lean beef production are frequently fattened on concentrate diets, which may be unfavorable for the n-6/n-3 polyunsaturated fatty acids ratio in meat. The reason for this is the fact that the fat in concentrates contains higher levels of C18:2n-6. Introducing forage in the diet of beef cattle should enhance the n-3 fatty acid concentrations since forages are a good source of C18:3n-3 [43].

4. Conclusions

The results from this research suggested that a fattening period of around 400 days is more than sufficient to eliminate differences which can be caused by the different rearing system and farm management for calves before the fattening. Therefore, more uniform values for most of the examined traits were achieved within the first group where the cattle for the fattening were from the same herd. So, the group had significant effect on the age of slaughter, where the cattle from the second group spent significantly longer time there and were older than the cattle from the first group. This can be explained by the fact that the calves from different herds took a longer period to adapt, especially the female population. Rearing practices and the production system might modify some of the characteristics of commercial beef, especially those associated with fat. Moreover, variability of rearing factors could make difficulties to simultaneously analyze their impacts on the carcass and the meat. Slaughter traits such as quality of meat samples may vary depending on the combinations of rearing practices utilised. For the future investigation in addition, it would be necessary to collect breeding data on female and male cattle with more detailed rearing practices before the fattening period to refine the characterization of management system with a shorter period of fattening.

Author Contributions: Conceptualization and methodology, D.K. and T.P.; software, D.K., D.S.; validation, D.K., M.P., T.P., S.K., I.J., and D.S.; formal analysis, D.K., V.T., and I.J.; investigation, T.P., D.K., and I.J.; resources, V.T. and T.P.; data curation, T.P., D.K., and V.T.; writing—original draft preparation, D.K., T.P., and S.K.; writing—review and editing, T.P., D.K., M.P., S.K., and D.S.; visualization, T.P., D.K., and S.K.; supervision, D.K., M.P., and S.K.; project administration, D.K. and I.J.; funding acquisition, D.K., I.J., and M.P.

Funding: This research received no external funding.

Acknowledgments: Research was financially supported by the Ministry of Science and Technological Development, the Republic of Serbia, project TR—31086.

Conflicts of Interest: The authors declare no conflict of interest. The funders had no role in the design of the study; in the collection, analyses, or interpretation of data; in the writing of the manuscript, or in the decision to publish the results.

References

1.   Albertí, P.; Panea, B.; Sañudo, C.; Olleta, J.L.; Ripoll, G.; Ertbjerg, P.; Christensen, M.; Gigli, S.; Failla, S.; Concetti, S.; et al. Live weight, body size and carcass characteristics of young bulls of fifteen European breeds. *Livest. Sci.* **2008**, *114*, 19–30.

2.   Bureš, D.; Bartoň, L.; Zahrádková, R.; Teslík, V.; Krejčová, M. Chemical composition, sensory characteristics, and fatty acid profile of muscle from Aberdeen Angus, Charolais, Simmental, and Hereford bulls. *Czech J. Anim. Sci.* **2006**, *51*, 279–284.

3.   Bureš, D.; Bartoň, L. Growth performance, carcass traits and meat quality of bulls and heifers slaughtered at different ages. *Czech J. Anim. Sci.* **2012**, *57*, 34–43.

4.   Weglarz, A. Meat quality defined based on pH and colour depending on cattle category and slaughter season. *Czech J. Anim. Sci.* **2010**, *55*, 548–556. [CrossRef]

5. Avilés, C.; Martínez, A.L.; Domenech, V.; Peña, F. Effect of feeding system and breed on growth performance, and carcass and meat quality traits in two continental beef breeds. *Meat Sci.* **2015**, *107*, 94–103.

6. Karolyi, D.; Đikić, M.; Salajpal, K.; Jurić, I. Fatty acid composition of muscle and adipose tissue of beef cattle. *Ital. J. Anim. Sci.* **2009**, *8*, 264–266. [CrossRef]

7. Loudon, K.M.; Tarr, G.; Pethick, D.W.; Lean, I.J.; Polkinghorne, R.; Mason, M.; Dunshea, F.R.; Gardner, G.E.; McGilchrist, P. The Use of Biochemical Measurements to Identify Pre-Slaughter Stress in Pasture Finished Beef Cattle. *Animals* **2019**, *9*, 503. [CrossRef]

8. Monteils, V.; Sibraa, C.; Ellies-Ourya, M.-P.; Botreau, R.; De la Torre, A.; Laurent, C. A set of indicators to better characterize beef carcasses at the slaughterhouse level in addition to the EUROP system. *Livest. Sci.* **2017**, *202*, 44–51. [CrossRef]

9. Ripoll, G.; Albertí, P.; Casasús, I.; Blanco, M. Instrumental meat quality of veal calves reared under three management systems and color evolution of meat stored in three packaging systems. *Meat Sci.* **2013**, *93*, 336–343. [CrossRef]

10. Petrič, N.; Levart, A.; Čepon, M.; Žgur, S. Effect of production system on fatty acid composition of meat from Simmental bulls. *Ital. J. Anim. Sci.* **2005**, *4*, 125–127.

11. McIntyre, B.L.; Tudor, G.D.; Read, D.; Smart, W.; Della Bosca, T.J.; Speijers, E.J.; Orchard, B. Effects of growth path, sire type, calving time and sex on growth and carcass characteristics of beef cattle in the agricultural area of Western Australia. *Anim. Prod. Sci.* **2009**, *49*, 504–514. [CrossRef]

12. Senk, I.; Ostojic, G.; Jovanovic, V.; Tarjan, L.; Stankovski, S. Experiences in Developing Labs for a Supervisory Control and Data Acquisition Course for Undergraduate Mechatronics Education. *Comput. Appl. Eng. Educ.* **2015**, *23*, 54–62. [CrossRef]

13. Stankovski, S.; Ostojic, G.; Senk, I.; Rakic-Skokovic, M.; Trivunovic, S.; Kucevic, D. Dairy cow monitoring by RFID. *Sci. Agric.* **2012**, *69*, 75–80. [CrossRef]

14. McKean, J. The importance of traceability for public health and consumer protection. *Rev. Sci. Tech.* **2001**, *20*, 363–371. [CrossRef] [PubMed]

15. Kukolj, D.; Ostojić, D.; Stankovski, S.; Nemet, S. Technology Status Visualisation Using Patent Analytics: Multi-Compartment Refrigerators Case. *J. Mechatron. Autom. Identif. Technol.* **2019**, *4*, 1–8.

16. Litwińczuk, Z.; Barłowska, J.; Florek, M.; Tabała, K. Slaughter value of heifers, cows and young bulls from commercial beef production in the central-eastern region of Poland. *Anim. Sci. Pap. Rep.* **2006**, *24*, 187–194.

17. Nuernberg, K.; Dannenberger, D.; Nuernberg, G.; Ender, K.; Voigt, J.; Scollan, N.D.; Wood, J.D.; Nute, G.R.; Richardson, R.I. Effect of a grass-based and a concentrate feeding system on meat quality characteristics and fatty acid composition of *longissimus* muscle in different cattle breeds. *Livest. Prod. Sci.* **2005**, *94*, 137–147. [CrossRef]

18. National Research Council. *Nutrient Requirements of Beef Cattle*, Seventh ed.; National Acdemies Press: Washington, DC, USA, 2000.

19. Commission Regulation (EC). Commission Regulation (EC) No 1249/2008 of 10 December 2008 laying down detailed rules on the implementation of the Community scales for the classification of beef, pig and sheep carcasses and the reporting of prices thereof. Available online: https://op.europa.eu/en/publication-detail/-/publication/9716803a-8887-4956-9877-629031ec7723/language-en (accessed on 31 October 2019).

20. International Organisation for Standardisation. *ISO 2917: Meat and meat products, Measurement of pH (Reference method)*; International Organisation for Standardisation: Geneva, Switzerland, 1999.

21. Tomović, V.M.; Petrović, L.S.; Džinić, N.R. Effects of rapid chilling of carcasses and time of deboning on weight loss and technological quality of pork semimembranosus muscle. *Meat Sci.* **2008**, *80*, 1188–1193.

22. Honikel, K.O. Reference methods for the assessment of physical characteristics of meat. *Meat Sci.* **1998**, *49*, 447–457. [CrossRef]

23. De LEclairage, C.I. Colorimetry, official recommendations of the international commission on illumination. *Publ. CIE* **1976**, No. (E-1.31).

24. Hughes, J.M.; Oiseth, S.K.; Purslow, P.P.; Warner, R.D. A structural approach to understanding the interactions between colour, water-holding capacity and tenderness. *Meat Sci.* **2014**, *98*, 520–532. [CrossRef] [PubMed]

25. American Meat Science Association. *Meat Color Measurement Guidelines*; American Meat Science Association: Campaign, IL, USA, 2012.

26. Grau, R.; Hamm, R. Eine einfache Methode zur Bestimmung der Wasserbindung im Muskel. *Naturwissenschaften* **1953**, *40*, 29–30. [CrossRef]

27. Van Oeckel, M.J.; Warnants, N.; Boucqué, C.V. Comparison of different methods for measuring water holding capacity and juiciness of pork versus on-line screening methods. *Meat Sci.* **1999**, *51*, 313–320. [CrossRef]

28. Tomović, V.M.; Šević, R.; Jokanović, M.; Šojić, B.; Škaljac, S.; Tasić, T.; Ikonić, P.; Lušnic Polak, M.; Polak, T.; Demšar, L. Quality traits of *longissimus lumborum* muscle from White Mangalica, Duroc x White Mangalica and Large White pigs reared under intensive conditions and slaughtered at 150 kg live weight: A comparative study. *Arch. Anim. Breed* **2016**, *59*, 401–415.

29. Van Oeckel, M.J.; Warnants, N.; Boucqué, C.V. Pork tenderness estimation by taste panel, Warner-Bratzler shear force and on-line methods. *Meat Sci.* **1999**, *53*, 259–267. [CrossRef]

30. United States Department of Agriculture (USDA). *Official Marbling Photographs*; National Cattlemen's Beef Association United States Department of Agriculture: Fort Collins, CO, USA, 2007.

31. International Organisation for Standardisation. *ISO 1442: Meat and Meat Products, Determination of Moisture Content (Reference Method)*; International Organisation for Standardisation: Geneva, Switzerland, 1997.

32. International Organisation for Standardisation. *ISO 937: Meat and Meat Products, Determination of Nitrogen Content (Reference Method)*; International Organisation for Standardisation: Geneva, Switzerland, 1978.

33. International Organisation for Standardisation. *ISO 1443: Meat and Meat Products, Determination of Total Fat Content*; International Organisation for Standardisation: Geneva, Switzerland, 1973.

34. International Organisation for Standardisation. *ISO 936: Meat and Meat Products, Determination of Total Ash*; International Organisation for Standardisation: Geneva, Switzerland, 1998.

35. Tomović, V.M.; Petrović, L.S.; Tomović, M.S.; Kevrešan, Ž.S.; Džinić, N.R. Determination of mineral contents of semimembranosus muscle and liver from pure and crossbred pigs in Vojvodina (northern Serbia). *Food Chem.* **2011**, *124*, 342–348.

36. International Organisation for Standardisation. *ISO 13730: Meat and Meat Products, Determination of Total Phosphorus Content—Spectrometric Method*; International Organisation for Standardisation: Geneva, Switzerland, 1996.

37. International Organisation for Standardisation. *ISO 1444: Meat and Meat Products, Determination of Free Fat Content*; International Organisation for Standardisation: Geneva, Switzerland, 1996.

38. Arneth, W. Über die Bestimmung des intramuskulären Fettes. *Fleischwirtschaft* **1998**, *78*, 218–220.

39. Yurchenko, S.; Sats, A.; Poikalainen, V.; Karus, A. Method for determination of fatty acids in bovine colostrum using GC-FID. *Food Chem.* **2016**, *212*, 117–122. [CrossRef]

40. Kaminiecki, H.; Wójcik, J.; Pilarczyk, R.; Lachowicz, K.; Sobczak, M.; Grzesiak, W.; Błaszczyk, P. Growth and carcass performance of bull calves born from Hereford, Simmental and Charolais cows sired by Charolais bulls. *Czech J. Anim. Sci.* **2009**, *54*, 47–54. [CrossRef]

41. Scollan, N.; Hocquette, J.F.; Nuernberg, K.; Dannenberger, D.; Richardson, I.; Moloney, A. Innovations in beef production systems that enhance the nutritional and health value of beef lipids and their relationship with meat quality. *Meat Sci.* **2006**, *74*, 17–33. [CrossRef]

42. Herva, T.; Huuskonen, A.; Virtala, A.M.; Peltoniemi, O. On-farm welfare and carcass fat score of bulls at slaughter. *Livest. Sci.* **2011**, *138*, 159–166. [CrossRef]

43. Monteils, V.; Sibra, C. Rearing practices in each life period of beef heifers can be used to influence the carcass characteristics. *Ital. J. Anim. Sci.* **2019**, *18*, 734–745. [CrossRef]

44. Chambaz, A.; Scheeder, M.R.L.; Kreuzer, M.; Dufey, P.-A. Meat quality of Angus, Simmental, Charolais and Limousin steers compared at the same intramuscular fat content. *Meat Sci.* **2003**, *63*, 491–500. [CrossRef]

45. Pilarczyk, R. Concentrations of toxic and nutritional essential elements in meat from different beef breeds reared under intensive production systems. *Biol. Trace Elem. Res.* **2014**, *158*, 36–44. [CrossRef] [PubMed]

46. Florek, M.; Domaradzki, P.; Skałecki, P.; Stanek, P.; Litwińczuk, Z. Longissimus lumborum quality of Limousine suckler beef in relation to age and post-mortem vacuum ageing. *Ann. Anim. Sci.* **2015**, *15*, 785–797. [CrossRef]

47. Bureš, D.; Bartoň, L. Performance, carcass traits and meat quality of Aberdeen Angus, Gascon, Holstein and Fleckvieh finishing bulls. *Livest. Sci.* **2018**, *214*, 231–237. [CrossRef]

48. Sami, A.S.; Augustini, C.; Schwarz, F.J. Effects of feeding intensity and time on feed on performance, carcass characteristics and meat quality of Simmental bulls. *Meat Sci.* **2004**, *67*, 195–201. [CrossRef]

49. Nogalski, Z.; Pogorzelska-Przybyłek, P.; Sobczuk-Szul, M.; Purwin, C.; Modzelewska-Kapituła, M. Effects of rearing system and feeding intensity on the fattening performance and slaughter value of young crossbred bulls. *Ann. Anim. Sci.* **2018**, *18*, 835–847. [CrossRef]

50. Marti, S.; Realini, C.E.; Bach, A.; Pérez-Juan, M.; Devant, M. Effect of castration and slaughter age on performance, carcass, and meat quality traits of Holstein calves fed a high-concentrate diet. *J. Anim. Sci.* **2013**, *91*, 1129–1140. [CrossRef]

51. Domaradzki, P.; Florek, M.; Staszowska, A.; Litwińczuk, Z. Evaluation of the Mineral Concentration in Beef from Polish Native Cattle. *Biol. Trace Elem. Res.* **2016**, *171*, 328–332. [CrossRef]

52. Nogalski, Z.; Pogorzelska-Przybyłek, P.; Sobczuk-Szul, M.; Nogalska, A.; Modzelewska-Kapituła, M.; Purwin, C. Carcass characteristics and meat quality of bulls and steers slaughtered at two different ages. *Ital. J. Anim. Sci.* **2017**, *17*, 279–288. [CrossRef]

53. De Smet, S.; Webb, E.C.; Claeys, E. Effect of dietary energy and protein levels on fatty acid composition of intramuscular fat in double-muscled Belgian Blue bulls. *Meat sci.* **2000**, *56*, 73–79. [CrossRef]

Article

Weaning Holstein Calves at 17 Weeks of Age Enables Smooth Transition from Liquid to Solid Feed

Sarah Schwarzkopf [1], Asako Kinoshita [1], Jeannette Kluess [2], Susanne Kersten [2], Ulrich Meyer [2], Korinna Huber [1], Sven Dänicke [2] and Jana Frahm [2,*]

[1] Institute of Animal Science, Faculty of Agricultural Sciences, University of Hohenheim, 70599 Stuttgart, Germany; Sarah.Schwarzkopf@uni-hohenheim.de (S.S.); asakokinoshita@googlemail.com (A.K.); Korinna.Huber@uni-hohenheim.de (K.H.)
[2] Institute of Animal Nutrition, Friedrich-Loeffler-Institute, 38116 Braunschweig, Germany; jeannette.kluess@fli.de (J.K.); Susanne.Kersten@fli.de (S.K.); Ulrich.Meyer@fli.de (U.M.); Sven.Daenicke@fli.de (S.D.)
* Correspondence: Jana.Frahm@fli.de

Received: 23 October 2019; Accepted: 9 December 2019; Published: 12 December 2019

Simple Summary: Weaning calves from liquid to solid feed can be a stressful event in their life and can affect growth, development and welfare. It is commonly done at the age of 7 to 8 weeks on dairy farms, but weaning at a greater age could potentially reduce the associated stress. Therefore, it might improve growth rates and enable a smooth transition to an adult liver metabolism. To confirm this hypothesis this study evaluated the effect of two different weaning ages (7 vs. 17 weeks of age) on female Holstein calves. Furthermore, the effect of mothers' parity was analyzed (primiparous vs. multiparouos). Primiparous cows were often immature and still developing during their first pregnancy. This can lead to negative intrauterine conditions and result in long-term changes in the calf's metabolism. Late-weaned calves consumed high amounts of concentrate feed before weaning despite their high milk replacer intake, indicating the maturation of their rumen. In addition, they experienced a smooth transition to an adult liver metabolism as reflected by steady plasma glucose and cholesterol concentrations. Later weaning corrected the reduced growth of calves born to primiparous cows as well, indicating that those particularly benefitted from late weaning. All benefits were indicated by slower changes of blood metabolites and higher growth rates, which might lead to better health and productivity in their subsequent lifetime.

Abstract: Development of calves depends on prenatal and postnatal conditions. Primiparous cows were still maturing during pregnancy, which can lead to negative intrauterine conditions and affect the calf's metabolism. It is hypothesized that weaning calves at higher maturity has positive effects due to reduced metabolic stress. We aimed to evaluate effects of mothers' parity and calves' weaning age on growth performance and blood metabolites. Fifty-nine female Holstein calves (38.8 ± 5.3 kg birth weight, about 8 days old) were used in a 2 × 2 factorial experiment with factors weaning age (7 vs. 17 weeks) and parity of mother (primiparous vs. multiparous cows). Calves were randomly assigned one of these four groups. Live weight, live weight gain and morphometry increased over time and were greater in calves weaned later. Metabolic indicators except total protein were interactively affected by time and weaning age. Leptin remained low in early-weaned calves born to primiparous cows, while it increased in the other groups. The results suggest that weaning more mature calves has a positive effect on body growth, and calves born to primiparous cows particularly benefit from this weaning regimen. It also enables a smooth transition from liquid to solid feed, which might reduce the associated stress of weaning.

Keywords: weaning age; Holstein calves; growth; milk replacer; metabolism; development

---

1. Introduction

Calves are born as functional monogastric animals that rely on nutrients from milk or milk replacer (MR) [1,2]. Therefore, weaning is a vital event in the young ruminant's life, as it means that lactose and milk fat are no longer available as main sources for energy metabolism. The change from functional monogastric to ruminant not only relies on volatile fatty acids (VFA) production in the rumen to supply energy, but also on well-functioning endocrine and biochemical features such as ruminant-specific insulin homeostasis and hepatic gluconeogenesis. Thus, weaning causes stress [3,4], and could affect animal welfare, growth, development and future performance [5–7]. Calf mortality was high (5%–9%) in the last decades [8,9]. For economic reasons the preweaning period was substantially shortened in dairy cow production systems [10]. In the USA and Canada, the typical weaning age was 6–8 weeks [9,11]. Early weaning was introduced to promote the early intake of concentrate feed and hay, which are cheaper feeds than milk or MR. Therefore, feeding strategies for dairy calves focused on increasing the capacity for solid feed and accelerating the development of the forestomach system. In scientific studies, only a few variables such as beta-hydroxybutyrate in the plasma and rumen epithelium growth are used to indicate this [1,2]. A suggested potential benefit of early weaning was a faster rumen development [12], whereas the maturity of other parts of the gastrointestinal tract and organs like the liver was not considered. Therefore, little research was done about weaning calves older than 14 weeks [13].

Weaning calves which are more mature was discussed to have many benefits for the growth and development of dairy calves. It has been associated with greater live weight gain (LWG) and improved gastrointestinal development at the time of weaning [5]. For the first 56 days of life, feed efficiency (gain: feed ratio) tended to be greater for calves that were fed milk compared to grain [14]. The utilization of solid feed (corn silage, wheat straw, concentrate) for growth increased until 27 weeks of age [15].

Considering the great impact of weaning on dairy calves, it is crucial to find an optimal age for it. In this study, the optimal age denotes sufficient maturity in all organ and tissue functions, and not only in the ruminal digestion of solid feed. In fact, we define maturity for weaning as the capacity of all organs to fulfill the digestive and metabolic needs for changing to a ruminant status.

In the present study, early weaning was conducted at the age of 7 weeks, as this is a common management decision taken on dairy farms [9,11]. As opposed to that, late weaning was executed at the age of 17 weeks, because the reticulorumen volume of calves reaches adult proportions of 23 to 36 L/100 kg of ingesta-free body weight at 12 to 16 weeks [16].

The prenatal period as well as the early postnatal period are critical stages of development at which metabolic imprinting may occur and have great impact on health and performance in adult life [17]. Opsomer et al. [18] concluded in their review article that the parity of the mother could have a major impact, as older cows are lactating and heifers are still growing during pregnancy. Older cows tended to give birth to larger calves [19]. As birth weight was associated with improved glucose metabolism in humans in adulthood [20], mother's parity can influence the calf in the long-term. In most studies on calf development, the authors did not consider this parity of the mother as a potential influencing factor of development. Furthermore, few studies examined post-weaning development in female calves for a longer period.

The present study aimed to determine the influence of mother's parity and calves' weaning age on growth, energy and protein metabolism and on endocrine regulators. Growth performance and weight gain was evaluated by morphometric measures, and metabolic maturity was assessed by insulin, leptin and adiponectin as regulators of organ maturation. Since energy and protein metabolism are crucial for growth and development, indicators such as glucose, beta-hydroxybutyrate (BHB), non-esterified fatty acids (NEFA), cholesterol, urea and total protein were measured by spectrophotometric methods.

To assess metabolic imprinting as well as economic aspects, these animals were monitored in an ongoing observational study.

2. Materials and Methods

In accordance with the German Animal Welfare Act, pertaining to the protection of experimental animals and approved by The Lower Saxony State Office for Consumer Protection and Food Safety (LAVES), Oldenburg, Germany, the present trial was carried out at the experimental station of the Institute of Animal Nutrition, Friedrich-Loeffler-Institute (FLI), Brunswick, Germany (file No.: 33.19-42502-04-15/1858).

2.1. Animals, Housing, and Diets

Female German Holstein calves (n = 59) were studied from day of birth until day 149 ± 2 (mean ± standard deviation (SD)) of life. All calves originated from one established herd of Holstein cows and were born within a seasonal calving period of three months (October–December). They were all vaccinated with inactive *Mannheimia haemolytica* serotype A1 and A6, parainfluenza3 vaccine and bovine respiratory syncytial virus (Bovigrip® RSP plus, MSD Animal Health, Unterschleißheim, Germany) in weeks 5 and 9 of age, against *Trichophyton* in weeks 6 and 8 of age with live attenuate vaccine (Bovilis® Ringvac, MSD Animal Health, Unterschleißheim, Germany) and against blue tongue disease (BTV) (Zulvac 8, Zoetis Belgium SA, Louvain-la-Neuve, Belgium) in weeks 11 and 15 of age.

Calves were weighed with an electronic scale directly after birth and received 3 L of colostrum through a nipple bucket within 2 h after birth. The quality of colostrum was evaluated using a colostrum densimeter (Wahl GmbH, Dietmannsried, Germany) and had to be greater than 1035 g/L, otherwise colostrum from another cow from the same herd was used. They were moved 2–3 h after birth into straw-bedded single hutches and were fed twice with three liters of pooled herd milk each day. The pre-experimental feeding period for neonatal calves was done according to standard dairy management practice at the experimental station. In detail, starting at the age of three days, milk replacer (MR) (NOLAC GmbH, Zeven, Germany, Table 1) was mixed with the pooled herd milk, with gradually increasing amounts from 0.3 kg MR powder/d (third day of life) to 0.9 kg MR/d (fifth day of life), while the maximum of 6 L liquid feed with a concentration of 150 g/L MR was available (Table 2). Calves entered the study at a mean age of 8 ± 1.9 days and 44.5 ± 5.2 kg of live weight and moved into straw-bedded stables with MR and concentrate self-feeding systems (Förster-Technik GmbH, Engen, Baden-Württemberg, Germany). Differential feeding and monitoring of feed intake were achieved using a transponder in the calf's ear. They were randomly allocated to either early weaning at 7 weeks of age (early-weaned calves from multiparous cows (earlyMC)/early-weaned calves from primiparous cows (earlyPC)) or late weaning at 17 weeks of age (late-weaned calves from multiparous cows (lateMC)/late-weaned calves from primiparous cows (latePC)) group considering an equal allocation of calves from primiparous cows (PC) and calves from multiparous cows (MC). Our experimental trial started with 0.9 kg MR powder/d, which were available for all calves for the first five experimental days. MR was increased gradually within the next five days (experimental days 6 to 10) from 0.9 kg MR powder/d to 1.35 kg MR powder/d, and remained at this level until the beginning of the weaning period (early-weaned group = day 28, late-weaned group = day 98). Concentration of MR was continuously at 150 g MR powder/L over the complete experimental time, and a maximum of 9 L liquid feed was available (Table 2). Over the entire trial, all calves received hay and water ad libitum and had access to a maximum of 2 kg concentrate feed per day until weaning. With the start of weaning at experimental day 98, the amount of concentrate feed was reduced to 1 kg/d according to standardized dairy management practice at the experimental station. During weaning, the milk replacer was gradually reduced within 14 days from 1.35 kg/d to 0.3 kg/d. Post-weaning calves were moved to another barn and received hay and a total mixed ration (TMR) consisting of 48% grass, 32% maize silage and 20% concentrate feed.

The ingredients of MR powder and concentrate feed are shown in Table 1. Composition of concentrate feed, roughage, milk replacer and TMR were determined according to the suggestions of the Association of German Agricultural Analysis and Research Centers [21] (Table 3).

Table 1. Ingredients of milk replacer (MR) powder and concentrate feed.

Ingredients of MR Powder		Ingredients of Concentrate Feed	
Component	g/kg	Component	g/kg
Skimmed milk powder	320	Soybean meal	300
Sweet whey powder	198	Oat	305
Vegetable fat	140	Barley	180
Whey powder	102	Wheat	170
Whole milk powder	100	Soy bean oil	15
Buttermilk powder	100	Minerals and vitamins *	20
Minerals and vitamins	40	Calcium carbonate	10

* Ingredients per kg feed: 160 g Ca; 80 g P; 100 g Na; 30 g Mg; 1000 mg Fe; 800 mg Cu; 6000 mg Zn; 50 mg I; 50 mg Se; 30 mg Co; 800,000 IU vitamin A; 80,000 IU vitamin D3; 1000 mg vitamin E.

Table 2. Feeding regimen before and during experiment.

Experi-mental Day	Age in Days	MR Powder (g/L)		Available Volume of Liquid Feed per Day (L)		Available MR Powder per Day (kg)	
		Early Weaned	Late Weaned	Early Weaned	Late Weaned	Early Weaned	Late Weaned
	1–3 *	0	0	6	6	0	0
	3–5	150	150	6	6	Gradually increased from 0.3 to 0.9	Gradually increased from 0.3 to 0.9
	6 until start of experiment **	150	150	6	6	0.9	0.9
1 to 5		150	150	6	6	0.9	0.9
6 to 10		150	150	Gradually increased from 6 to 9	Gradually increased from 6 to 9	Gradually increased from 0.9 to 1.35	Gradually increased from 0.9 to 1.35
11 to 28		150	150	9	9	1.35	1.35
29 to 42		150	150	Gradually decreased from 9 to 2	9	Gradually decreased from 1.35 to 0.3	1.35
42 to 98		0	150	0	9	0	1.35
99 to 112		0	150	0	Gradually decreased from 9 to 2	0	Gradually decreased from 1.35 to 0.3
113 to 140		0	0	0	0	0	0

* 3 L of colostrum within 2 h after birth; pooled herd milk in the first three days of life. ** start of experiment at mean age of 8 ± 1.9 days, ranging from 6–12 days, one animal of earlyMC was 18 days old when entering the experiment.

Table 3. Ingredients of concentrate feed, roughage, milk replacer and total mixed ration (TMR).

Feed	DM %	XA g/kg T	XP g/kg T	XL g/kg T	XF g/kg T	NDF g/kg T	ADF g/kg T	Starch g/kg T	Sugar g/kg T
concentrate feed	86.84	63.21	232.31	47.64	68.63	199.09	82.61	363.67	48.32
roughage	86.06	66.61	98.28	22.05	326.74	660.53	365.18		
milk replacer	96.98	79.12	248.56	182.63					
TMR	39.29	43.75	81.22	35.35	217.76	426.36	244.81	301.51	

All ingredients were assessed by Weender analysis. Dry matter (DM), crude ash (XA), crude protein (XP) and crude fat (XL) were analyzed in all feedstuff. Crude fiber (XF), neutral detergent fiber (NDF) and acid detergent fiber (ADF) were analyzed in the solid feed. In concentrate feed and TMR starch was analyzed and in concentrate feed additionally sugar was analyzed.

2.2. Morphometry of Calves

Concerning morphometry, the hip height, withers height, back length, body length and heart girth were determined as shown in Table 4 at days 1, 7, 14, 28, 42, 56, 70, 84, 98, 112, 126 and 140 of this trial. Hip and withers height were measured with a folding rule, the other measurements were taken with a measuring tape. Live weight (LW) was recorded on day of birth, and also on days 1, 28, 42, 70, 98, 112 and 140 with an electronic scale. Live weight gain (LWG) in kg per day was calculated from this data by dividing the weight gain between our sample days through the number of days between sampling.

Table 4. Morphometry of calves.

Measure	Definition
Withers height	From floor to dorsal process of first thoracic vertebra
Hip height	From floor to sacrum
Back length	From dorsal process of first thoracic vertebra to sacrum
Body length	From shoulder joint to ischium
Heart girth	Behind front leg

Definition of morphometry.

2.3. Collection and Analysis of Blood Samples

Blood samples of each individual animal were taken on experimental days 1, 28, 42, 70, 98, 112 and 140 by jugular venipuncture and collected in serum and ethylenediaminetetraacetic acid (EDTA) plasma tubes (10 mL tubes; Sarstedt, Nuembrecht, Germany). Serum tubes were incubated for 30 min at 30 °C. After centrifugation at 3000× g for 15 min at 15 °C, serum and plasma aliquots were stored at −80 °C for subsequent analyses. Serum leptin concentrations were determined using a competitive enzyme immunoassay according to Sauerwein et al. [22]. Adiponectin concentrations were analyzed in serum with an indirect competitive bovine specific enzyme-linked immunosorbent assay (ELISA) according to Mielenz et al. [23]. Analyses of serum concentrations of beta-hydroxybutyrate (BHB), non-esterified fatty acids (NEFA), cholesterol, urea, total protein and glucose were done by an automatic analyzing system, based on spectrometric measures (Eurolyser, Type VET CCA, Salzburg, Austria). Insulin concentration in plasma was analyzed with a bovine insulin ELISA (Mercodia, Sweden).

2.4. Statistical Analysis

Live weight (LW), live weight gain (LWG), hip and withers height, body length, heart girth, back length, serum glucose, beta-hydroxybutyrate (BHB), non-esterified fatty acids (NEFA), leptin concentrations and plasma insulin concentrations were presented as least squares means (LSMeans) and standard errors (SEs) which were evaluated by repeated measures using the PROC MIXED procedure in SAS (V 9.4., SAS Institute Inc., Cary, NC, USA), and employing a restricted maximum likelihood model (REML). The model included a fixed factor of time, weaning age, parity of the mother and their interactions while the time was taken into consideration by a "REPEATED" statement. Best fitting covariance structures (compound symmetry, autoregressive and unstructured) was tested and used, based on the Akaike Information Criterion (AICC). Significant effects were further tested with the Tukey–Kramer procedure using the piecewise differentiable (PDIFF) procedure. Visualization and correlations computed as Pearson correlation coefficients were done using GraphPad Prism 6.0 (GraphPad software, San Diego, CA, USA). For all statistical tests, $p < 0.05$ was the level of significance. For visualization, the measurements on serial time points were interpolated linearly.

3. Results

Multiparous cows were 1592 ± 805 days (Mean ± SD) old when they gave birth to MC. Their mean lactation number was 1.875 ± 0.074 lactations (Mean ± SD). The age of primiparous cows was 710 ± 67 days (Mean ± SD) at calving. Birth weight of PC was 37.9 ± 4 kg (Mean ± SD) and birth weight of MC was 39.6 ± 6 kg (Mean ± SD).

3.1. Feed Intake

There was no difference in feed intake between calves from multiparous (MC) and primiparous cows (PC) in both weaning groups. Therefore, all data from calves of one weaning group were combined for the visualization of feed intake patterns in early and late weaned calves (Figure 1). Both groups had the same MR intake for the first 28 days of trial. Early-weaned calves consumed 11,288 g MR DM on average during their weaning period (days 28–42) whereas late-weaned calves consumed 7182 g MR DM on average during their weaning (days 98–112). Thus, the MR intake during weaning was lower for late-weaned calves compared to early-weaned calves. Early-weaned calves consumed their whole MR allowance until weaning, whereas late-weaned calves reduced their MR intake earlier than they had to. Both weaning groups started to consume concentrate feed around day 21 of trial. Early-weaned calves increased their concentrate feed intake during their weaning period. Late-weaned calves also increased their intake until day 63 of the trial and then consumed between 1500 and 1700 g concentrate DM/day until weaning. Late-weaned calves increased their roughage intake when the MR supply was reduced, whereas early-weaned calves started to consume roughage when weaning was already done (data not shown).

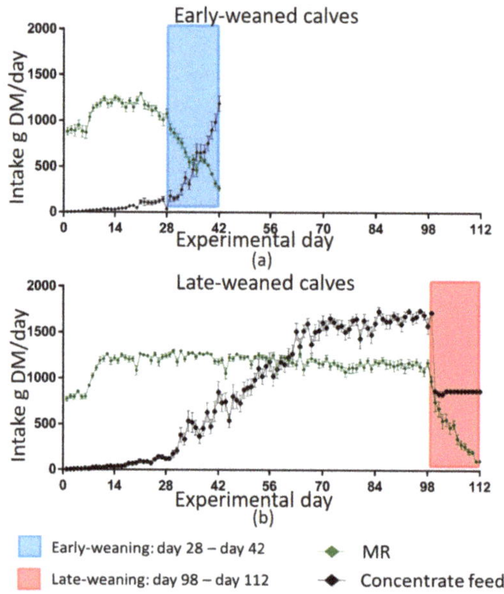

Figure 1. Milk replacer (MR) and concentrate feed intake in g dry matter/d for early-weaned calves (**a**) and late-weaned calves (**b**). Early-weaned calves ($n = 24$) were weaned gradually between days 28 and 42 of trial. Late-weaned calves ($n = 28$) were weaned gradually between days 98 and 112 of trial. Amount of concentrate feed was limited to 1 kg/d after weaning for early-weaned calves and during weaning period of late-weaned calves. Data shown as means ± the standard error of the mean (SEM) evaluated with GraphPad Prism 6. Due to technical problems, not all calves could be included in monitoring feed intake.

3.2. Morphometry

Morphometric variables increased over time ($p < 0.001$) and were greater in late-weaned calves. Interactions between time and weaning group were also observed to be highly significant for LW, withers and hip height ($p < 0.001$), heart girth ($p < 0.01$) and body length ($p < 0.05$; Figure 2). LW was greater for all calves in the late-weaned group from day 70 until the end of trial (Figure 2a). On day 140 the mean live weight of early-weaned calves was 164.1 kg ± 3.65 kg, whereas that of late-weaned calves was 186.1 ± 3.88 kg ($p = 0.009$). LWG (Figure 2b) was strongly influenced by the weaning age ($p < 0.001$). Additionally, a significant interaction between time and weaning age was found ($p < 0.001$). Late-weaned calves had a significantly higher LWG from day 42 until day 98 of our trial ($p < 0.05$). Withers height differed significantly between late- and early-weaned calves on day 84 ($p = 0.014$) and day 140 ($p < 0.001$). Hip height differed significantly between the weaning groups from day 56 until day 84 ($p < 0.05$) and on day 140 ($p < 0.001$; Figure 2c,d). Late-weaned calves had a greater heart girth from day 84 onwards ($p < 0.05$; Figure 2f). Body length was significantly lower for early-weaned calves at the end of trial on day 126 and day 140 ($p < 0.01$). Back length was the only morphometric variable influenced by parity of the mother as indicated by an interaction between parity and weaning age ($p = 0.011$). EarlyPC had significantly lower back length than earlyMC ($p < 0.001$), whereas it did not differ in the parity groups of late-weaned calves (Figure 2e).

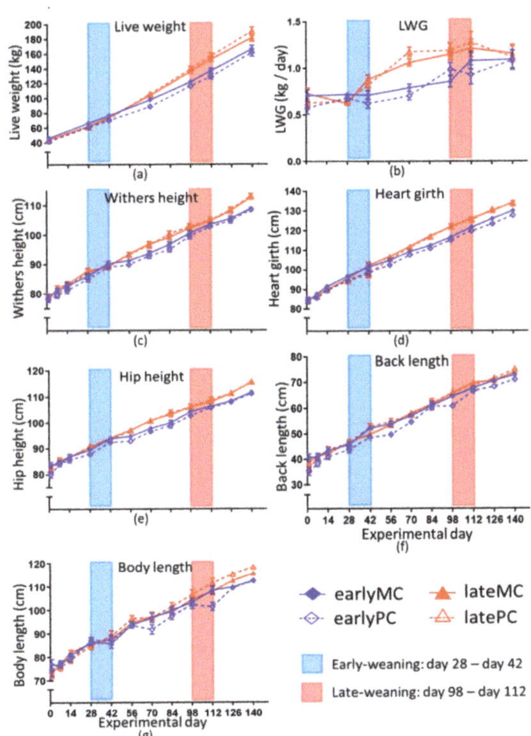

| *p*-Values | | | | | | | |
Parameters	Time (T)	Parity (P)	Weaning Age (W)	T × P	T × W	P × W	T × P × W
Live weight	**<0.001**	0.396	**<0.001**	0.685	**<0.001**	0.099	0.453
LWG	**<0.001**	0.422	**<0.001**	0.221	**<0.001**	0.325	0.556
Withers height	**<0.001**	0.419	**0.005**	0.444	**<0.001**	0.436	0.659
Hip height	**<0.001**	0.365	**0.002**	0.233	**<0.001**	0.364	0.985
Back length	**<0.001**	**0.035**	**0.008**	0.481	0.560	**0.011**	0.129
Heart girth	**<0.001**	0.122	**0.004**	0.899	**<0.001**	0.537	0.530
Body length	**<0.001**	0.791	**0.023**	0.140	**0.005**	0.116	0.217

Figure 2. Morphometry of calves. Shown are live weight (**a**), live weight gain (LWG) (**b**), withers (**c**), heart girth (**d**), hip height (**e**) and back length (**f**). Early-weaned calves were weaned gradually between days 28 and 42 of the trial. Late-weaned calves were weaned gradually between days 98 and 112 of the trial. Data shown as LSmeans ± SEM, early-weaned calves from multiparous cows (earlyMC) $n = 16$, late-weaned calves from multiparous cows (lateMC) $n = 16$, early-weaned calves from primiparous cows (earlyPC) $n = 15$, late-weaned calves from primiparous cows (latePC) $n = 12$.

3.3. Blood Parameters

Time had a significant effect on all measured variables in the blood ($p < 0.001$) and there was an interaction of time and weaning age observed for all variables except total protein (Figures 3 and 4). On day 70, which was between the two weaning periods, the two weaning groups differed highly significant in their serum glucose concentration ($p < 0.001$, Figure 3a). Blood glucose concentration increased significantly from day 70 to day 112 in the early-weaned calves ($p = 0.014$). Insulin concentration dropped on days 98 and 112 in early-weaned calves and stayed below those insulin concentrations of late-weaned calves during the rest of the trial until day 140 (Figure 3b). Total protein concentration in calves (Figure 3c) was influenced by mother's parity ($p = 0.021$) and was higher in MCs. Urea concentration (Figure 3d) in late-weaned calves increased constantly up to day 98 and started to drop to the initial level during weaning. In early-weaned calves, it increased during weaning and decreased afterwards. Therefore, they reached lower urea concentrations after weaning than late-weaned calves from day 42 until day 98 ($p < 0.001$).

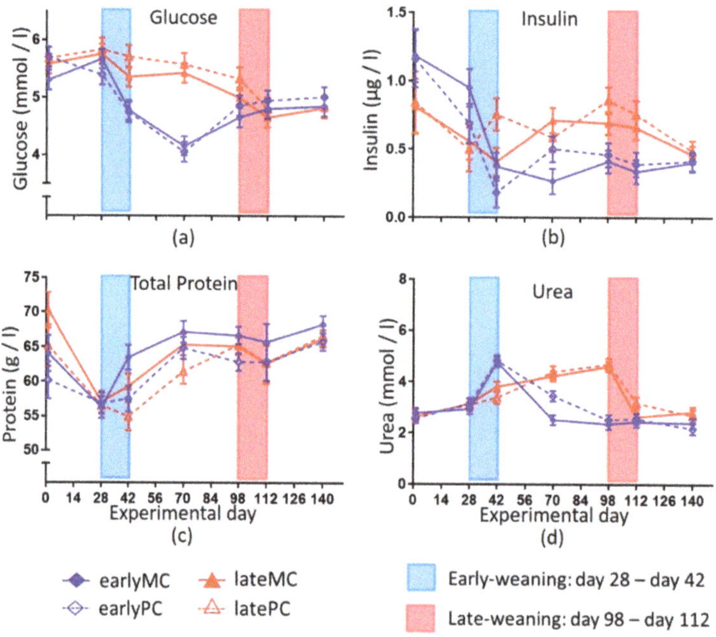

p-Values							
Parameters	Time (T)	Parity (P)	Weaning Age (W)	T × P	T × W	P × W	T × P × W
insulin	**<0.001**	0.654	0.089	0.790	**<0.001**	0.448	0.080
glucose	**<0.001**	0.224	**<0.001**	0.538	**<0.001**	0.620	0.781
total protein	**<0.001**	**0.021**	0.779	0.229	0.140	0.597	0.714
urea	**<0.001**	0.318	**<0.001**	0.064	**<0.001**	0.532	0.464

Figure 3. Blood concentrations of the glucose (**a**), insulin (**b**), total protein (**c**) and urea (**d**) of calves. Early-weaned calves were weaned gradually between days 28 and 42 of the trial. Late-weaned calves were weaned gradually between days 98 and 112 of this trial. Data shown as LSmeans ± SEM, early-weaned calves from multiparous cows (earlyMC) $n = 16$, late-weaned calves from multiparous cows (lateMC) $n = 16$, early-weaned calves from primiparous cows (earlyPC) $n = 15$, late-weaned calves from primiparous cows (latePC) $n = 12$.

Cholesterol concentrations increased similarly in all groups from day 1 to day 28. After weaning, it decreased in the early-weaned calves from day 28 to day 70 ($p < 0.001$). Therefore, they showed lower cholesterol concentrations than late-weaned calves until day 112 ($p < 0.01$). From day 70 to day 140 it increased significantly in early-weaned calves ($p = 0.011$; Figure 4a). NEFA concentrations decreased with weaning in early-weaned calves. Therefore, late-weaned calves had higher NEFA concentrations on day 70 ($p = 0.001$; Figure 4b). The serum BHB concentration increased after weaning for the early-weaned calves (day 42–70, $p < 0.001$), whereas it remained low in the late-weaned calves and increased after their weaning period (day 112–140; $p < 0.001$). As a result, a significant difference in serum BHB concentration was observed between the weaning groups on day 70 ($p = 0.001$; Figure 4c). After day 70, BHB concentration decreased until day 140 in early-weaned calves ($p < 0.001$). BHB concentration was negatively correlated with glucose concentration when all treatments and time points were considered collectively ($p = 0.0001$; r = −0.1895). Serum leptin concentration showed a significant interaction of weaning age and time ($p < 0.001$), and was also influenced by parity of the mother ($p < 0.05$). Serum leptin concentration increased from day 28 to day 140 in late-weaned calves ($p = 0.008$), whereas no significant increase was found in early-weaned calves (Figure 4d). Calves' plasma leptin concentrations correlated positively with the lactation number of the mother ($p = 0.0015$; r = 0.4177; data not shown). All measured blood metabolites except insulin (Figure 3a), total protein (Figure 3c) and NEFA (Figure 4b) were affected by weaning age. Weaning age did not affect Adiponectin concentrations, but there were significant effects of time ($p < 0.001$) and interaction between time and parity ($p = 0.031$; data not shown).

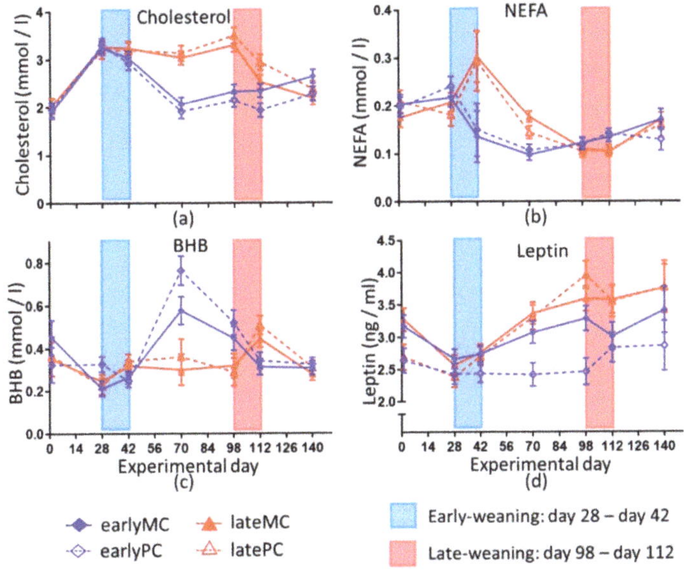

p-Values							
Parameters	**Time (T)**	**Parity (P)**	**Weaning Age (W)**	**T × P**	**T × W**	**P × W**	**T × P × W**
cholesterol	**<0.001**	0.995	**<0.001**	0.990	**<0.001**	0.327	0.344
NEFA	**<0.001**	0.766	0.072	0.737	**<0.001**	0.735	0.440
BHB	**<0.001**	0.225	**0.026**	0.540	**<0.001**	0.685	0.511
leptin	**<0.001**	**0.025**	**<0.001**	0.448	**<0.05**	0.100	0.115

Figure 4. Indicators for lipid metabolism of calves. Shown are cholesterol (**a**), non-esterified fatty acids (NEFA) (**b**), beta-hydroxybutyrate (BHB) (**c**) and leptin (**d**). Early-weaned calves were weaned gradually between days 28 and 42 of this trial. Late-weaned calves were weaned gradually between days 98 and 112 of this same trial. Data shown as LSmeans ± SEM, early-weaned calves from multiparous cows (earlyMC) *n* = 16, late-weaned calves from multiparous cows (lateMC) *n* = 16, early-weaned calves from primiparous cows (earlyPC) *n* = 15, late-weaned calves from primiparous cows (latePC) *n* = 12.

4. Discussion

This study assessed the impact of two different weaning ages. Furthermore, calves were grouped according to their mother's parity. Precisely, calves born to primiparous and born to multiparous cows were allocated to both weaning groups. To assess the effect of weaning age, a high-quality MR was used, consisting mostly of milk components (Table 2). One liter of MR contained the same amount of protein as whole milk (36.13 g XP/L MR vs. 35 g XP/L whole milk). The amount of MR powder used in literature ranges from 0.383 kg per day to 1.49 kg per day [24], thus the calves in the present study received a high amount of MR, which was quite similar to an ad libitum intake [25]. Voluntary DM intake/day from MR was lower than 1300 g/day in the first 8 weeks of life [26], which was the maximum allowance in the current study. Therefore, the effects of weaning age were assessed under sufficient milk-derived energy and nutrient supply, and not negatively influenced by a low amount and quality of MR.

4.1. Feed Intake

Data from computer-controlled mangers for roughage were not shown as there were technical problems with recognizing the individual calf, precluding data collection. As expected, MR intake was not different in the four groups during the first 28 days of trial when MR allowance was the same for all calves. Afterwards, MR intake followed the regimen of weaning. However, the pattern of voluntary solid feed intake over time varied between the weaning groups. Late-weaned calves consumed concentrate feed—despite consumption of the full amount of MR and even before MR supply was reduced, which led to a greater concentrate feed intake at the beginning of their weaning compared to the early-weaned group. This was in line with previous findings [5,13]. Later weaning (12.7 weeks) resulted in concentrate feed intake before weaning (close to 0.5 kg/d), whereas early-weaned calves (6.7 weeks) did not increase their concentrate intake before weaning started [13]. Eckert et al. [5] observed the same feed intake pattern for calves that were weaned with 6 or 8 weeks, with calves weaned later consuming more concentrate feed 1 week pre- and post-weaning. Male calves permitted to choose their preferred feed between MR and different solid feed components started to consume concentrate feed at the age of 49 days [26]. Despite the high MR allowance, calves consumed solid feed as observed in the present study. As milk production in the first lactation was associated with a higher intake of grain and forage at weaning [7], voluntary solid feed intake before weaning can be a potential benefit for later life production. Higher starter intake was related to a higher weight gain during weaning [13]. Late-weaned calves probably consumed more energy until their weaning, as they had a higher concentrate feed intake and still consumed MR. This was associated with the higher growth rates of late-weaned calves (Figure 2). This also indicates that even though the late-weaned calves had a high MR supply for 15 weeks, they started to consume solid feed and their rumen probably started to develop and to maturate. They even restricted their MR intake during the weaning period voluntarily more than they had to, which might indicate the ability of mature organs to function ruminant-specifically.

It is proposed that solid feed was digested in the rumen as indicated by several rumen development parameters (Schwarzkopf et al., unpublished data). This indicated a development of rumen digestive functions despite high MR intake. Besides gastrointestinal development, however, liquid feeding in addition to voluntary solid feed intake over 17 weeks in the early life of calves might also be of advantage for other body functions and endocrine regulatory processes, as demonstrated in the following sections.

4.2. Morphometry of Calves

Weaning calves at a more mature developmental stage (17 weeks of life) resulted in increased LWG and higher LW (Figure 2). This was also demonstrated by several other studies [5,27,28]. The reticulorumen reached its adult proportions at the age of 12 to 16 weeks [16]. Additionally, the utilization of solid feed for body growth increased with age. Berends et al. [15] adjusted the quantity of MR for male calves to achieve the same weight gain across different solid feed levels. This way, they measured an increasing utilization of solid feed until the age of 27 weeks. If this was a sign of maturity and a proper function, it could be an explanation why later-weaned calves had a greater live weight and were able to maintain this at least 4 weeks after weaning. It could mean that the early-weaned young ruminants were physiologically unable to use all the energy provided with solid feed, and their preferred sources of energy remained lactose and milk fat. As promoted by early-weaning, the intake of solid feed instead of MR also resulted in a greater need of energy for ruminal activity. Elevated muscle work in the rumen needs energy. Furthermore, heat production through fermentation in the rumen [29] increased the energetic need for thermoregulation. Thus, the metabolic rate for maintenance might increase by solid feed intake. This could also be an explanation for the lower LWG in early-weaned calves compared to the late-weaned groups. Early-weaned calves were unable to compensate the reduced growth and could not catch up with weight, wither and hip height, body length and heart girth at least until the age of 5 months (Figure 2, $p < 0.05$). Furthermore, back length showed a significant interaction of mother's parity and weaning age ($p = 0.011$), showing shorter back lengths in earlyPC. Reduced back length is a well-known symptom of prenatal imprinting by intrauterine malnutrition in rodents, sheep and humans [30]. Weaning late, however, was advantageous to correct the imprinted change in body proportions, as latePC did not express shorter back lengths.

4.3. Blood Metabolites

Collection of blood samples was always in the morning between eight and ten, but it was not controlled for feed intake. Part of the variance of serum insulin and glucose concentration could be a result of different times of feed intake relative to sampling. In milk-fed calves, blood glucose increased 1 h after feeding and then decreased rapidly during the next 2 h [28]. A high capacity for hepatic gluconeogenesis is an essential metabolic feature for a ruminant due to low intestinal glucose availability. Weaning late at 17 weeks of age resulted in a smooth transition of glucose metabolism from lactose to endogenous glucose production by the liver without any signs of dysregulation. In early-weaned calves, weaning led to a strong decrease in blood glucose concentration (Figure 3a), as lactose is the most important source of glucose for young calves. Obviously, early-weaned calves were not able to compensate the lack of dietary glucose (lactose) by hepatic gluconeogenesis. The glucose gap was closed slowly. There was a significant increase in blood glucose concentration from day 70 to day 112 in the early-weaned calves ($p = 0.014$), which indicated that liver gluconeogenetic function matured slowly until 4 months of age. This is in accordance with findings of other authors [27]. Calves that were weaned at the age of 5 weeks had lower blood glucose concentrations than calves that still received MR, even 7 weeks after weaning [28].

Low blood glucose concentrations are detrimental for a developing young animal, since related endocrine status is concomitantly changed. The decrease in glucose concentrations at early-weaning resulted in a decrease in serum insulin concentrations. In more mature calves, the endocrine system was smoothly adapting when weaning was done with only marginal changes in hormone concentrations. The decrease of insulin was steeper for early-weaned calves and resulted in more abrupt changes (Figure 3b). Therefore, weaning stress of calves was attenuated when weaned later, because it has more time to mature and adapt. In accordance, other studies also showed that a lower intake of MR decreased insulin and insulin like growth factor 1 (IGF1) concentrations in calves [24,31,32]. Therefore, a catabolic status was most likely established in early-weaned calves, as insulin and IGF1 are the strongest anabolic hormones.

Consequently, lipolysis and proteolysis most likely were promoted to gain energy and produce precursors for gluconeogenesis. The catabolic state, however, was not able to increase glucose concentrations for several weeks in early-weaned calves. Both pathways led to an increase in ketone bodies in blood when the oxidative capacity of mitochondria was limited [33]. The negative correlation between glucose and BHB concentrations supported the hypothesis that BHB originated from lipolysis and proteolysis because of glucose shortage. NEFA derived from lipolysis were used in beta-oxidation.

Hence, the blood concentrations decreased in early-weaned calves (Figure 4b). Simultaneously, the blood BHB concentration increased, reflecting a low capacity of hepatic oxidative phosphorylation. Furthermore, an incomplete oxidation of amino acids could also result in a higher BHB concentration. The decrease in cholesterol concentrations might also be linked to higher ketone body production, as the precursor metabolite (3-Hydroxy-3-methyl-glutaryl-CoA) for cholesterol production was used for BHB production [34]. Low insulin concentration led to a lower activation of HMG-CoA-reductase, which is vital for cholesterol biosynthesis [35]. This might also be a reason for lower cholesterol production. There was a significant increase in cholesterol concentration in the early-weaned groups from day 70 to day 140 of this trial ($p = 0.011$), which reflected that the liver was unable to produce as much cholesterol in the young, early-weaned calves on account of later production when they were more mature. A higher energy supply enhanced cholesterol biosynthesis in 16-week-old bull calves [36], which is in line with our findings in late-weaned female calves that received more energy through MR and a higher concentrate feed intake (Figure 1). Moreover, the MR is a nutritional source for cholesterol, which is no longer available after weaning. Therefore, late-weaned calves had a constant supply of cholesterol until day 98 of the trial. The liver of early-weaned calves was not able to produce enough cholesterol and glucose to compensate the dietary lack through weaning for several weeks. Possibly, these are not the only metabolic pathways that did not mature in early-weaned calves, and other important ones were impaired, as well.

Blood BHB concentration was used as a marker for rumen development as it originated from rumen wall ketogenesis [37]. Many authors observed a rise in blood BHB at weaning and with starter intake [14,38]; while lower concentrations were detected in ad libitum MR fed calves (0.14 ± 0.01 mmol/L) compared to calves fed with restricted MR (0.17 ± 0.01 mmol/L) [31]. A similar pattern was observed in this study, as serum BHB concentrations were higher in early-weaned calves compared to late-weaned calves still on MR feeding on day 70 (Figure 4c). But these higher concentrations declined again with age ($p < 0.001$). Thus, their relevance as a marker for rumen development could be questioned. It is likely that to some extent BHB was derived from an incomplete oxidation of nutrients, such as fatty acids and amino acids as described above. Parts of higher NEFA concentration in late-weaned calves on day 70 might also be explained by NEFA content in MR (Figure 4). Other authors had also seen that weaning resulted in lower NEFA concentration compared to calves that received MR [14,39]. Low glucose and high BHB after weaning indicated that capacity of liver functions was less developed in early-weaned calves. The increase in plasma urea concentration during weaning could have resulted from elevated proteolysis as well. After weaning the urea concentration likely decreased because the rumen used it in the ruminohepatic recycling of nitrogen. Another explanation might be a lack of microbial protein and low protein sustenance. Hence, plasma urea concentrations were lower in early-weaned calves than in late-weaned calves that still received MR and thus got enough protein for body protein turn-over.

Influence of Mother's Parity on Early- and Late-weaned Calves

As discussed before, heifers have to allocate nutrients and energy between their own body needs and the requirements of the fetus. This could create unfavorable conditions such as intrauterine malnutrition for the unborn calf, which could affect them for their whole life [18]. Besides reduced back length, a lower leptin concentration appeared to be another sign of intrauterine imprinting by malnutrition (Figure 4d). Furthermore, the lactation number of the dams and serum leptin concentrations in their 1-week old calves were positively correlated ($p = 0.0015$; $r = 0.4177$). Leptin plays an important role in the onset of puberty and regulation of the immune system [40,41].

Low leptin concentration in early postnatal life was associated with a leptin resistance in later life in rats [42]. Thus, the lower leptin concentrations in PC could indicate a potential risk factor for a dysregulated energy metabolism and development also in later life [43]. The main effect of leptin is a decrease in feed intake. In general, leptin concentration changes during a long-term negative energy balance in mammalians [44]. Hence, the lower serum leptin concentrations in the PC can be hypothesized as a sign of hunger and lack of adipose deposition.

4.4. Implications and Perspectives

Most of the evaluated blood metabolites did not differ among the groups at the age of 5 months, but changes occurred more abruptly for early-weaned, and slower for later-weaned calves. Hence, early weaning may cause more postnatal metabolic stress. This experience could also lead to metabolic imprinting and affect health and productivity in later life [20]. Kenéz et al. [6] examined a reduced amount of MR supply during the rearing period, which sustainably affected the development and altered the metabolism. These changes could still be seen at first lactation. So, it is possible that, weaning age or mother's parity will affect these animals in later life through metabolic imprinting. The existence and quality of long-term consequences are examined in an ongoing study with the same animals during their first and second lactation. Further research is needed to identify an optimal age to transition from MR to solid feed and an optimal amount of MR. It is challenging to distinguish the effect of older weaning age from the effect of an overall higher amount of MR that is consumed. Thus, it would be advisable to conduct further research on both factors and their influence on the growth and development of calves. Voluntary MR and solid feed intake are different in individual calves [26,45], and should therefore be considered in further research about optimal weaning age. From this study, a later weaning regimen can be considered as advantageous in early life with potential consequences for later health and metabolic performance. Naturally, calves suckling their dam were weaned at the age of 8–9 months [46,47]. This behavior might have been established through evolution, creating the best possible outcome for calves, and should therefore be considered in re-thinking weaning regimens in dairy calves.

5. Conclusions

Increasing weaning age to 17 weeks enables a smooth transition of physiological functions from the pseudomonogastric status to full ruminant status in dairy calves. However, weaning at 17 weeks of age is not only influenced by milk. The calves ingested up to 2 kg of concentrate feed, despite a high intake of MR. Thus, body maturation was supported by both sources of energy.

Author Contributions: Conceptualization and Methodology, K.H., S.D. and J.F.; Investigation, J.F., S.S., A.K., J.K., S.K., U.M.; Formal analysis, J.F.; Visualization, S.S.; Writing—original draft, S.S.; Writing—review & editing, J.F., K.H., S.D., S.S.

Funding: This research received no external funding.

Acknowledgments: Many thanks to the technical staff of the experimental station of the Institute of Animal Nutrition, Friedrich-Loeffler-Institute (FLI), Brunswick, Germany. We also want to thank the workgroup of Sauerwein (Institute of Animal Science, Physiology and Hygiene Unit of the University of Bonn) for leptin and adiponectin analyzes.

Conflicts of Interest: The authors declare no conflict of interest.

References

1. Baldwin, R.L.; McLeoad, K.R.; Klotz, J.L.; Heitmann, R.N. Rumen development, intestinal growth and hepatic metabolism in the pre- and postweaning ruminant. *J. Dairy Sci.* **2004**, *87* (Suppl. E.), E55–E65. [CrossRef]
2. Khan, M.A.; Bach, A.; Weary, D.M.; von Keyserlingk, M.A.G. Invited review: Transitioning from milk to solid feed in dairy heifers. *J. Dairy Sci.* **2016**, *99*, 885–902. [CrossRef] [PubMed]
3. Kim, M.-H.; Yang, J.-Y.; Upadhaya, S.D.; Lee, H.-J.; Yun, C.-H.; Ha, J.K. The stress of weaning influences serum levels of acute-phase proteins, iron-binding proteins, inflammatory cytokines, cortisol, and leukocyte subsets in Holstein calves. *J. Vet. Sci.* **2011**, *12*, 151–157. [CrossRef] [PubMed]
4. Steele, M.A.; Penner, G.B.; Chaucheyras-Durand, F.; Le Guan, L. Development and physiology of the rumen and the lower gut: Targets for improving gut health. *J. Dairy Sci.* **2016**, *99*, 4955–4966. [CrossRef] [PubMed]
5. Eckert, E.; Brown, H.E.; Leslie, K.E.; DeVries, T.J.; Steele, M.A. Weaning age affects growth, feed intake, gastrointestinal development and behavior in Holstein calves fed an elevated plan of nutrition during the preweaning stage. *J. Dairy Sci.* **2015**, *98*, 6315–6326. [CrossRef] [PubMed]

6. Kenéz, Á.; Koch, C.; Korst, M.; Kesser, J.; Eder, K.; Sauerwein, H.; Huber, K. Different milk feeding intensities during the first 4 weeks of rearing dairy calves: Part 3: Plasma metabolomics analysis reveals long-term metabolic imprinting in Holstein heifers. *J. Dairy Sci.* **2018**, *101*, 8446–8460. [CrossRef] [PubMed]

7. Heinrichs, A.J.; Heinrichs, B.S. A prospective study of calf factors affecting first-lactation and lifetime milk production and age of cows when removed from the herd. *J. Dairy Sci.* **2001**, *94*, 336–341. [CrossRef]

8. Compton, C.W.R.; Heuer, C.; Thomsen, P.T.; Carpenter, T.E.; Phyn, C.V.C.; McDougall, S. Invited review: A systematic literature review and meta-analysis of mortality and culling in dairy cattle. *J. Dairy Sci.* **2017**, *100*, 1–16. [CrossRef]

9. USDA. Dairy 2007. Heifer Calf Health and Management Practices on U.S. Dairy Operations 2010. Available online: https://www.aphis.usda.gov/animal_health/nahms/dairy/downloads/dairy07/Dairy07_ir_CalfHealth.pdf (accessed on 8 October 2019).

10. Spiekers, H.; Potthast, V. *Erfolgreiche Milchviehfütterung*, 4th ed.; DLG-Verlags-GmbH: Frankfurt am Main, Germany, 2004; ISBN 3-7690-0573-2.

11. Vasseur, E.; Borderas, F.; Cue, R.I.; Lefebvre, D.; Pellerin, D.; Rushen, J. A survey of dairy calf management practices in Canada that affect animal welfare. *J. Dairy Sci.* **2010**, *93*, 1307–1315. [CrossRef]

12. Berends, H.; van Reenen, C.G.; Stockhofe-Zurwieden, N.; Gerrits, W.J.J. Effects of early rumen development and solid feed composition on growth performance and abomasal health in veal calves. *J. Dairy Sci.* **2012**, *95*, 3190–3199. [CrossRef]

13. Passillé, A.M.d.; Borderas, T.F.; Rushen, J. Weaning age of calves fed a high milk allowance by automated feeders: effects on feed, water and energy intake, behavioral sings of hunger and weight gains. *J. Dairy Sci.* **2011**, *94*, 1401–1408. [CrossRef] [PubMed]

14. Quigley, J.D.; Bernard, J.K. Effects of nutrient source and time of feeding on changes in blood metabolites in young calves. *J. Anim. Sci.* **1992**, *70*, 1543–1549. [CrossRef] [PubMed]

15. Berends, H.; van den Borne, J.J.G.C.; Mollenhorst, H.; van Reenen, C.G.; Bokkers, E.A.M.; Gerrits, W.J.J. Utilization of roughages and concentrates relative to that of milk replacer increases strongly with age in veal calves. *J. Dairy Sci.* **2014**, *97*, 6475–6484. [CrossRef] [PubMed]

16. Huber, J.T. Development of the digestive and metabolic apparatus of the calf. *J. Dairy Sci.* **1969**, *52*, 1303–1315. [CrossRef]

17. Fowden, A.L.; Giussani, D.A.; Forhead, A.J. Intrauterine programming of physiological systems: Causes and consequences. *Physiology* **2006**, *21*, 29–37. [CrossRef]

18. Opsomer, G.; van Eetvelde, M.; Kamal, M.; van Soom, A. Epidemiological evidence for metabolic programming in dairy cattle. *Reprod. Fertil. Dev.* **2017**, *29*, 52–57. [CrossRef]

19. Heinrichs, A.J.; Heinrichs, B.S.; Harel, O.; Rogers, G.W.; Place, N.T. A prospective study of calf factors affecting age, body size, and body condition score at first calving of holstein dairy heifers. *J. Dairy Sci.* **2005**, *88*, 2828–2835. [CrossRef]

20. Waterland, R.A.; Garza, C. Potential mechanisms of metabolic imprinting that lead to chronic disease. *Am. J. Clin. Nut.* **1999**, *69*, 179–197. [CrossRef]

21. Verband Deutscher Landwirtschaftlicher Untersuchungs- und Forschungsanstalten (VDLUFA). Volume III Die Chemische Untersuchung von Futtermitteln. In *Handbuch der Landwirtschaftlichen Versuchs- und Untersuchungsmethodik (VDLUFA-Methodenbuch*; VDLUFA-Verlag: Darmstadt, Germany, 1993; ISBN 978-3-941273-14-6.

22. Sauerwein, H.; Heintges, U.; Hennies, M.; Selhorst, T.; Daxenberger, A. Growth hormone induced alterations of leptin serum concentrations in dairy cows as measured by a novel enzyme immunoassay. *Livest. Prod. Sci.* **2004**, *87*, 189–195. [CrossRef]

23. Mielenz, M.; Mielenz, B.; Singh, S.P.; Kopp, C.; Heinz, J.; Häussler, S.; Sauerwein, H. Development, validation, and pilot application of a semiquantitative western blot analysis and an ELISA for bovine adiponectin. *Domest. Anim. Endocrinol.* **2013**, *44*, 121–130. [CrossRef]

24. Abdelsamei, A.H.; Fox, D.G.; Tedeschi, L.O.; Thonney, M.L.; Ketchen, D.J.; Stouffer, J.R. The effect of milk intake on forage intake and growth of nursing calves. *J. Anim. Sci.* **2005**, *83*, 940–947. [CrossRef] [PubMed]

25. Korst, M.; Koch, C.; Kesser, J.; Müller, U.; Romberg, F.-J.; Rehage, J.; Eder, K.; Sauerwein, H. Different milk feeding intensities during the first 4 weeks of rearing in dairy calves: Part 1: Effects on performance and production from birth over the first lactation. *J. Dairy Sci.* **2017**, *100*, 3096–3108. [CrossRef] [PubMed]

26. Webb, L.E.; Engel, B.; Berends, H.; van Reenen, C.G.; Gerrits, W.J.J.; De Boer, I.J.M.; Bokkers, E.A.M. What do calves choose to eat and how do preferences affect behaviour? *Appl. Anim. Behav. Sci.* **2014**, *161*, 7–19. [CrossRef]

27. Greenwood, R.H.; Morrill, J.L.; Titgemeyer, E.C. Using dry feed intake as a percenage of initial body weight as a weaning criterion. *J. Dairy Sci.* **1997**, *80*, 2542–2546. [CrossRef]

28. Ndumbe, R.D.; Runcie, K.V.; McDonald, P. The effect of early weaning on the blood sugar and rumen acid levels of the growing calf. *Br. J. Nutr.* **1964**, *18*, 29–38. [CrossRef]

29. Czerkawski, J.W. A novel estimate of the magnitude of heat produced in the rumen. *Br. J. Nutr.* **1980**, *43*, 239–243. [CrossRef]

30. Symonds, M.E.; Stephenson, T.; Gardner, D.S.; Budge, H. Long-term effects of nutritional programming of the embryo and fetus: Mechanisms and critical windows. *Reprod. Fertil. Dev.* **2007**, *19*, 53–63. [CrossRef]

31. Kesser, J.; Korst, M.; Koch, C.; Romberg, F.-J.; Rehage, J.; Müller, U.; Schmicke, M.; Eder, K.; Hammon, H.M.; Sadri, H.; et al. Different milk feeding intensities during the first 4 weeks of rearing dairy calves: Part 2: Effects on the metabolic and endocrine status during calfhood and around the first lactation. *J. Dairy Sci.* **2017**, *100*, 3109–3125. [CrossRef]

32. Breier, B.H.; Gluckman, P.D.; Bass, J.J. Plasma concentrations of insulin-like growth factor-I and insulin in the infant calf: Ontogeny and influence of altered nutrition. *J. Endocrinol.* **1988**, *119*, 43–50. [CrossRef]

33. Hegardt, F.G. Mitochondrial 3-hydroxy-3-methylglutaryl-CoA synthase: A control enzyme in ketogenesis. *Biochem. J.* **1999**, *338*, 569–582. [CrossRef]

34. Cerqueira, N.M.F.S.A.; Oliveira, E.F.; Gesto, D.S.; Santos-Martins, D.; Moreira, C.; Moorthy, H.N.; Ramos, M.J.; Fernandes, P.A. Cholesterol biosynthesis: A mechanistic overview. *Biochemistry* **2016**, *55*, 5483–5506. [CrossRef] [PubMed]

35. Lakshmanan, M.R.; Nepokroeff, C.M.; Ness, G.C.; Dugan, R.E.; Porter, J.W. Stimulation by insulin of rat liver β-hydroxy-β-methylglutaryl coenzyme A reductase and cholesterol-synthesizing activities. *Biochem. Biophys. Res. Commun.* **1973**, *50*, 704–710. [CrossRef]

36. Johnson, C.; Dance, A.; Kovalchuk, I.; Kastelic, J.; Thundathil, J. Enhanced early-life nutrition upregulates cholesterol biosynthetic gene expression and Sertoli cell maturation in testes of pre-pubertal Holstein bulls. *Sci. Rep.* **2019**, *9*, 6448. [CrossRef] [PubMed]

37. Walker, D.M.; Simmonds, R.A. The development of the digestive system of the young animal. VI. The metabolism of short-chain fatty acids by the rumen and caecal wall of the young lamb. *J. Agric. Sci.* **1962**, *59*, 375–379. [CrossRef]

38. Haga, S.; Fujimoto, S.; Yonezawa, T.; Yoshioka, K.; Shingu, H.; Kobayashi, Y.; Takahashi, T.; Otani, Y.; Katoh, K.; Obara, Y. Changes in hepatic key enzymes of dairy calves in early weaning production systems. *J. Dairy Sci.* **2008**, *91*, 3156–3164. [CrossRef]

39. Quigley, J.D.; Caldwell, L.A.; Sinks, G.D.; Heitmann, R.N. Changes in blood glucose, nonesterified fatty acids, and ketones in response to weaning and feed intake in young calves. *J. Dairy Sci.* **1991**, *74*, 250–257. [CrossRef]

40. Zieba, D.A.; Amstalden, M.; Williams, G.L. Regulatory roles of leptin in reproduction and metabolism: A comparative review. *Domest. Anim. Endocrinol.* **2005**, *29*, 166–185. [CrossRef]

41. La Cava, A. Leptin in inflammation and autoimmunity. *Cytokine* **2017**, *98*, 51–58. [CrossRef]

42. Attig, L.; Solomon, G.; Ferezou, J.; Abdennebi-Najar, L.; Taouis, M.; Gertler, A.; Djiane, J. Early postnatal leptin blockage leads to a long-term leptin resistance and susceptibility to diet-induced obesity in rats. *Int. J. Obes.* **2008**, *32*, 1153–1160. [CrossRef]

43. Granado, M.; Fuente-Martín, E.; García-Cáceres, C.; Argente, J.; Chowen, J.A. Leptin in early life: A key factor for the development of the adult metabolic profile. *Obes. Facts* **2012**, *5*, 138–150. [CrossRef]

44. Graaf, C.d.; Blom, W.A.M.; Smeets, P.A.M.; Stafleu, A.; Hendriks, H.F.J. Biomarkers of satiation and satiety. *Am. J. Clin. Nut.* **2004**, *79*, 946–961. [CrossRef] [PubMed]

45. Dennis, T.S.; Suarez-Mena, F.X.; Hill, T.M.; Quigley, J.D.; Schlotterbeck, R.L.; Klopp, R.N.; Lascano, G.J.; Hulbert, L. Effects of gradual and later weaning ages when feeding high milk replacer rates on growth, textured starter digestibility, and behavior in Holstein calves from 0 to 4 months of age. *J. Dairy Sci.* **2018**, *101*, 1–13. [CrossRef] [PubMed]

46. Reinhardt, V.; Reinhardt, A. Natural sucking performance and age of weaning in zebu cattle (Bos indicus). *J. Agric. Sci.* **1981**, *96*, 309–312. [CrossRef]
47. Walker, D.E. Suckling and grazing behaviour of beef heifers and calves. *N. Z. J. Agric. Res.* **1962**, *5*, 331–338. [CrossRef]

Article

Appropriate Dairy Calf Feeding from Birth to Weaning: "It's an Investment for the Future"

Laura J. Palczynski [1,*], Emma C. L. Bleach [1], Marnie L. Brennan [2] and Philip A. Robinson [1,3]

1   Animal Production, Welfare and Veterinary Sciences Department, Harper Adams University, Newport, Shropshire TF10 8NB, UK; ebleach@harper-adams.ac.uk (E.C.L.B.); Philip.Robinson.2@glasgow.ac.uk (P.A.R.)
2   School of Veterinary Medicine and Science, University of Nottingham, Sutton Bonington Campus, Leicestershire LE12 5RD, UK; marnie.brennan@nottingham.ac.uk
3   School of Veterinary Medicine, College of Medical, Veterinary and Life Sciences, University of Glasgow, Garscube Campus, Bearsden Road, Glasgow G61 1QH, UK
*   Correspondence: lpalczynski@harper-adams.ac.uk

Received: 28 November 2019; Accepted: 3 January 2020; Published: 10 January 2020

Simple Summary: Research has indicated that dairy farms often do not feed calves according to recommended best practice, despite legislation and industry advisory efforts. This study used interviews with dairy farmers and their advisors to investigate why farmers feed calves the way they do. Various calf feeding practices were used by participating farmers, largely based on perceived convenience and calf performance. Advisors were concerned that calves were commonly underfed, which may be partly due to farmers receiving inadequate instructions for calf feeding. Our results highlight the need for more consistent and effective recommendations for farmers regarding calf feeding and weaning. Standard guidelines for calf milk replacers should be improved to ensure that calves are fed enough to support basic biological functions and growth. Further research is needed to establish best practices for weaning calves whilst supporting rumen development, health and weight gain. All recommendations for calf feeding should facilitate the achievement of standard industry targets including rearing replacement dairy heifers to calve by 24 months of age.

Abstract: Dairy calves must be fed appropriately to meet their nutritional needs, supporting optimal growth and development to achieve the recommended target age at first calving (AFC) of 24 months. Traditional restricted milk feeding practices suppress growth, contribute to negative welfare states and may result in malnutrition and immunosuppression. Despite more recent recommendations to increase milk allowances for pre-weaned calves, restricted feeding remains a common practice. This study explored the rationales behind the calf feeding protocols used by dairy farmers in England. Forty qualitative interviews (26 farmers, 14 advisors) were conducted between May 2016 and June 2017, transcribed in full, then coded into themes. Results indicate that a variety of calf feeding regimes are used on farms, largely determined by farmers' attitudes regarding ease of management and the wellbeing of calves. Advisors were concerned about widespread underfeeding of calves, which may be partially due to insufficiently clear recommendations for calf milk replacer (CMR) feeding rates. There was also evidence of uncertainty regarding best practices for weaning calves. Collaboration between academic research and industry is essential to establish a consensus on calf feeding standards which support physiological function, facilitate weaning, support growth targets and ensure calf health and welfare is protected.

Keywords: dairy calf feeding; health; welfare; nutrition; stakeholder attitudes

---

1. Introduction

Dairy calves must be fed appropriately to meet their nutritional needs for optimal growth and development. Diet must also support and reflect the development of calves' digestive function from the liquid-fed pre-ruminant phase through the transition into a functional ruminant [1]. There are also financial implications since milk feeding accounts for 40% of total rearing costs from birth to weaning, the most expensive phase of rearing replacement dairy heifers [2,3]. Calf growth rates at least partly determine their age at first calving (AFC), with heifers calving at 23–24 months being more cost-efficient than later calving animals [2]. The recommended target AFC of 24 months achieves optimal economic efficiency resulting from increased lifetime fertility, survival and milk production compared to later calving heifers [4–6].

A typical Holstein-type heifer must maintain a growth rate of about 750 g/day from birth to achieve adequate body weight and stature to calve at 24 months [5]. The optimal protein to energy ratio for growth in pre-weaned calves has been estimated to be approximately 11.5 g of crude protein per MJ of metabolisable energy (ME) [7]. Approximately 325 g/day whole milk solids (2.5 L/day) or 380 g/day calf milk replacer (CMR) (3 L/day), which contain about 22.5 MJ ME/kg and 19.5 MJ ME/kg respectively, provide sufficient ME to meet the maintenance requirements of a 45 kg calf under thermoneutral conditions with surplus nutrients supporting growth [1].

Traditional feeding practices provide daily milk allowances of approximately 10% of calf bodyweight, primarily to increase solid-feed intakes to facilitate rumen development for earlier weaning. These restricted feeding practices limit the growth potential of calves [8] and are likely to provide insufficient energy in temperatures below 15 °C [9]. When calves are malnourished, particularly in cases of insufficient energy intakes, their immunity is impaired and they are more susceptible to disease (e.g., [10–12]). The effect of feeding higher planes of nutrition, above maintenance requirements, on the immunocompetence of calves is less clear cut as intensive milk feeding does not appear to affect the health and immune status of calves in a consistent manner [12,13].

However, calves will voluntarily consume over 9 L/day of milk [8,14], indicating that larger milk meals are required to satiate calves and improve their welfare. Indeed, restricted milk feeding causes calves to experience persistent hunger, as indicated by higher numbers of unrewarded visits to milk feeders [14,15], more frequent and higher pitched vocalisations [16] and reduced play behaviour [17]. More recent recommendations suggest daily milk or CMR feeds should equate to 20% of calf bodyweight to support calf growth and health [18] and a common target is to have doubled the birth weight of calves by the time of weaning at 8 weeks of age [19]. Increasing the amount of milk or CMR fed per day supports higher growth rates, with the weight advantage persisting post-weaning [20,21], and is linked to developmental effects which positively affect future milk yield [22].

Despite these recommendations, once-a-day milk feeding is sometimes used on farms to reduce labour requirements whilst achieving adequate gains in calf bodyweight [23,24]. In England, The Welfare of Farmed Animals (England) Regulations 2007 and EU Directive 2008/119/EC on the minimum standards for the protection of calves require calves to be fed at least twice-a-day up to six months of age. European legislation also requires that all calves over two weeks of age must be provided with sufficient fresh drinking water to satisfy their needs and have access to water at all times in hot weather or if they are ill. The national legislation in England requires that all calves are provided with sufficient fresh drinking water each day from birth. Once-a-day milk feeding in the first month of life may contribute to abomasal disorders (abomasitis and/or bloat) in calves [25] and is illegal since the limited intakes of solid feed during early life do not constitute a meal. Twice daily milk feeding is necessary to meet calves' nutritional requirements prior to 28 days of age [25,26].

Water is a key nutrient and plays a critical role in calf growth and rumen development [1] and calves should be provided free access to clean drinking water from birth. Although calves obtain the majority of their water intake through consumption of milk or CMR [27], this water from feed goes directly to the abomasum. Drinking water enters and supports the development of the rumen [28] and encourages greater intakes of starter concentrates [29], milk consumption and growth performance [30].

Despite the research outlined above evidencing the benefits of feeding calves greater milk allowances and offering drinking water from birth, many farms feed a restricted milk diet, and some do not provide access to water prior to weaning [3,31]. Restricted calf feeding has been highlighted as an area of concern in the scientific literature [31–33], suggesting that legislation and current industry advisory efforts may have failed to assert best practice on farms. Very few studies have explored the rationale behind the calf feeding systems adopted by farmers. The present study used qualitative interviews to explore the practices, experiences and perspectives of participant dairy farmers and advisors. Such social science approaches are advocated by a growing proportion of the animal health and welfare research community e.g., [34–38]. This paper aims to explore the nuanced reasoning behind the different pre-weaning calf feeding protocols used on English dairy farms to provide greater holistic understanding of the wider context which might influence on-farm decisions.

2. Materials and Methods

This study employed a critical realist paradigm which asserts that subjective experiences of phenomena and objective facts are equally important in understanding a topic within its wider context [39]. This epistemology enabled the exploration of different perspectives regarding dairy calf management, providing a more holistic understanding of pre-weaning calf feeding.

2.1. Data Collection

Calf management on English dairy farms was investigated through 40 in-depth semi-structured interviews (26 with farmers, 14 with advisors) conducted between May 2016 and June 2017. All interviews were conducted by the first author, a doctoral student who sought to investigate human influences on calf health and welfare regarding rearing practices from birth to first calving. Presented here are findings relating to calf feeding following the provision of colostrum, which has been addressed in a previous paper [40].

Purposive and snowball sampling [41] was used to recruit participants from existing contacts, veterinary practices, dairy events and conferences, and individuals suggested by interviewees. This method yielded farmers who managed a range of dairy herd sizes and production systems (Table 1) and advisors who tended to have a specific interest in dairy youngstock (Table 2). Interviews were grouped according to geographical location with participants from areas of England with high densities of dairy farms (Southwest and Midlands) and from a north-easterly area with less dairy focus in Yorkshire.

Interviewees included 37 dairy farmers (farm managers (n = 17), farm workers (n = 9), calf rearers (n = 8) and herd managers (n = 3)) and 14 advisors (veterinarians (n = 11), feed (n = 2) and a veterinary pharmaceutical company representative (n = 1)). One of three interview formats were used according to participants' preferences: all advisors and nine farmers were interviewed individually in a seated setting; 20 farmers participated in nine joint interviews where two to three participants were interviewed together; and eight farmers were interviewed whilst walking around the farm.

Two separate interview topic guides were used, one for farmer interviews, the other for advisor interviews. These guides included open-ended questions which ensured interviews remained relevant to calf rearing whilst allowing flexibility to explore areas of most importance to participants [42] rather than being predefined by the researchers. Farmers were asked questions about the practices used on their farm and their opinions about how calves are reared elsewhere, whereas advisors were asked about their main areas of concern regarding calf rearing and their role in providing information and advice. Seven pilot interviews were conducted, four with farmers (F1, F2, F3, F4) and three with advisors (V1, V2, N1) to ensure topic guides were suitable. Responses were useful to the research project and only minor refinements were made to the topic guides so the pilot interviews were included in the overall dataset.

Table 1. Farmer participant demographics.

Interview Code, Style	Interviewee Details: Job, Gender, Age Estimate	Farm Details: Calving Pattern, Herd Size, Farm System	Location within UK
F1, Mobile	Calf rearer, f, 20–30	AYR, 380, conventional	Midlands
F2, Sit-down	Calf rearer, f, 40–50	AB, 350, conventional	Midlands
F3, Sit-down	Farm hand/calf rearer, m, 20–30	AYR, 350, conventional	Midlands
F4, Joint	Farm manager, m, >50 Farm hand, f, 20–30 Son/trainee vet, m, 20–30	AYR, 120, conventional	Midlands
F5, Sit-down	Farm manager, m, >50	AB/SB, 70, conventional	Midlands
F6, Sit-down	Calf rearer, f, 30–40	SB, 300, organic	Midlands
F7, Mobile	Farm manager/calf rearer, m, 30–40	AYR, 280, conventional	Midlands
F8, Joint	Farm manager, m, 40–50 Farm wife, f, 40–50	Dairy bull calf rearer, batches of 20 calves	Yorkshire
F9, Mobile	Farm manager, m, 40–50	AYR, 250, conventional	Yorkshire
F10, Mobile	Farm manager, m, >50	AB, 90, conventional	Yorkshire
F11, Mobile	Farm administrator, f, 30–40	AYR, 400, conventional	Yorkshire
F12, Joint	Farm manager, m, 40–50 Herd manager, m, 20–30	AB, 370, conventional	Yorkshire
F13, Sit-down	Farm manager, m, >50	SB, 600, conventional	Southwest
F14, Joint	Farm manager, m, >50 Calf rearer, m, 40–50	AB, 420, organic	Southwest
F15, Joint	Farm manager, m, 30–40 Calf rearer, m, 30–40	AYR, 120, conventional	Southwest
F16, Joint	Calf rearer, f, 30–40 Farm manager, m, 30–40	SB, 250, organic	Southwest
F17, Joint	Farm manager, m, >50 Farm hand, m, 20–30 Farm hand, f, 20–30	Dairybull/beef calf rearer, 1400 calf places	Southwest
F18, Sit-down	Calf rearer, f, 20–30	AYR, 180, conventional	Southwest
F19, Sit-down	Farm manager, m, 30–40	AYR, 160, conventional	Southwest
F20, Sit-down	Farm manager, m, 30–40	AB, 330, conventional	Southwest
F21, Mobile	Farm manager, m, 40–50	AYR, 1200, conventional	Yorkshire
F22, Mobile	Herd manager, f, 20–30	AYR, 130, conventional	Yorkshire
F23, Mobile	Farm hand/calf rearer, m, 30–40	AB, 250, organic	Southwest
F24, Sit-down	Herd manager, m, 20–30	AYR, 200, conventional	Southwest
F25, Joint	Farm manager, m, >50 Calf rearer, m, 20–30	AYR, 350, organic	Southwest
F26, Joint	Farm manager, m, >50 Calf rearer, f, >50	AB, 500, conventional	Southwest

Abbreviations: male (m), female (f), all-year-round calving pattern (AYR), autumn block calving pattern (AB), and spring block calving pattern (SB).

Table 2. Advisor participant demographics.

Interview Code, Style	Interviwee Details: Job, Gender, Age Estimate	Location within UK
N1, Sit-down	Feed company salesperson, m, 40–50	Midlands
N2, Sit-down	Feed company calf specialist, f, 30–40	Midlands
DR1, Sit-down	Pharmaceutical company veterinary advisor, f, 30–40	Midlands
GA1, Sit-down	Government veterinary advisor, f, 40–50	Southwest
V1, Sit-down	Veterinary specialist in cattle health, m, 30–40	Midlands
V2, Sit-down	Youngstock veterinarian, f, 20–30	Midlands
V3, Sit-down	Veterinarian starting a youngstock discussion group, m, 20–30	Yorkshire
V4, Sit-down	Farm veterinarian, works on beef calf rearing unit, m, 20–30	Yorkshire
V5, Sit-down	Practice director and youngstock veterinarian, m, 30–40	Southwest
V6, Sit-down	Youngstock veterinarian, m, 30–40	Southwest
V7, Sit-down	Practice partner and farm veterinarian, f, 40–50	Southwest
V8, Sit-down	Practice partner and farm veterinarian, m, >50	Southwest
V10, Sit-down	Out of practice veterinarian, now feed consultant, m, 40–50	Midlands
V11, Sit-down	Youngstock veterinarian, f, 30–40	Southwest

Abbreviations: male (m), female (f).

Data collection and analysis overlapped in an iterative approach so that topics raised in earlier interviews could be further examined with later interviewees [43]. Interviews were audio recorded with consent and subsequently manually transcribed in full using f4transkript software (Version 6.2.5 Edu, audiotranskription.de, Marburg, Germany). Data collection ceased when it was judged that thematic saturation was established [43], i.e., the main concepts and range of opinions relevant to calf rearing had been identified, and no new themes were emerging.

2.2. Data Analysis

Transcripts were analysed using thematic coding which involved reading and re-reading the data and grouping extracts into common themes [44]. Transcripts were coded in NVivo 11 for Windows (Version 11.4.1.1064 Pro, QSR International Pty Ltd., Victoria, Australia). In first cycle coding excerpts were arranged according to topic, personal values, and processes [43] to inform ongoing interviews and indicate focal subjects including calf feeding. Coding was repeated to explore the topic of calf feeding in-depth and relevant interview extracts were chosen to represent the perceptions of participants relevant to the themes and explanations being constructed.

2.3. Ethical Approval

Prior to participation in the study, all participants gave their informed consent—specifically for interviews to be conducted, audio recorded, transcribed, securely stored and for anonymised interview excerpts to be used when reporting findings. The study was conducted in accordance with the Declaration of Helsinki, and the protocol was approved by the Harper Adams University Research Ethics Committee on 13 January 2016 under project number 75-201511.

3. Results

Average interview length was 56 min (rage 26–90 min). Most results within the theme of calf feeding pertained to liquid feeds, with some reference to the provision of water and solid feeds in preparation for weaning.

3.1. Milk Feeding: Amount Fed

Participating farmers fed their calves 4–8 L milk per day (10 fed whole milk, 16 fed CMR) (Table 3) and the mixing rates, brands and composition of CMR varied. Few farmers could recollect basic details of their CMR, including the protein and fat content. Most farmers provided the weight of CMR fed, since *"water is just the carriage to get [nutrients from CMR] into the calves"* (F8 male, farm manager); the total CMR provided ranged from 500 g–996 g per day, though some farmers referred only to the volume of CMR fed. The majority of milk was fed in two daily feeds unless calves had access to an automatic feeder throughout the day. One organic farm fed cold whole milk ad libitum to calves after the first week. Two farms fed once-a-day milk to calves from 1 to 2 weeks of age and F7 used a particularly concentrated 3 L feed once a day with a mixing rate of 300 g/L, believing that increasing the feeding rate in this manner had improved calf health:

> *"Prior to the feeding regime we're on now I generally tended to restrict milk to 4 L of milk a day, 750 g of milk solids over two feeds, and I would get a lot more enteric disease. I'd get a lot more of all calf health issues".*

(F7, male farm manager)

Table 3. Information given by farmer participants regarding heifer calf milk feeding.

Farm	Colostrum	Milk Feeding			
		Type	Amount Per Day	Feeding Method	Temperature
F1	1 feed of 4 L	CMR	2.8 L twice daily	Teat bottles filled from mixer	40 °C set on equipment
F2	2–3 days: 4 L first feed then 2.5 L twice daily	CMR (26% CP)	3.5 L twice daily (2.5 L twice daily first week)	Multi-teat bucket feeder filled from mixer	40 °C set on equipment
F3	4 days: 2 L twice daily	CMR	3 L twice daily (166 g/L) (2 L over 2 weeks)	Multi-teat bucket feeder	Warm (not measured)
F4	3–4 days: 3 L first feed, then amount not stated	Waste WM	Not stated	Multi-teat bucket feeder	Warm (not measured)
F5	4 days: amount not stated	CMR (26% CP)	400 g milk solids twice daily	Multi-teat bucket feeder	Warm (not measured)
F6 [2]	3–4 days: 3–5 L first feed, left with dam for 24 h then 3–4 L twice daily	WM (Johne's-free only)	3–4 L twice daily	Via teat	Warm, straight from parlour
F7	"As much colostrum as I can get it to drink"	CMR (26% CP, 20% oil, skim-based)	3 L once daily (300 g/L) (3 L twice daily 150 g/L until day 7–14)	Teat bottles filled from mixer	Not stated
F8 [1]	Calves not on farm at this point	CMR (whey-based)	Total amount not stated, 150 g/L	Automated feeders with teat	Warm, set on feeder
F9	2–3 feeds of 3 L	WM, soon CMR again [3]	Not stated	Multi-teat bucket feeder	Warm, straight from parlour
F10	2 feeds of 3–4 L	CMR (skim-based)	3.5 L twice daily (125 g/L)	Not stated	Warm (measured on thermometer)
F11	1 feed of 3 L	CMR (skim-based)	6 L over the day (150 g/L)	Automated feeders with teat	Warm, set on feeder
F12	2 feeds, amount not stated	WM [3]	Not stated	Multi-teat bucket feeder	Warm, straight from parlour
F13	1 feed of 2 L	Pasteurised waste WM [3]	Not stated	Multi-teat trailer feeder	40 °C from pasteuriser
F14 [2]	One feed then left with dam for 24–48 h	Pasteurised WM	3 L twice daily	Multi-teat bucket feeder	Warm from pasteuriser
F15	One feed of 2–4 L then left with dam for 3–4 days.	CMR	2.5 L twice daily (100 g/L) (2 L twice daily, 125 g/L until day 9)	Multi-teat bucket feeder	38–40 °C
F16 [2]	Left with dam for 24 h	WM	*Ad libitum* (3 L twice daily first week)	Multi-teat buckets, barrels or trailer feeder according to group size	Warm for first week, then cold
F17 [1]	Calves not on farm at this point	WM	3L once daily (125 g/L) from arrival date (14 days of age)	Trough (no teats) filled from mixer	Not stated
F18	6 L within six hours of birth	CMR	Not stated	Teat bottle for first couple of weeks then bucket (no teat)	Not stated
F19	Left with dam for 24–48 h. Two 3 L feeds if necessary	CMR	3 L twice daily (150 g/L)	Not stated	38–40 °C measured using thermometer by interviewee, but not others
F20	2 feeds of 2.5–3L	CMR (50% skim)	Not stated, but decrease to once daily feeds at 3 weeks	Multi-teat bucket feeder	35 °C
F21	1 feed of 4 L	CMR	6 L over the day (150 g/L (increased from 4.5 L first couple of weeks)	Bucket fed for 10 days then automated feeders with teat	Warm, set on equipment
F22	Left with dam for 3 days. Will feed if necessary	WM	2 L twice daily	Bottle fed for first few days then bucket fed	Warm (not measured)
F23 [2]	Left with dam for a week	Waste WM	3 L twice daily	Multi-teat bucket feeder	Warm from parlour
F24	1 feed of 2.5–3 L within six hours	CMR	3 L tiwce daily (166 g/L) (increased fro 2 L first week)	Multi-teat bucket feeder	Not stated
F25 [2]	2 feeds of 2–3 L within 24 h	Waste WM	2 L until 3–4 weeks, then 2.5 L twice daily	Multi-teat bucket feeder	Warm
F26	2–3 feds within 24 h, amount not stated	CMR	Up to 7 L over the day (137 g/L)	Automated feeders with teat	Warm, set on feeder

Abbreviations: calf milk replacer (CMR), whole milk (WM), crude protein (CP). [1] Rears dairy bull or beef cross calves. [2] Organic. [3] Price driven decision. Any details not included in the table mean those aspects were not covered during the interview.

Most farmers appreciated that a higher rate of nutrition could contribute to improved calf health and recognised the high feed conversion efficiency for calf growth and potential impacts on future performance. However, several participants believed that on some farms calves were not prioritised as the focus was centred on the milking herd, and advisor participants were concerned that underfeeding of calves was commonplace:

"The amount of people that feed once a day cold milk to calves despite the fact it's illegal is still quite high".

(V2, female youngstock vet)

"I think these calves are starved […] The number of people that feed two litres twice a day—which is not even maintenance growth rates, especially considering the [cold] weather".

(V3, male youngstock vet)

Farmers seemed less concerned by legislation and calf growth requirements, focusing instead on what suited their management routine and whether calves *"looked well"* (F22, female herd manager). Reasons for restricted feeding included maintenance of traditional practices, following instructions on CMR packaging, and attempts to save money. Calf feeding protocols were usually only changed in response to problems:

"[On the packaging] 250 g was what was recommended, so that's what [the calves] got, but they weren't really doing well on it. You think "it's disease", or "it's the [starter] feed" […] it was actually the lack of a decent amount of milk […] You can't hide behind saying "I'll save a bit of money on milk powder" […] it's an investment for the future".

(F5, male farm manager)

That CMR guidelines on commercial product packaging did not provide sufficient nutrition to meet recommended growth targets, e.g., to double the birth weight by weaning, was raised by a veterinarian-turned-feed-consultant (V10):

"Current recommendations often to a farmer are only about 750 g of milk powder a day […] Even if they're being as efficient as they possibly could, you're only gonna get 750 g a day of growth […] and that's before you factor in any cold or draughty conditions."

Furthermore, one farmer (F15) admitted finding instructions to be unclear and fed the same milk solids as a more dilute milk solution when attempting to increase the amount fed to calves (Table 3):

"Generally it's just water I've been adding […] because reading the instructions on the bag, it doesn't actually say if you're supposed to give more powder."

3.2. Milk Feeding: Type of Milk Fed

The majority of participant farmers (16/26) fed CMR while all participating organic farmers (n = 5) and five conventional farmers fed whole milk (Table 3). Three participants stated that they fed calves unpasteurised non-saleable milk, two fed pasteurised whole milk and five did not specify. Three participants had started feeding whole milk to reduce feed costs during the 2014 downturn in milk prices:

"I did fall out with my powder milk supplier because the price didn't come down when milk price came crashing down […] so I put a pasteuriser in. It was expensive […] but the calves are so much better on whole milk than they are on powdered milk".

(F13, male farm manager)

Some farmers were very positive about the information and support provided by their feed company representative, and most were willing to invest in *"a feed that's right"* (F17, male farm hand)—CMR, which was cost-effective rather than the cheapest available. However, what constitutes a 'good' CMR was not specified, though some referred to the protein and oil content of their milk powder. Other farmers were distrustful of salespeople and one youngstock veterinarian questioned both farmers' knowledge of feed components and the ethics of feed companies:

"If you look at milk powders, some of them, particularly when money was getting very tight, their vitamin E levels suddenly crashed. I think that's a bit naughty of them [the feed companies] because a lot of farmers won't really know what's in their milk powder".

(V11, female youngstock vet)

Several participants, particularly organic farmers, perceived feeding whole milk to be more natural and suggested that it resulted in better calf performance, having been *"designed"* (F13, male farm manager) for calf feeding. Feeding whole milk was also considered beneficial in terms of consistency in feeding if more than one person was responsible for feeding calves. Dairy-bred bull and beef-cross calves were either fed the same as dairy replacement heifer calves for ease of management in dual dairy-beef systems or considered to be low-priority *"milk thieves"* (F10, male farm manager) which would be quickly removed from the farm. In these cases, dairy-bred bull calves received poorer-quality feeds, largely due to a poor market value for those calves:

"I'm rearing a calf, and it's margin with me […] If they put another £20 worth of milk powder into that calf and get that heifer in-calf three months quicker that's cheap, but for me it's £20 directly off".

(F8, male farm manager, rears dairy bull calves)

Although feeding waste milk may be standard practice for replacement calves on some farms, unpasteurised non-saleable milk was more commonly fed to bull or beef-cross calves on dairy enterprises.

"The bull calves and any beef calves, they get […] antibiotic milk, […] high cell count milk, anything really because they're not going to be around for long enough to pick up anything serious".

(F5, male farm manager)

These non-saleable milk feeds often included milk from cows treated with antimicrobials, an area of concern acknowledged by this farm manager:

"If you're feeding milk from cows which have been treated with [antibiotics], you're feeding that antibiotic to those calves. So what problems are you creating? What resistance do you create?".

(F19, male farm manager)

3.3. Milk Feeding: Preparation and Feeding Method

In addition to what was fed to calves, many farmers emphasised the importance of how milk was prepared and delivered to calves. Farmers using automatic machine feeders believed calves benefited from being able to feed throughout the day:

"If you're bottle feeding a calf twice a day, when you feed it it's always starving and it guzzles it really fast. You don't get that when they're on machine because they're doing it in a more natural way, as if they were on a cow".

(F8, female calf rearer)

Automated feeders could also help to ensure consistency of milk feeding, a fundamental principle according to farmer participants. They could also provide farmers with flexible time as they could check the calves when it was convenient rather than being tied to a specific feeding time.

"If you're really busy, you don't have to tend the machines, two or three hours either way, it's really flexible […] The milk's always there at the right temperature, it's well mixed, should be [hygienic] if they've kept the machines clean".

(F21, male farm manager)

However, the cost of machine-feeders prevented many farms from installing them.

Several participants stressed the need for staff to have the time and equipment available to make calf feeding easy and simple to facilitate proper feeding. However, mixing CMR involves several variables, including water temperature, mixing rates and timings, and if the person responsible for calf feeding does not use measuring implements or if several people feed the calves, consistency may suffer and affect calf performance.

"I use a thermometer and I mix at 40 °C and I feed at about 38 °C. Dad uses his finger and I couldn't tell you what [temperature] he feeds at […] Then concentration, I've given him a scoop that's pretty failsafe, but when I was doing it myself I did get better results".

(F19, male farm manager)

Teat feeding was considered beneficial by most farmers. Some had made the change from bucket feeding and were impressed with the results, or acted on external information:

"One journal said that teat feeding over bucket feeding actually helps them grow a little quicker […] I'm not sure if it does, but I tried doing it anyway".

(F3, male farm hand and calf rearer)

"[I visited a farm with stunning calves, the farmer] said whatever you do, do not feed a calf on a bucket. It gulps it down, it gets into the wrong stomach. He said, when a calf suckles, it produces saliva, you can see it around its mouth, that aids digestion".

(F8, male farm manager)

However, one farm veterinarian indicated that the feeding position resulting from the angle of teats on bar feeders may contribute to respiratory disease:

"I think calves on a bar feeder get a certain degree of aspiration pneumonia from the teats being horizontal […] I can't understand why no one's invented a calf bucket that's got like a corner cut off and the teat coming out on the 45° angle so that it forces them into a neck down, head up position which is more natural".

(V4, male farm vet)

Hygiene of the feeding equipment was considered important by both farmer and advisor participants to foster good calf health.

"[Calves] are babies. You have to keep your bottles clean, disinfect everything in-between feeding each calf on a bottle […] even if they're healthy calves, I always disinfect the teat".

(F18, female calf rearer)

However, cleaning may not be done to a high standard on farms and may not be recognised as a problem by farmers:

"[I recommended increasing] everyone's milk that they were feeding, and everyone would say "oh no, if I do that they scour!" […] I think it was just general hygiene of the milk preparation and the buckets. So when they cleaned that, adding more milk wasn't the problem".

(V11, female youngstock vet)

Advisors tended to attribute lack of hygiene to farm facilities and poor availability of hot water. Reasons given by farmers for a lack of hygiene in calf feeding included lack of perceived efficacy in disease control and a perception that sanitation hinders the acquisition of immunity:

"Some people say you should disinfect between [feeding groups of calves], but I never have done. If one lot gets [an infection], they usually all get it anyway".

(F14, male calf rearer)

"Everything should be washed and sterilised with hot water after every calf's fed. With that you're not giving the calf the chance to build up any immunity".

(F16, male farm manager)

3.4. Solid Feed, Weaning and Water

A range of weaning methods were implemented by farmers, although the majority were weaning calves at around 7–8 weeks (Table 4). Some based weaning decisions on age alone whilst others considered calf weight or starter intakes. There was generally a negative view of early weaning practices:

"It seems to me there's this race to wean the calves as quickly as you can. "We wean all calves at six weeks old." It's unnatural. […] You're gonna grow better animals by just feeding them milk for longer".

(F16, male farm manager)

Farmers fed calves different starter feeds and forage, and used different methods for gradual weaning. Some decreased the volume or concentration of milk fed, others decreased the number of daily milk feeds. One farm veterinarian (V4) admitted being unsure of the 'best' weaning technique:

"Weaning, I don't think there's a right answer with that. I certainly haven't found it yet […] How you reduce the milk? Some people will do it by going from two times a day to once a day. Some people will continue twice a day, feeding smaller amounts. Some people will continue twice a day, feeding the same amount but a lower concentration and I don't know what the right answer is to be honest with you."

Table 4. Information given by farmer participants regarding weaning practices.

Farm	Water	Solid Feed	Weaning Process	Calf Weight Recording
F1	From birth	Rearing pellets from birth	Gradual when calve weighs 80 kg and consume 1 kg starter	Weekly from birth using weigh-crate. Aim for 0.8–0.9 kg/d growth
F2	From birth	Corn and straw from birth	Decrease to one daily milk feed at 7–8 weeks. Weaned when consuming 2 kg starter	At turn-out (6–7 months). Plan to improve weigh system
F3	From birth	Rearing pellets from birth	Group housed at 6 weeks to begin weaning by decreasing volume or concentration of milk	No. lacks time. Mental record of intakes and growth
F4	Not stated	Straw and concentrates from a week old	Gradual decrease in milk concentration between 6–10 weeks depending on availability of milk and intakes of concentrates	No. Judge by end product (target AFC 24 months)
F5	Not stated	Corn	Decrease to one daily milk feed at 6 weeks, weaned at 8 weeks depending on availability of milk and intakes of concentrates	No. Judge by end product (target AFC 24 months)
F6 [2]	Not stated	Rearing nuts, oats, straw from birth	Gradual decrease in volume of milk at each feed. Weaned at 12 weeks (organic standard)	At movements between accommodation and vaccinations. Aim for 0.8 kg/d growth.
F7	From birth	Rearing pellets (18% CP) from birth	Decrease volume of milk according to intakes of dry starter feed not based on age. Weaned when consuming 2 kg starter for one week	Used to. Established regime that achieved desired growth rates. Aim for >850 g liveweight gain by calving (target AFC 24 months)
F8 [1]	Not stated	Rearing pellets, home mix (barley, distillers grains, soya, rape meal and minerals), straw	Automated feeders programmed to decrease volume of milk allowance	No. Intends to start
F9	From birth	Rearing pellets from birth and straw from three weeks	Weaned at 8–10 weeks, later if calf is small	No. Labour intensive. Plan to incorporate automated weigh system
F10	From birth	Rearing pellets and straw from birth	Weaned over the course of a week at 7–8 weeks when calf weighs 80–85 kg	Girth measurements at birth and before weaning at 7 weeks. Aim to double birth weight by weaning
F11	Not stated	Concentrates, home mix	Automated feeders programmed to reduce milk allowance by 0.2 L/d day 40–65	Girth measurement at birth Weigh scale output manually recorded periodically. Aim to double birth weight by weaning.
F12	Not stated	Minimal concentrates, grass	Weaned at about 12 weeks when calf weighs 100 kg	Weighed when approaching weaning and about a month after weaning. Compare annual average values.
F13	Not stated	Minimal concentrates, barley, grass	"we probably keep them on milk a little bit longer than we need to"	No. New employee to take groups of calves over local weighbridge
F14 [2]	First week	Rearing pellets	Decrease milk from 7–12 weeks	Monthly weights taken to calculate growth rate
F15	First week	Rearing pellets, barley straw or hay	Decrease to one daily milk feed at 6–7 weeks for one week	Not stated (Targed AFC > 24 months)
F16 [2]	Four weeks	Straw, grass, no concentrates	Decrease to one daily milk feed of decreasing volume to wean at 12 weeks	Not stated
F17 [1]	From arrival	Concentrates, straw	Start weaning when calf weighs about 80 kg	Weighed on arrival and departure over local weighbridge
F18	From birth	Rearing nuts, barley straw	Decrease to one daily milk feed at 6–7 weeks for one week before weaning at 7–8 weeks, depending how calf is doing	No. Intends to start
F19	From birth	Concentrates and straw first week	Weaned at 12 weeks	Girth measurements taken throughout rearing period
F20	From birth	Rearing pellets, chopped wheat straw	Weaned at 8–9 weeks	No. Wants a simple, easy system to use
F21	Not stated	Rearing pellets, straw	Automated feeders programmed to reduce milk allowance by 0.6 L/d day 49–59	Periodically. Would like vet-tech service to reduce labour cost
F22	Four weeks	Rearing nuts, hay	Not stated	No. Does not seem feasible or small farms
F23 [2]	Three weeks	Rearing pellets, straw	Weaned at 12 weeks	No. Would like to start but can judge by eye
F24	Not stated	Concentrates	Weaned at 8–10 weeks	No. Intends to start
F25 [2]	Not stated	Rearing pellets	Decrease to one daily milk feed from 10–12 weeks	Regular use of weigh-crate
F26	From birth	Concentrates, straw	Weaned at 7–9 weeks. Automated feeders programmed to decrease volume of milk allowance.	Not stated

Abbreviations: age at first calving (AFC), crude protein (CP). [1] Rears dairy bull or beef cross calves. [2] Organic. Since no quantitative survey of farm practices was conducted, some details were not included in the interviews—this does not necessarily indicate that calves were not provided with components e.g., straw, water. Straw is stated where it is provided as a feed substrate rather than as bedding.

Participants were aware that calves should be consuming solid feed and forage to aid rumen development, and milk feeding practices sometimes needed to be altered to facilitate intakes of dry starter.

"We do struggle to get roughage in them […] We've had the odd post-mortem done on calves which have been poor and we've had poor rumen development so it's something we're trying to improve on".

(F9, male farm manager)

"We tried a kilogram [of CMR] a day, but we found that although the calves looked great at weaning time, they didn't wean as well. I don't think they had room to eat as many pellets. This way [875 g/day], they eat more pellets and it's a more seamless weaning".

(F10, male farm manager)

Problems encountered at weaning time included pot-bellied calves, growth checks and diarrhoea. Some farmers had changed their practices and improved weaning, whereas others struggled to prevent problems, despite trying several alterations in a trial-and-error approach:

"I used to wean everything at six weeks. We'd go once a day milk at five weeks and they'd be weaned at six. But now we do twice a day feeding until six weeks and then once a day for another two weeks, monitoring how much corn they're eating. By eight weeks old they're taking a lot of corn, and then we wean them. That's made quite a difference to the calves in that they used to be pot bellied and horrible after weaning, but they're not now".

(F5, male farm manager)

"[The calves] do get very loose [at weaning] and that's mostly when the coccidiosis kicks in […] I know you shouldn't do everything all at once. They're trying to be weaned, they're changing the ration, they're introduced onto silage—that's when they get loose. I've tried not giving them silage, I've tried keeping them on pellets, I've tried putting them on rearing nuts [sooner] and they still get loose, so it doesn't really seem to make a lot of difference".

(F14 male calf rearer)

Water affects calf consumption of concentrates, plays an important role in rumen development and its provision is required under UK and EU law. However, many advisors were frustrated that calves on many farms did not have access to fresh water.

"You can walk around quite a lot of dairy farms in the UK that the calves don't have access to water. The fact that it's illegal let alone detrimental to growth rates … ".

(V2, female youngstock vet)

"[Farmers will] complain to you "oh, they're not eating much dry starter feed, your feed's rubbish"—you're not really gonna want to eat dry crackers without a drink of water, are you? They don't realise that [calves] need fresh water for rumen development. Their milk feeds twice a day—it doesn't constitute free water. It doesn't go to the rumen for rumen development—it goes to the abomasum".

(N2, female feed company calf specialist)

Some farmers who did not provide water to young calves believed that calves would reject their milk feed after gorging on water, particularly if both were provided in buckets rather than milk via a teat. Others did not realise that calves required access to free water in addition to their liquid feeds.

"One thing is that they don't fill up on water, so when you feed them they're hungry enough to drink the milk. They shouldn't really need it. It's like a newborn baby, you don't give them water. Apart from warm milk, they don't need anything else".

(F16, female calf rearer)

"Milk when you feed it is a fixed dry matter content and fixed fat and protein content, so you haven't got the element of a thirst-quenching feed for the baby calf".

(GA1, female government veterinary advisor)

If calves seem to be doing well, often practices are not altered and farm staff may not have control over management decisions.

"This is a source of contest between me and the bosses because I think they should have water all the time, but they only feed water when they get to about a month old […] that's how they've always done it, and the calves look really well so I can't really tell them to do otherwise".

(F22, female herd manager)

4. Discussion

Our results indicate that a wide variety of calf feeding regimes, primarily to rear replacement heifers, are used on English dairy farms. Whilst participant farmers reported providing concentrates and forage to calves, discussion in our interviews was focused on liquid feeding, particularly CMR. Farmers' actions concerning calf feeding practices were largely determined by their attitudes regarding the ease of management and wellbeing of calves. Some farmers made proactive changes seeking to achieve optimal calf performance, with several noting the benefits of feeding programmes which promote accelerated growth. Most participants maintained the status quo, continuing historic practices, including limiting liquid feed allowances and only making alterations in response to perceived problems with calf health or growth rates. However, farmers may struggle to accurately assess calf performance due to a lack of calf monitoring data [45], possibly resulting in failure to identify problems. Calf feeding is also often regarded as a simple, childhood task that does not require discussion or deliberation, particularly if calves are perceived to be performing well [46].

In the present study, advisors, particularly veterinarians, were concerned about widespread underfeeding of dairy calves. Sumner and von Keyserlingk [33] found that Canadian dairy cattle veterinarians were also concerned about calf hunger and malnutrition, suggesting that underfeeding calves is potentially a global problem in the dairy industry in developed countries. This may, at least in part, be due to the long-established industry standard for restricted milk feeding which has only relatively recently been challenged to favour greater milk allowances for improved calf performance [18,20–22] and better welfare standards [8,14]. However, it has also recently been argued that increasing intakes of solid feed during the pre-weaning period alongside appropriate liquid feeding (as opposed to accelerated liquid feeding programmes) offers a more cost-effective route to achieving greater growth rates whilst also supporting rumen development and future lactation performance [47]. This lack of consensus in the research literature is reflected by the range of milk allowances provided by participant farmers. Farmers were providing approximately 5–6 L/day of liquid feed to calves on average, with most feeding above the historically-favoured daily rate of 4 L/day. However, the traditional practice of restricted milk feeding persists on many farms [3,31], including a minority of those participating in this study. Several farmers had increased the milk allowance for calves and perceived the change positively, largely pertaining to improved calf health. This indicates that their previous milk ration did not provide calves with sufficient nutrition, impairing their immune function [12,13], and increasing liquid feed allowances covered this nutritional deficit.

Contrary to the legislative requirements, once-a-day milk feeding for young calves was used on two farms in this study. One farm was a rearing unit for dairy bull calves seeking the most time- and cost-effective feeding method for their calves. The other farmer provided the recommended daily milk solids to replacement heifer calves in one highly concentrated feed (30% CMR solution) and observed improved calf health as a result. However, these perceived health benefits are again likely due to the provision of increased nutrition compared to the previous restricted feeding programme rather than the provision of a single, concentrated daily feed. Calves can digest large milk meals of up to 6.8 L

(13.2% of bodyweight) without evidence of abdominal discomfort or milk entering the rumen [48]. However, large, infrequent milk meals can cause negative metabolic changes including impaired insulin sensitivity which may negatively affect animals long-term [49]. Despite the legal requirement to provide two liquid meals per day to calves under 28 days of age, some CMR products have been marketed as being suitable for once-a-day feeding [25], thereby encouraging it as an acceptable protocol on farms.

The ethics or technical competency of some animal feed companies was questioned by some of the participants in this study. In particular, concerns were raised that recommended feeding rates from manufacturers of CMR may not facilitate optimal growth efficiency. Calves fed high rates of CMR can achieve growth rates of 1 kg/day [8], but a recent study showed that normal pre-weaning feeding practices on commercial farms resulted in 70% of calves failing to achieve the recommended growth rate of 0.7 kg/day, and 20% of those calves grew at less than 0.5 kg/day [50]. That study did not report how the participating farms established their feeding protocols, but it is likely that current industry standards which may not be based on the optimal physiological requirements of calves [50] contribute to the consistent failure to meet the recommended AFC of 24 months [51]. It is imperative that recommended feeding rates are sufficient to meet calf nutritional requirements and support growth rates which are compatible with industry targets, and that product packaging is updated to reflect these recommendations.

The current study also raises concerns about the clarity of the instructions provided on CMR product packaging, as written instructions for mixing CMR with water to obtain the correct concentration for calf feeding were misunderstood by at least one farmer in the present study. Farmers respect the advice given by trusted feed company representatives who are familiar with their farm and the farms of others [52] so in-person advice which can account for farm-specific rearing targets may be the best way to facilitate optimal feeding protocols on farms. Regardless, written instructions for preparing liquid feeds to pre-weaned calves should be easy to follow in order to support farmers who do not accept in-person advice, and to act as a reference or reminder when mixing CMR at calf feeding.

Few participant farmers accurately measured the temperature of the liquid mix or the amount of CMR included in the feed provided to calves. A consistent liquid diet is important for calf performance; inconsistent provision of milk solids hinders growth, starter intake and feed efficiency [53]. Whilst most farmers appreciated the need for consistency in calf feeding systems, it could be difficult to achieve in practice, largely affected by the values and priorities held by the person responsible for calf feeding, but also the time, equipment and facilities available. Despite the importance of stockmanship [54], most studies have focused on the feeding systems employed by farms, rather than the individuals employing them (e.g., [3,55]). This study indicated that designated calf rearers tended to be most diligent regarding calf feeding, prioritising attention to detail including measuring the variables affecting CMR feeding consistency. On farms where calf feeding was carried out by persons with other responsibilities on the farm, feeding processes were more variable, possibly stemming from a lack of time dedicated to calves and a sense of diminished responsibility compared to designated calf rearers. Automated milk feeders were useful calf management aids for the farms that had them, and can improve welfare due to calf socialisation and constant access to feed which is consistently mixed and at an appropriate pre-set temperature. However, machine feeders have high upfront capital costs, require suitable accommodation for grouping calves, and may contribute to increased disease incidence due to the hygiene challenges presented by calves sharing a single teat [56].

Good hygiene regarding food preparation was prioritised to varying degrees on farms; some diligently disinfected equipment between feeding each calf or pen, others did not. This was sometimes due to pessimistic perceptions that hygiene was ineffectual in disease control, but management problems including uncleanliness have been shown to contribute to increased rates of diarrhoea [57,58]. Others believed sterilisation hindered the acquisition of immunity, similar to misunderstandings previously reported in areas of colostrum management [40] and biosecurity [37]. Indifference or negative attitudes towards ensuring good hygiene are problematic since sanitary feeding equipment and accommodation are critical to maintaining good calf health [18,56]. Furthermore, such attitudes may compound the restricted feeding of calves, as indicated in the literature [18] and by a youngstock veterinarian in the present study, who revealed that farmers often associated increased milk allowances with increased incidences of diarrhoea in calves, but cases of calf scour were more likely to stem from poor hygiene.

In addition to the contribution of poorly sanitised feeding equipment to calf ill-health, one veterinarian in the current study believed the angle of artificial teats on bar feeders could cause aspiration pneumonia in calves. The authors are not aware of research investigating this issue, since aspiration pneumonia is more commonly associated with incorrect oesophageal feeding [59,60] but if proven, calf feeders may need to be adapted and their design improved to encourage correct feeding position and reduce the risk of aspiration. Artificial teat feeding is recommended to allow expression of natural sucking behaviour and aid digestion [58] through activation of the oesophageal groove reflex which bypasses the rumen for milk to enter the abomasum. Farmer participants appreciated this, referencing milk entering 'the wrong stomach' in the absence of a teat and saliva.

Feeding unpasteurised whole milk, or non-saleable milk, can also contribute to pathogenic risk [1]. Of the nine participating farmers feeding whole milk to calves, only two stated that they pasteurised whole milk before feeding it to calves, one of whom was using waste milk, and a further two participants fed unpasteurised non-saleable milk. The practice of feeding milk from cows treated with antimicrobials is also a key area of concern in relation to antibiotic resistance [61] as antibiotic residues cannot be decreased through pasteurisation. Also, feeding milk containing antimicrobial residues causes microbial imbalance in the gut microbiome of pre-weaned calves [62]. These issues appear to be most common in relation to bull or beef-cross calves from dairy enterprises due to the cost of feeding CMR or saleable milk, but some farms also fed their dairy heifers non-saleable milk as standard practice. This could be because the up-front cost of installing a pasteuriser is considered prohibitive or the benefits of pasteurisation and the risks of feeding non-saleable milk are not well understood by farmers, suggesting a need for proactive advice from veterinarians.

The information interviewees provided regarding their CMR lacked detail. Whilst farmers would refer to the need to use a 'good' feed, they did not provide definition. This suggests that farmers require further guidance on calf nutrition, and it is likely that they relied heavily upon the information provided by their feed merchant or product packaging. The current study relied only on interviewee accounts which limited our ability to precisely assess what was fed to calves. However, detailed analyses of feed packaging or written records were beyond the scope of the study. The interviews did provide a useful overview of calf feeding and highlighted a potential disconnect between current recommendations and information provided on CMR packaging as outlined above. The interviews also showed that participants were most focused on liquid feeding of calves, with limited discussion of concentrate and forage feeding for milk-fed calves beyond ensuring adequate intakes of dry feed prior to weaning. Young calves are most at risk of diarrhoea and mortality [63], and there are arguably more variables and effort involved in providing milk or CMR to calves (temperature, consistency, timing, feeding method, hygiene) compared to providing calf starter and roughage. Participants said very little about the post-weaning feeding of calves, attitudes which are reflected in the lack of coverage of the post-weaned period to approximately 4–5 months of age in the research literature [64].

Participants' main focus regarding dry feed for calves was ensuring adequate intakes to prepare calves for weaning. All producers in this study used some form of gradual weaning, and none weaned earlier than six weeks of age. Farmers mainly based weaning decisions on calf age, with some also considering calf bodyweight or starter intake, recognising that calves should be consuming over 1 kg/day of dry calf starter before weaning to indicate sufficient rumen development and prevent growth checks [1]. These practices should support gastrointestinal growth and development in dairy calves [65]. However, not all farmers provided calves with access to water from birth, which may negatively affect rumen development, restricting pre-weaning feed-efficiency and impeding growth both pre- and post-weaning [30]. This could be related to the poorly described water requirement for calves and few published research articles which include calf water intakes [64].

Furthermore, the range of weaning practices used on farms indicates that there is a lack of consistent guidance regarding the best way to wean calves, or if there is, it is not being consistently implemented at farm level. Research has largely focused on the positive effects of gradual weaning based on concentrate intakes [66] and the effect of pre-weaning milk or CMR allowances on the weaning and post-weaning period [67]. However, participants were unsure of the best weaning methods, largely pondering whether transition should be done by diluting milk feeds, reducing the number of feeds, or reducing the quantity fed at each meal. Even a veterinarian who would be expected to have a good understanding of the developing bovine digestive physiology was unsure which weaning method was most effective. This suggests the industry requires further evidence-based recommendations for practical methods to wean calves, particularly *how* to reduce milk provision to best transition calves onto solely solid feed. Several participant farmers also reported that calf health status and growth rates were most problematic at weaning time, suggesting their calves did not have sufficiently-developed rumens when transitioned from milk to solid feed, or that forage intakes are insufficient to mitigate ruminal acidosis [68] and support the establishment of diverse rumen bacteria [69]. Our results indicate a need for further research to establish a consensus on optimal weaning techniques so that farmers can be more effectively advised.

In summary, there is considerable variation in the calf feeding practices used on UK dairy farms, possibly reflecting the current lack of consensus in the scientific literature regarding the most cost-effective feeding protocols to promote growth and future performance. Although now outdated, restricted milk feeding was the predominant recommendation for decades, and advice must be consistent and have evident benefits at the farm level to shift mindsets away from restricted milk feeding. Some CMR feed manufacturers may need to review their feeding recommendations in order to better ensure calves' nutritional needs are met. More consistent advice, for example, about the importance of drinking water and hygiene practices regarding milk feeding, have also not stimulated all farmers to implement best practice. In these cases, it is possible that more effective calf performance monitoring and peer-to-peer learning may help to show farmers that their methods may not be as efficient as they could be, thus motivating them to make improvements [46].

Farmers would also likely benefit from more input from their advisors to counter the variation and confusion about what to feed calves and how to do it. However, it appears that the area of calf nutrition is somewhat of a grey area in terms of advice. Veterinarians may not be focused on the calf rearing of their dairy farm clients [33] and are often not asked by the farmers about calf feeding. It might seem more appropriate to seek advice from trusted animal nutritionists or feed merchants [70], though some participants in this study indicated they would be distrustful of receiving a sales pitch rather than honest information about the best way to feed their calves. Collaboration between veterinarians and the feed industry could help to improve the consistency of recommendations for ensuring suitable calf nutrition. Working together, veterinarians, feed merchants and nutritionists could offer farmers high-quality, bespoke advice about the most cost-effective nutrition and feeding systems that would provide for the health and wellbeing of calves on individual farms.

5. Conclusions

Feeding practices on dairy farms tended to be based on perceived calf performance, and the simplicity, efficiency and cost- or time-effectiveness of their feeding practices versus potential alternatives. However, farmers cannot be expected to implement best practice if the recommendations for standard feeding provide insufficient nutrition and guidance regarding weaning protocols. The advice available to farmers on the subject of practical calf feeding needs to be improved and communicated by advisors. In particular, the animal feed industry should make a more concerted effort to ensure guidelines are compatible with the physiological needs of calves, facilitate weaning and support growth targets to achieve earlier AFC.

Author Contributions: The individual contributions of each author are as follows. Conceptualization, L.J.P., P.A.R., E.C.L.B., M.L.B.; methodology, L.J.P., P.A.R., M.L.B., E.C.L.B.; software, N/A; validation, N/A; formal analysis, L.J.P.; investigation, L.J.P.; resources, L.J.P.; data curation, L.J.P.; writing—original draft preparation, L.J.P.; writing—review and editing, L.J.P., E.C.L.B., P.A.R., M.L.B.; visualization, L.J.P.; supervision, P.A.R., E.C.L.B., M.L.B.; project administration, L.J.P.; funding acquisition, P.A.R. All authors have read and agreed to the published version of the manuscript.

Funding: This research was funded by Barham Benevolent Foundation.

Acknowledgments: The authors are most grateful for the participation of our interviewees and the assistance of those who supported our calls for participants for this study.

Conflicts of Interest: The authors declare no conflict of interest. The funders had no role in the design of the study; in the collection, analyses, or interpretation of data; in the writing of the manuscript, or in the decision to publish the results.

References

1. Drackley, J.K. Calf Nutrition from Birth to Breeding. *Vet. Clin. N. Am. Food Anim. Pract.* **2008**, *24*, 55–86. [CrossRef]
2. Boulton, A.C.; Rushton, J.; Wathes, D.C. An empirical analysis of the cost of rearing dairy heifers from birth to first calving and the time taken to repay these costs. *Animal* **2017**, *11*, 1372–1380. [CrossRef]
3. Boulton, A.C.; Rushton, J.; Wathes, D.C. A Study of Dairy Heifer Rearing Practices from Birth to Weaning and Their Associated Costs on UK Dairy Farms. *Open J. Anim. Sci.* **2015**, *5*, 185–197. [CrossRef]
4. Cooke, J.S.; Cheng, Z.; Bourne, N.E.; Wathes, D.C. Association between growth rates, age at first calving and subsequent fertility, milk production and survival in Holstein-Friesian heifers. *Open J. Anim. Sci.* **2013**, *3*, 1–12. [CrossRef]
5. Wathes, D.C.; Pollott, G.E.; Johnson, K.F.; Richardson, H.; Cooke, J.S. Heifer fertility and carry over consequences for life time production in dairy and beef cattle. *Animal* **2014**, *8*, 91–104. [CrossRef]
6. Eastham, N.T.; Coates, A.; Cripps, P.; Richardson, H.; Smith, R.; Oikonomou, G. Associations between age at first calving and subsequent lactation performance in UK Holstein and Holstein-Friesian dairy cows. *PLoS ONE* **2018**, *13*, e0197764. [CrossRef]
7. Hill, T.M.; Bateman, H.G.; Quigley, J.D.; Aldrich, J.M.; Schlotterbeck, R.L.; Heinrichs, A.J. REVIEW: New information on the protein requirements and diet formulation for dairy calves and heifers since the Dairy NRC 2001. *Prof. Anim. Sci.* **2013**, *29*, 199–207. [CrossRef]
8. Bleach, E.; Gould, M.; Blackie, N.; Beever, D. Growth Performance of Holstein-Friesian Heifer Calves Reared using Three Milk Replacer Feeding Regimes. In *Recent Advances in Animal Nutrition*; Garnsworthy, P.C., Wiseman, J., Eds.; Nottingham University Press: Nottingham, UK, 2005; pp. 347–357.
9. National Research Council. Chapter 10: Nutrient Requirements of the Young Calf. In *Nutrient Requirements of Dairy Cattle: Seventh Revised Edition*; The National Academies Press: Washington, WA, USA, 2001; pp. 214–233. ISBN 978-0-309-06997-7.
10. Godden, S.M.; Fetrow, J.P.; Feirtag, J.M.; Green, L.R.; Wells, S.J. Economic analysis of feeding pasteurized nonsaleable milk versus conventional milk replacer to dairy calves. *J. Am. Vet. Med. Assoc.* **2005**, *226*, 1547–1554. [CrossRef]
11. Ollivett, T.L.; Nydam, D.V.; Linden, T.C.; Bowman, D.D.; Van Amburgh, M.E. Effect of nutritional plane on health and performance in dairy calves after experimental infection with Cryptosporidium parvum. *J. Am. Vet. Med. Assoc.* **2012**, *241*, 1514–1520. [CrossRef]
12. Gerbert, C.; Frieten, D.; Koch, C.; Dusel, G.; Eder, K.; Stefaniak, T.; Bajzert, J.; Jawor, P.; Tuchscherer, A. Effects of ad libitum milk replacer feeding and butyrate supplementation on behavior, immune status, and health of Holstein calves in the postnatal period. *J. Dairy Sci.* **2018**, *101*, 7348–7360. [CrossRef]
13. Hengst, B.A.; Nemec, L.M.; Rastani, R.R.; Gressley, T.F. Effect of conventional and intensified milk replacer feeding programs on performance, vaccination response, and neutrophil mRNA levels of Holstein calves. *J. Dairy Sci.* **2012**, *95*, 5182–5193. [CrossRef] [PubMed]
14. Rosenberger, K.; Costa, J.H.C.; Neave, H.W.; von Keyserlingk, M.A.G.; Weary, D.M. The effect of milk allowance on behavior and weight gains in dairy calves. *J. Dairy Sci.* **2017**, *100*, 504–512. [CrossRef] [PubMed]
15. De Paula Vieira, A.; Guesdon, V.; de Passillé, A.M.; von Keyserlingk, M.A.G.; Weary, D.M. Behavioural indicators of hunger in dairy calves. *Appl. Anim. Behav. Sci.* **2008**, *109*, 180–189. [CrossRef]

16. Thomas, T.J.; Weary, D.M.; Appleby, M.C. Newborn and 5-week-old calves vocalize in response to milk deprivation. *Appl. Anim. Behav. Sci.* **2001**, *74*, 165–173. [CrossRef]

17. Krachun, C.; Rushen, J.; de Passillé, A.M. Play behaviour in dairy calves is reduced by weaning and by a low energy intake. *Appl. Anim. Behav. Sci.* **2010**, *122*, 71–76. [CrossRef]

18. Khan, M.A.; Weary, D.M.; von Keyserlingk, M.A.G. Invited review: Effects of milk ration on solid feed intake, weaning, and performance in dairy heifers. *J. Dairy Sci.* **2011**, *94*, 1071–1081. [CrossRef]

19. Soberon, F.; Raffrenato, E.; Everett, R.W.; Van Amburgh, M.E. Preweaning milk replacer intake and effects on long-term productivity of dairy calves. *J. Dairy Sci.* **2012**, *95*, 783–793. [CrossRef]

20. Khan, M.A.; Lee, H.J.; Lee, W.S.; Kim, H.S.; Kim, S.B.; Ki, K.S.; Ha, J.K.; Lee, H.G.; Choi, Y.J. Pre- and Postweaning Performance of Holstein Female Calves Fed Milk Through Step-Down and Conventional Methods. *J. Dairy Sci.* **2007**, *90*, 876–885. [CrossRef]

21. Silper, B.F.; Lana, A.M.Q.; Carvalho, A.U.; Ferreira, C.S.; Franzoni, A.P.S.; Lima, J.A.M.; Saturnino, H.M.; Reis, R.B.; Coelho, S.G. Effects of milk replacer feeding strategies on performance, ruminal development, and metabolism of dairy calves. *J. Dairy Sci.* **2014**, *97*, 1016–1025. [CrossRef]

22. Soberon, F.; Van Amburgh, M.E. Lactation biology syposium: The effect of nutrient intake from milk or milk replacer of preweaned dairy calves on lactation milk yield as adults: A meta-analysis of current data. *J. Anim. Sci.* **2013**, *91*, 706–712. [CrossRef]

23. Galton, D.M.; Brakel, W.J. Influence of Feeding Milk Replacer Once Versus Twice Daily on Growth, Organ Measurements, and Mineral Content of Tissues. *J. Dairy Sci.* **1976**, *59*, 944–948. [CrossRef]

24. Jenny, B.F.; van Dijk, H.J.; Grimes, L.W. Performance of Calves Fed Milk Replacer Once Daily at Various Fluid Intakes and Dry Matter Concentrations. *J. Dairy Sci.* **1982**, *65*, 2345–2350. [CrossRef]

25. Van der Burgt, G.; Hepple, S. Legal position on 'once a day' feeding of artificial milk to calves. *Vet. Rec.* **2013**, *172*, 371–372.

26. Farm Animal Welfare Committee (FAWC). *Opinion on the Welfare Implications of Nutritional Management Strategies for Artificially-Reared Calves from Birth to Weaning*; Farm Animal Welfare Committee (FAWC): London, UK, 2015.

27. Thomas, L.C.; Wright, T.C.; Formusiak, A.; Cant, J.P.; Osborne, V.R. Use of flavored drinking water in calves and lactating dairy cattle. *J. Dairy Sci.* **2007**, *90*, 3831–3837. [CrossRef]

28. Govil, K.; Yadav, D.S.; Patil, A.K.; Nayak, S.; Baghel, R.P.S.; Yadav, P.K. Feeding management for early rumen development in calves. *J. Entomol. Zool. Stud.* **2017**, *5*, 1132–1139.

29. Kertz, A.F.; Reutzel, L.F.; Mahoney, J.H. Ad Libitum Water Intake by Neonatal Calves and Its Relationship to Calf Starter Intake, Weight Gain, Feces Score, and Season. *J. Dairy Sci.* **1984**, *67*, 2964–2969. [CrossRef]

30. Wickramasinghe, H.K.J.P.; Kramer, A.J.; Appuhamy, J.A.D.R.N. Drinking water intake of newborn dairy calves and its effects on feed intake, growth performance, health status, and nutrient digestibility. *J. Dairy Sci.* **2019**, *102*, 377–387. [CrossRef]

31. Vasseur, E.; Borderas, F.; Cue, R.I.; Lefebvre, D.; Pellerin, D.; Rushen, J.; Wade, K.M.; de Passillé, A.M. A survey of dairy calf management practices in Canada that affect animal welfare. *J. Dairy Sci.* **2010**, *93*, 1307–1315. [CrossRef]

32. Lorenz, I.; Mee, J.F.J.; Earley, B.; More, S.S.J.; Donovan, G.; Dohoo, I.; Montgomery, D.; Bennett, F.; More, S.S.J.; McKenzie, K.; et al. Calf health from birth to weaning. I. General aspects of disease prevention. *Ir. Vet. J.* **2011**, *64*, 10. [CrossRef]

33. Sumner, C.L.; Keyserlingk, M.A.G. Von Canadian dairy cattle veterinarian perspectives on calf welfare. *J. Dairy Sci.* **2018**, *101*, 10303–10316. [CrossRef]

34. Kauppinen, T.; Vainio, A.; Valros, A.; Rita, H.; Vesala, K.M. Improving animal welfare: Qualitative and quantitative methodology in the study of farmers' attitudes. *Anim. Welf.* **2010**, *19*, 523–536.

35. Escobar, M.P.; Buller, H. *Projecting Social Science into Defra's Animal Welfare Evidence Base: A Review of Current Research and Evidence Gaps on the Issue of Farmer Behaviour with Respect to Animal Welfare*; Department for Environment, Food and Rural Affairs: London, UK, 2014.

36. Ruston, A.; Shortall, O.; Green, M.; Brennan, M.; Wapenaar, W.; Kaler, J. Challenges facing the farm animal veterinary profession in England: A qualitative study of veterinarians' perceptions and responses. *Prev. Vet. Med.* **2016**, *127*, 84–93. [CrossRef] [PubMed]

37. Brennan, M.; Wright, N.; Wapenaar, W.; Jarratt, S.; Hobson-West, P.; Richens, I.; Kaler, J.; Buchanan, H.; Huxley, J.; O'Connor, H. Exploring Attitudes and Beliefs towards Implementing Cattle Disease Prevention and Control Measures: A Qualitative Study with Dairy Farmers in Great Britain. *Animals* **2016**, *6*, 61. [CrossRef]

38. Robinson, P.A. Farmers and bovine tuberculosis: Contextualising statutory disease control within everyday farming lives. *J. Rural Stud.* **2017**, *55*, 168–180. [CrossRef]

39. Maxwell, J.A. *A Realist Approach to Qualitative Research*; Sage Publications: Thousand Oaks, CA, USA, 2012.

40. Palczynski, L.; Bleach, E.; Brennan, M.; Robinson, P. Giving calves "the best start": Perceptions of colostrum management on dairy farms in England. *Anim. Welf.* **2020**, in press.

41. Cohen, L.; Manion, L.; Morrison, K. *Research Methods in Education*, 6th ed.; Routledge: London, UK, 2007; ISBN 0-203-02905-4.

42. Turner, D.W. Qualitative Interview Design: A Practical Guide for Novice Investigators. *Qual. Rep.* **2010**, *15*, 754–760.

43. Miles, M.B.; Huberman, A.M.; Saldana, J. *Qualitative Data Analysis: A Methods Sourcebook*, 3rd ed.; Sage Publications: Thousand Oaks, CA, USA, 2014.

44. Braun, V.; Clarke, V. Using thematic analysis in psychology. *Qual. Res. Psychol.* **2006**, *3*, 77–101. [CrossRef]

45. Bach, A.; Ahedo, J. Record Keeping and Economics of Dairy Heifers. *Vet. Clin. N. Am. Food Anim. Pract.* **2008**, *24*, 117–138. [CrossRef]

46. Sumner, C.L.; von Keyserlingk, M.A.G.; Weary, D.M. How benchmarking motivates farmers to improve dairy calf management. *J. Dairy Sci.* **2018**, *101*, 3323–3333. [CrossRef]

47. Heinrichs, A.J.; Gelsinger, S.L. Economics and Effects of Accelerated Calf Growth Programs. In Proceedings of the 28th Annual Florida Ruminant Nutrition Symposium, Gainesville, FL, USA, 6–8 February 2017; pp. 160–168.

48. Ellingsen, K.; Mejdell, C.M.; Ottesen, N.; Larsen, S.; Grøndahl, A.M. The effect of large milk meals on digestive physiology and behaviour in dairy calves. *Physiol. Behav.* **2016**, *154*, 169–174. [CrossRef]

49. Bach, A.; Domingo, L.; Montoro, C.; Terré, M. Short communication: Insulin responsiveness is affected by the level of milk replacer offered to young calves. *J. Dairy Sci.* **2013**, *96*, 4634–4637. [CrossRef] [PubMed]

50. Johnson, K.F.; Chancellor, N.; Burn, C.C.; Wathes, D.C. Analysis of pre-weaning feeding policies and other risk factors influencing growth rates in calves on 11 commercial dairy farms. *Animal* **2018**, *12*, 1413–1423. [CrossRef] [PubMed]

51. Hanks, J.; Kossaibati, M. *Key Performance Indicators for the UK National Dairy Herd: A Study of Herd Performance in 500 Holstein/Friesian Herds for the Year Ending 31st August 2018*; National Milk Records: Reading, UK, 2018.

52. Croyle, S.L.; Belage, E.; Khosa, D.K.; LeBlanc, S.J.; Haley, D.B.; Kelton, D.F. Dairy farmers' expectations and receptivity regarding animal welfare advice: A focus group study. *J. Dairy Sci.* **2019**, *102*, 7385–7397. [CrossRef]

53. Hill, T.M.; Bateman, H.G.; Aldrich, J.M.; Schlotterbeck, R.L. Effect of Consistency of Nutrient Intake from Milk and Milk Replacer on Dairy Calf Performance. *Prof. Anim. Sci.* **2009**, *25*, 85–92. [CrossRef]

54. Tucker, C.B.; Verkerk, G.A.; Small, B.H.; Tarbotton, I.S.; Webster, J.R. Animal welfare in large dairy herds: A survey of current practices. *Proc. N. Z. Soc. Anim. Prod.* **2005**, *65*, 127–131.

55. Medrano-Galarza, C.; LeBlanc, S.J.; DeVries, T.J.; Jones-Bitton, A.; Rushen, J.; Marie de Passillé, A.; Haley, D.B. A survey of dairy calf management practices among farms using manual and automated milk feeding systems in Canada. *J. Dairy Sci.* **2017**, *100*, 6872–6884. [CrossRef]

56. Curtis, G.C.; Argo, C.M.; Jones, D.; Grove-White, D.H. Impact of feeding and housing systems on disease incidence in dairy calves. *Vet. Rec.* **2016**, *179*, 512. [CrossRef]

57. Appleby, M.C.; Weary, D.M.; Chua, B. Performance and feeding behaviour of calves on ad libitum milk from artificial teats. *Appl. Anim. Behav. Sci.* **2001**, *74*, 191–201. [CrossRef]

58. Jasper, J.; Weary, D.M. Effects of Ad Libitum Milk Intake on Dairy Calves. *J. Dairy Sci.* **2002**, *85*, 3054–3058. [CrossRef]

59. Poulsen, K.P.; McGuirk, S.M. Respiratory Disease of the Bovine Neonate. *Vet. Clin. N. Am. Food Anim. Pract.* **2009**, *25*, 121–137. [CrossRef]

60. Gorden, P.J.; Plummer, P. Control, management, and prevention of bovine respiratory disease in dairy calves and cows. *Vet. Clin. N. Am. Food Anim. Pract.* **2010**, *26*, 243–259. [CrossRef] [PubMed]

61. Ricci, A.; Allende, A.; Bolton, D.; Chemaly, M.; Davies, R.; Fernández Escámez, P.S.; Girones, R.; Koutsoumanis, K.; Lindqvist, R.; Nørrung, B.; et al. Risk for the development of Antimicrobial Resistance (AMR) due to feeding of calves with milk containing residues of antibiotics. *EFSA J.* **2017**, *15*, 4665.

62. Malmuthuge, N.; Guan, L.L. Understanding the gut microbiome of dairy calves: Opportunities to improve early-life gut health. *J. Dairy Sci.* **2017**, *100*, 5996–6005. [CrossRef] [PubMed]

63. Windeyer, M.C.; Leslie, K.E.; Godden, S.M.; Hodgins, D.C.; Lissemore, K.D.; LeBlanc, S.J. Factors associated with morbidity, mortality, and growth of dairy heifer calves up to 3 months of age. *Prev. Vet. Med.* **2014**, *113*, 231–240. [CrossRef]

64. Kertz, A.F.; Hill, T.M.; Quigley, J.D.; Heinrichs, A.J.; Linn, J.G.; Drackley, J.K. A 100-Year Review: Calf nutrition and management. *J. Dairy Sci.* **2017**, *100*, 10151–10172. [CrossRef]

65. Schäff, C.T.; Gruse, J.; Maciej, J.; Pfuhl, R.; Zitnan, R.; Rajsky, M.; Hammon, H.M. Effects of feeding unlimited amounts of milk replacer for the first 5 weeks of age on rumen and small intestinal growth and development in dairy calves. *J. Dairy Sci.* **2018**, *101*, 783–793. [CrossRef]

66. Roth, B.A.; Keil, N.M.; Gygax, L.; Hillmann, E. Influence of weaning method on health status and rumen development in dairy calves. *J. Dairy Sci.* **2009**, *92*, 645–656. [CrossRef]

67. Quigley, J.D.; Hill, T.M.; Dennis, T.S.; Schlotterbeck, R.L. Effects of feeding milk replacer at 2 rates with pelleted, low-starch or texturized, high-starch starters on calf performance and digestion. *J. Dairy Sci.* **2018**, *101*, 5937–5948. [CrossRef]

68. Laarman, A.H.; Oba, M. Short communication: Effect of calf starter on rumen pH of Holstein dairy calves at weaning. *J. Dairy Sci.* **2011**, *94*, 5661–5664. [CrossRef]

69. Kim, Y.H.; Nagata, R.; Ohtani, N.; Ichijo, T.; Ikuta, K.; Sato, S. Effects of dietary forage and calf starter diet on ruminal pH and bacteria in holstein calves during weaning transition. *Front. Microbiol.* **2016**, *7*, 1–12. [CrossRef]

70. Ellingsen, K.; Mejdell, C.M.; Hansen, B.; Grøndahl, A.M.; Henriksen, B.I.F.; Vaarst, M. Veterinarians' and agricultural advisors' perception of calf health and welfare in organic dairy production in Norway. *Org. Agric.* **2012**, *2*, 67–77. [CrossRef]

MDPI

St. Alban-Anlage 66

4052 Basel

Switzerland

Tel. +41 61 683 77 34

Fax +41 61 302 89 18

www.mdpi.com

Animals Editorial Office

E-mail: animals@mdpi.com

www.mdpi.com/journal/animals